Advances in FISH AND WILDLIFE ECOLOGY AND BIOLOGY

— Vol. 5 —

Advances in
FISH AND WILDLIFE ECOLOGY AND BIOLOGY

— Vol. 5 —

— Editor —

B.L. Kaul

Former Professor of Zoology and Principal,
Government SPMR, College of Commerce and Management,
Canal Road, Jammu – 180 001, J&K, India

2011

DAYA PUBLISHING HOUSE®

Delhi – 110 035

ISBN 81-7035-718-7
ISBN 978-81-7035-718-6

Published by : **Daya Publishing House®**
4760-61/23, Ansari Road, Darya Ganj,
New Delhi - 110 002
Phone: 23245578, 23244987
Fax: (011) 23260116
e-mail : dayabooks@vsnl.com
website : www.dayabooks.com

Laser Typesetting : **Classic Computer Services**
Delhi - 110 035

Printed at : **Chawla Offset Printers**
Delhi - 110 052

PRINTED IN INDIA

Dedicated to

Professor (Dr.) B.L. Kaw

Editorial Board

5. Dr. P.L. Koul

Professor of Zoology and Principal (Retd.),
Roop Nagar Enclave, Jammu, India
E-mail: plkoul@rediffmail.com

6. Dr. Anil K. Verma

Professor & Head, Department of Zoology
Government (P.G.) College, Rajouri, J&K, India
E-mail: anilverma.ak@gmail.com

Preface

The present volume of the series "Advances in Fish and Wildlife Ecology and Biology" is being published at a time when the whole world is severely experiencing the effects of global warming. While some parts of the world are reeling under extreme heat and drought conditions other parts are flooded with rains and water from melting glaciers. Cyclones and typhoons are causing destruction of life and property. Unmindful activities such as wanton destruction of forests, excessive dependence on fossil fuels, too much use of insecticides and fertilizers in agriculture and many other anthropogenic actions are adversely affecting lives of humans, domestic animals, aquatic plants and animals, wildlife and forests. It is increasingly being felt that to reverse the trend immediate remedial measures must be taken. Kyoto Protocol which entered into force on 16 February 2005 was ratified by 183 parties in January 2009. Under Kyoto Protocol, the Industrialized countries agreed to reduce their collective Green House Gas (GHG) emissions by 5.2 per cent compared to the year 1990. But to date nothing on ground seems to have been done except lip service paid to the cause of environment and prevention of global warming.

The earth summit held at Rio De Janeiro (Brazil) in June 1992 had recognized the need to redirect international and national plans and policies to ensure that all economic decisions fully took into account any environmental impact. The Kyoto Protocol went a step ahead and an international treaty was produced with the goal of "achieving stabilization of GHG concentrations in the atmosphere at a level that would prevent dangerous anthropogenic interference with the climate system".

It is heartening that besides discussing global recession and terrorism G8 summit in July 2009 at L'Aquila, Italy climate change and food security were also on the agenda. It is also heartening that the developed countries have agreed to take remedial steps to reduce GHG emissions to half present levels by 2050. US President Mr. Barack Obama said emphatically at the summit that carbon emissions must be reduced to stop climate change. It is but natural that a carbon free atmosphere and an improved climate will help conserve biodiversity.

The present volume is dedicated to the late Professor (Dr.) B.L. Kaw, a distinguished teacher, researcher and authority on *Acanthocephala*. It is my earnest hope that this volume too will be well received like the earlier volumes.

I am thankful to Er. Brij Lal Kaw the younger son of Prof. B.L. Kaw for providing me details about his illustrious father.

In the end I heartily thank the contributors, my wife Promila and Mr. Anil Mittal (Publisher) of Daya Publishing House, for their cooperation and help in bringing out this volume.

B.L. Kaul

Prof. (Dr.) Bidh Lal Kaw– A Tribute

Professor Bidh Lal Kaw, a great teacher and pioneer researcher of the last century, was among the founding fathers of Zoology as a subject of study in the Jammu and Kashmir State of undivided India. He was held in great reverence as a talented teacher, a discreet researcher and an able administrator and is remembered even after having passed away more than three decades ago. I feel privileged to pen down the following as my tribute to him:

Prof. B.L. Kaw was born at the village Nadigaum, District Shopian of Jammu and Kashmir in a middle class family in 1907. He received his primary education at Shopian and had to shift to Srinagar in 1922 to matriculate from the Govt. High School Bagh-i-Dilavar Khan. He was admitted to the Sri Pratap College, Srinagar to pursue higher studies and passed Intermediate Examination in the year 1924. Since graduate courses in sciences were not available at the S.P. College he had to go to Lahore (now in Pakistan) for admission in Govt. College there to pursue B.Sc. course. He passed B.Sc. from Punjab University Lahore in 1926. Soon after he was admitted to the Post-graduation course in Botany at Allahabad University. He passed M.Sc. in Botany in 1928 and returned to Srinagar, where he was appointed a demonstrator in Biology at the S.P. College. In 1935 he joined Lucknow University to pursue M.Sc. Zoology and L.L.B. simultaneously and attained the degrees in 1937. On his return he was appointed a lecturer in the newly created Department of Biology at Prince of Wales College (Now Govt. Gandhi Memorial Science College), Jammu. He worked hard to establish the Department and earned laurels for the services he rendered. After the establishment of Jammu and Kashmir University in 1949 the Department of Education wanted to start graduation courses in selected subjects at the S.P. College Srinagar. In view of his experience and reputation Dr. Kaw was entrusted with the job of starting B.Sc. course in Zoology in 1950. Students had earlier to move outside the state to pursue B.Sc. course and now they could study in their own State. After heading the Zoology Department at S.P. College Srinagar. Prof. Kaw was promoted as Principal in 1959 and posted at the newly established Govt. Degree College Bhaderwah, in Doda district of Jammu Province. During his tenure as the Principal there, he created infrastructure and introduced new courses of study at Bhaderwah. He got a student's hostel constructed and set up science laboratories and also created library facilities. After serving as Principal of the college for three years he retired in 1962.

Prof. Kaw came under the influence of the eminent Zoologist and Helminthologist Prof. G.S. Thapar as a student of Lucknow University. He worked under his supervision on the *Acanthocephalan* parasites of fish and other vertebrates of Jammu and Kashmir and named some new genera and species of helminth parasites. For this work he was awarded Ph. D. degree in Zoology by Lucknow University in 1949. He was the first Kashmiri to have earned a Ph.D. in a science subject. Prof. Kaw was an authority on *Acenthocephala* and his work was acknowledged in classical treatises on Helminth and Acenthocephalan Parasites by Yamaguti and Skrjabin. His work on Helminth and Acenthocephalan parasites paved the way for further research in these and other related fields in Zoology in Jammu and Kashmir.

Prof. B.L. Kaw had the advantage of teaching Botany as well as Zoology at the undergraduate level and he had gained a thorough knowledge of the fauna and flora of the J&K State through field studies. He took his students on field trips for days together to study fauna and flora of Kashmir Himalayas and enthused them with love for nature and interest in natural history. Thus he was instrumental in creating a generation of researchers in the biological sciences in the forties and fifties of the last century. Most of them later became eminent doctors, teachers, taxonomists and researchers, and occupied key positions in many universities, research institutions and colleges in India.

In recognition of his meritorious work in the field of Zoology Prof. B.L. Kaw was made a Fellow of the Zoological Society of India (F.Z.S.I.), of which he had also been a founder member. He was also made a Fellow of National Academy of Science (F.N.A. Sc.). The Indian Science Congress regularly invited him to attend its sessions and deliver lectures.

I had the good fortune of getting associated with him in the year 1957 as a student at the S.P. College Srinagar and learned a lot under his influence. He was an excellent teacher and taught every topic in great detail. He was gifted with qualities of understanding and patience. Above all he was a wonderful teacher and a strict disciplinarian. Zoology and Botany Departments were without doubt the pride of the S.P. College in those days. Both these departments had the good fortune of having been developed by such stalwarts as Prof. Bahadur Singh, Prof. Sham Lal Raina, Prof. Bidh Lal Kaw, Prof. Shiban Krishan Kaul, Prof. Qazi Mubarak, Prof. T.N. Dhar, Prof. M.K. Munshi, Prof. Shafi-ud-Din, Prof. B.N. Kaul, Prof. Saleh Mustafa, Prof. K.L. Razdan and Prof. S.N. Tickoo. Prof. Dina Nath Fotedar, who had later joined the University of Kashmir established the School of Parasitology in the Post Graduate Department of Zoology there. Prof. Jagan Nath Wattal and Prof. Ghulam Qadir Bhat as Heads and as noted teachers for varying spells of time.

The museums, libraries and laboratories in these departments were well-equipped and fit not only to meet the needs and demands of graduate courses but also Post-graduate courses. It was on account of this fact that a research centre in Limnology was set up at the S. P. College, Srinagar under Dr. Durga Prashad Zutshi a reputed limnologist and Dr. Bashir Ahmed Subla in 1977–78.

Prof. Kaw remained active in research even after his retirement. He remained associated with the Board of Studies in Zoology and was a member of the Syndicate of the University of Jammu and Kashmir. Post-retirement he had also developed an interest in religion and philosophy. Hindu scriptures and spiritual practices attracted him the most. But there was a problem. He did not know the Sanskrit language. So even at an advanced age he undertook the study of Sanskrit which later helped him to read and understand first hand the philosophy advocated by the Vedas, Upanshids and other Hindu scriptures, including the epics Ramayana and Mahabharata. This changed his life and the scientist in him turned a spiritual aspirant.

In 1977 I functioned as the Head of the Zoology Department at Govt. College, Udhampur in Jammu province. One day, to my joy, I received a phone call from Prof. B.L. Kaw that he was on a visit to Udhampur to meet his elder son who was the Chief Medical Officer of the district. The next day I went to meet him at his son's residence. At first I could not recognise him for he had grown a beard and looked like a Sadhu (ascetic). He told me that the Sanskrit language had opened a new field of knowledge to him namely spiritual knowledge. He mentioned that the Hindu scriptures are a treasure house of philosophy and knowledge and advised me to learn the language and engage a Sanskrit teacher for my sons. He also revealed to me the spiritual experiences he had gone through by coming into contact with the enlightened gurus at Swargha Ashram at Haridwar which he frequently visited. He apprised me of what had made him at peace with himself and his surroundings. When I asked him how he had reconciled science and spirituality, he smiled and said that there was no conflict between the two. According to him science and spirituality complemented each other. He was of the view that a person with a scientific mind found it easy to understand the intricacies of spirituality and experienced spiritual bliss.

When Prof. B.L. Kaw passed away after a brief illness in 1977 he was highly praised for his role in developing the spirit of enquiry and scientific temper among his students. The legacy left behind by him has been carried on by generations of students who have successfully brought Jammu and Kashmir on the world map of research in Biological Sciences, especially in Zoology.

B.L. Kaul

Contents

List of Contributors

Ahmad, Shahnawaz

Department of Applied Zoology, Kevumpu University, Shankarghatta – 577 451, Shimoga, Karnataka, India

Araújo, A.

Department of Physiology, Bioscience Centre, Universidade Federal do Rio Grande do Norte, Av. Salgado Filho, 3000, Lagoa Nova, Natal, Rio Grande do Norte, Brazil, CEP 59 072-970

Barai, S.R.

Department of Zoology, Udai Pratap College, Varanasi – 221 002, U.P, India

Bohinder, K.

Department of Zoology, Utkal University, Vani Vihar, Bhubaneshwar – 751004, Orissa, India

Challappa, N.T.

Department of Oceanography and Limnology, Centre of Bioscience, Unversidade do Rio Grande do Norte (UFRN), Praia de Mae Luiza, s/n, Natal, R.N. Brazil-59014-100

Challappa, S.

Department of Oceanography and Limnology, Centre of Bioscience, Unversidade do Rio Grande do Norte (UFRN), Praia de Mae Luiza, s/n, Natal, R.N. Brazil-59014-100

Fatuma, A. Mohammad

Department of Zoology, Institute of Science, 15 Madam Cama Road, Mumbai – 400 032, India

Gupta, Subash C.

Department of Environmental Sciences, University of Jammu, Jammu – 1800 06 J&K, India

Jayson, E.A.

Division of Forest Ecology and Biodiversity Conservation, Kerala Forest Research Institute, Peechi – 680 653, Kerala, India

Kar, Sudhakar

Office of Chief Wildlife Warden, Prakruti Bhawan, Nilakantha Nagar, Nayapalli, Bhubaneshwar – 751 012, Orissa, India

Kaul, B.L.

186, Upper Laxmi Nagar, Sarwal, Jammu – 180 005, J&K, India
E-mail: blkaul@gmail.com

Kotwal, Deepti

Department of Zoology, University of Jammu, Jammu – 180 006, J&K, India

Koul, P.L.

37/2A, Roop Nagar Enclave, Jammu, J&K, India

Kulkani, B.G.

Institute of Science, 15 Madam Cama Road, Mumbai – 400 032, Maharashtra, India

Kumar, Anjani

Central Institute of Freshwater Aquaculture, Kausalyaganga, Bhubaneshwar – 751 002, Orissa, India

Kumar, Sanjeev

Department of Zoology, University of Jammu, Jammu – 180 006, J&K, India

Mishra, D.K.

Department of Zoology, Banki Autonomous College, Banki, Cuttack – 754 508, Orissa, India

Nandini, S.

UIICSE, Division of Research and Postgraduate Studies, National Autonomous University of Mexico, Campus Inztacala, A P 314, CP 54090, Tlalnepantla, State of Mexico, Mexico

Pandey, A.K.

Central Institute of Freshwater Aquaculture, Kausalyaganga, Bhubaneshwar – 751 002, Orissa, India

Raina, M.K.

174/5 Trikuta Nagar, Jammu, J&K, India

Roy, P.K.

Central Institute of Fisheries Education (ICAR), 32 G N Block, Sector-V, Salt Lake City, Kolkata – 700 091, India

Sahi, D.N.

Department of Zoology, University of Jammu, Jammu – 180 006, J&K, India

Sharma, S.S.S.

Laboratory of Aquatic Zoology, Division of Research and Postgraduate Studies, National Autonomous University of Mexico, Campus Iztacala, AP 314, CP 54090, Tlalnapantla, State of Mexico, Mexico

Sharma, O.P.

J&K Forest Research Institute, Srinagar – 190 001, J&K, India

Sharma, M.B.

Department of Zoology, D.A.V. College for Women, Karnal – 132 001, Haryana, India

Sharma, R.K.

Reproductive Physiology Lab, Department of Zoology, Kurukshetra University, Kurukshetra – 136 119, Haryana, India

Sudan, Madhu

Department of Zoology, University of Jammu, Jammu (Tawi) – 180 006, J&K, India

Suryawenshi, S.A.

Swami Ramanand Teerth Marathwada University, Nanded – 431 606, Maharashtra, India

Venkateshwerlu, M.

Department of Applied Zoology, Kuvempu University, Shankarghatta – 575 451, Shimoga, Karnataka, India

Verma, Anil K.

Department of Zoology, Government (P.G.) College, Rajouri, J&K, India

Wanganeo, Ashwani

Department of Limnology, Barkatullah University, Bhopal, M.P., India.

Wanganeo, Rajni

Department of Zoology, S.V. College, Bairagarh, M.P., India

Xiemenes-Lima, J.T.A.

Department of Physiology, Bioscience Centre, Universidade Federal do Rio Grando do Norte, A V. Salgado Filho, 3000, Lagoa Nova, Natal, Rio Grando do Norte, Brazil, CEP 59072-970

Zargar, Tauseef A.

Department of Zoology, University of Jammu, Jammu – 180 006, J&K, India

Section I

Fish and Limnology

Chapter 1

Reproductive Biology of *Scomberomorus brasiliensis* (Perciformes: Scombridae)

☆ S. Chellappa, J.T.A. Ximenes-Lima,
A. Araújo and N.T. Chellappa

Abstract

Scomberomorus brasiliensis is a commercially important species, which occurs in the coastal waters of the Western Atlantic, along the Caribbean and Atlantic coasts of Central and South America, from Belize to Rio Grande do Sul, Brazil. This study verified the macroscopic and histological characterization of gonads, body size, sex ratio, gonadosomatic index, fecundity and spawning season of this species. A total of 424 males (51.3 per cent) and 402 females (48.7 per cent) were collected, total body length of males ranged from 92 to 661 mm and of females from 93 to 805 mm. There was a balanced sex ratio, with females slightly bigger and heavier than males and the onset of sexual maturity in males occurred earlier. Total fecundity was 871,523 mature eggs while relative fecundity was 952 eggs female g^{-1}. Ovaries revealed five stages of gonadal maturation and testes showed four stages. Relative frequency distributions of the oocyte diameter sizes indicate total spawning. Monthly values of GSI, gonadal maturation pattern and period of reproductive activity suggest that reproduction is influenced by the rainy season.

Keywords: *Scomberomorus brasiliensis, Reproduction, Gonadal development, Histology of gonads.*

Introduction

Scomberomorus brasiliensis (Collette, Russo and Zavala-Camin) (Osteichthyes: Perciformes: Scombridae) occurs in the Western Atlantic, along the Caribbean and Atlantic coasts of Central and South America, from Belize to Rio Grande do Sul, Brazil (Collette *et al.*, 1978). *S. brasiliensis* is an important fishery resource of the western

Central Atlantic waters and of Northeastern Brazil. It is a major component of the Brazilian northeast artisanal fishery and has high commercial value (Lucena *et al.*, 2004; Lima *et al.*, 2005).

Although *S. brasiliensis* is an important food fish throughout most of its distributional range, limited details are available about the reproductive characteristics of this species (Gesteira and Mesquita, 1976; Fonteles-Filho, 1988). Description of reproductive characteristics was a major aspect of the current study, since this information is required for stock assessment and for management controls. Information on fecundity and spawning are also required for stock assessments, which is insufficiently described in the literature available for this species. Considering this plethora of factors, the main objective of this study was to provide a comprehensive description of the reproductive biology of *S. brasiliensis*. Furthermore, this work presents and extends information on gonad development based on macroscopic stages and histological characteristics, sex ratio, size at sexual maturity, fecundity, spatial and temporal patterns in gonadosomatic index (GSI), type of spawning and reproductive period of *S. brasiliensis*.

Materials and Methods

Sample Collection

During the period of August 2005 to July 2006, monthly samples of *S. brasiliensis* were collected from artisanal fishermen at various locations in the coastal waters situated between latitudes 05° 52′ 30″–05° 45′ 00″ S, and longitudes 35° 08′ 00″–35° 10′ 35″ W, northeastern Brazil. Fish were caught by local fishermen using beach-seines from the coastal waters of approximately <10 m depth. The beach-seines were 110 m in length, 3 m in height, with a mesh size of 1 cm in the central part and 7 cm in the extremities. Fish collected from the beach-seine fishing process were transported to the laboratory on ice. They were numbered, measured, weighed and samples of whole fish were used for morphometric analysis to confirm the taxonomical identification of the species based on Collette *et al.* (1978) and Carpenter (2002). Rainfall data of the region during the study period were obtained from the Meteorological Department of Natal, Brazil.

Measurements

A total of 826 fish was collected during the study period and the sample size was sufficiently large to allow accurate estimations. The total body lengths of fish sampled were measured (± 1 mm) and body mass recorded (± 1 g). Fish were dissected within a few hours of capture, and gonads were removed, weighed (± 0.1 mg) and examined to separate the sex. Sex ratio was verified based on the monthly distribution of relative frequency of males and females. The length and weight composition of the males and females were determined based on the mean distribution of their class frequencies and grouped with intervals of 100 mm and 500 g (Vazzoler, 1996).

Determination of Size at Maturity

Size at sexual maturity (l_{50} and l_{100}) was established by calculating the percentage of mature and immature gonads observed for fish of given size classes, using total length (mm), sex and stages of gonad development of each individual. Both l_{50} (when 50 per cent of individuals were with gonads in maturing stages) and l_{100} (when all individuals were ready to participate actively in reproduction) were determined. Logistic curves were fitted to data by the use of a non-linear least-squares procedure weighted by the number of fish in each length-class (Fonteles-Filho, 1989).

Macroscopic and Histological Examinations of Gonads

The location and general aspects of the ovaries and testes were noted and stage of reproductive maturity determined using a macroscopic staging system. The degree of turgidity, colour and presence of blood vessels of the gonads were observed (Vazzoler, 1996; Mackie and Lewis, 2001). In order to avoid possible variation in the developmental stage of oocytes due to their position in the ovaries, histological examinations were carried out on sections from the anterior (cephalic), middle (central), and posterior (caudal) regions of 20 ovaries in different developmental stages (Yoshida, 1964). Development of sperm tissue throughout the testes was compared by microscopic examination of sections taken from anterior, middle and posterior sections of each lobe (n = 20). These data were later compared in order to determine whether samples taken from mid-section of the gonad of either lobe were representative of gamete development throughout the ovaries.

The gonads were preserved in Bouin's solution, later embedded in paraffin, sectioned at 3–5 µm thickness, and stained with Harris hematoxylin and eosin (H&E). Ovarian developmental stages were assessed microscopically with the help of light microscope (Taimin, model TM 800), coupled with a video camera (Kodo Digital). The terminology used for stages of oogenesis followed that of West (1990) and Palmer *et al.* (1995).

Estimation of Gonadosomatic Index

Periodicity of gonadal development and seasonal reproductive activity were estimated by the gonadosomatic index for each fish. GSI was estimated by dividing the weight of gonads by its body weight and multiplying by 100 (Wootton *et al.*, 1978).

Analysis of Fecundity and Breeding

For each stage of development, the diameters of oocytes from different ovaries (n = 20) were measured with an ocular micrometer (± 1 µm). The diameter of 60 oocytes at different stages of development was done using fresh ovaries. Each oocyte was measured twice on perpendicular axes, and the mean of the two measurements was used to represent the average diameter of the oocyte. In addition, the relative proportion of oocytes sizes present were estimated and size frequency distribution of the oocytes was plotted (Palmer *et al.*, 1995).

Period of breeding was determined by the temporal relative frequency distribution of the different stages of maturation of gonads of males and females. This study focused more on ovaries since their developmental stages were easier to distinguish than in testes, and because ovarian development usually defines the spawning season and number of offspring produced during spawning (De Martini and Fountain 1981).

Data Analysis

Sex ratio (M:F) was tested using χ^2 test at 5 per cent level. Gonadosomatic indices of males and females during rainy and dry periods were compared at 5 per cent level using Kruskal-Wallis One Way Analysis of Variance on Ranks (Software Statistica, version 7.0 Windows).

Results

Sex Ratio

A total of 826 samples of *S. brasiliensis* were collected during the study period, out of which 424 were males (51.3 per cent) and 402 were females (48.7 per cent). The monthly distribution frequency of occurrence of males and females show that sex ratio for the total sample was equivalent to 1M:1F as expected, although with a slight predominance of males (Figure 1.1). During the drought period males occurred more (52 per cent, n =275) than females (48 per cent, n =249). On the other hand, females occurred more (51 per cent, n =153) than males during the rainy season (49 per cent, n =149). However, the overall difference was not significant statistically at 5 per cent level ($P>0.05$).

Total Body Length and Weight

Amplitude of total body length (Lt) of both males and females varied from 92 to 805 mm (mean 298.0 ± S.D. 146.8), that of males alone ranged from 92 to 661 mm (306 ± 139) and of females from 93 to 805 mm (289.2 ± 153.4) (Figure 1.2). During the drought period, total body length of males and females ranged from 101 to 598 mm (344.4 ± 136.4) and 107 to 805 (312.2 ± 176.5) respectively. During the rainy season, total length of males and females varied from 92 to 661 mm (285.9 ± 37.5) and from 93 to 565 mm (275.0 ± 135.9) respectively. A higher frequency of occurrence of total body length for males was registered in the class intervals between 100-200 mm and 400-500 mm during the wet and dry periods respectively, whereas the same for females was observed throughout the year in the class intervals between 100-200 mm.

The amplitude of total body weight (Wt) of males and females ranged from 7.7 to 3385 g (256.4 ± 358.9), that of males varied from 7.7 to 1493.5 g (250.0 ± 274.2) and of females varied from 8.1 to 3385 g (265.6 ± 430.9) (Figure 1.3). Thus females were significantly heavier than males. During the drought period, total body weight of males and females varied from 14.6 to 1310 g (313.4 ± 243.9) and 8.1 to 4390 g (382.4 ± 692.5) respectively. During the rainy season, total weight of males and females varied from 7.7 to 1493.5 g (215.6 ± 83.8) and 11 to 1015.8 g (206.5 ± 250.9) respectively. A higher frequency of occurrence of total body weight of males and females was observed in the class intervals between 7 to 500 g.

Size at Maturity (l_{50} and l_{100})

Fifty per cent maturity was attained by males and females of *S. brasiliensis* at 280.5 mm. In case of males 50 per cent maturity was attained at 246 mm and females at 315 mm. All males and females were mature (l_{100}) at 440 and 520 mm respectively. Onset of maturity for males were earlier during the dry period at 255 mm (l_{50}) and 450 mm (l_{100}) and during the rainy season they mature later at 246 (l_{50}) and 460 mm (l_{100}).

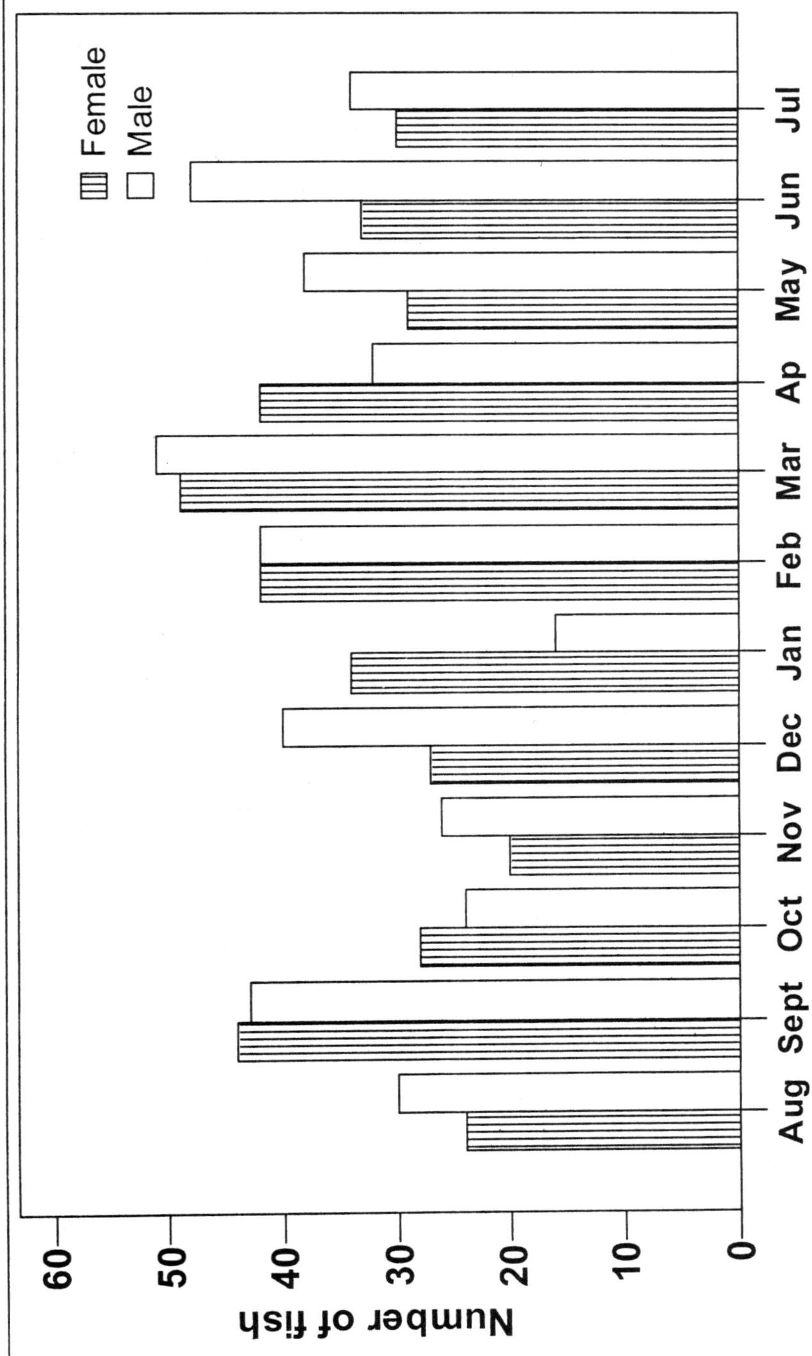

Figure 1.1: Monthly Frequency of Occurrence of Males and Females of *Scomberomorus brasiliensis* during September 2005 to August 2006

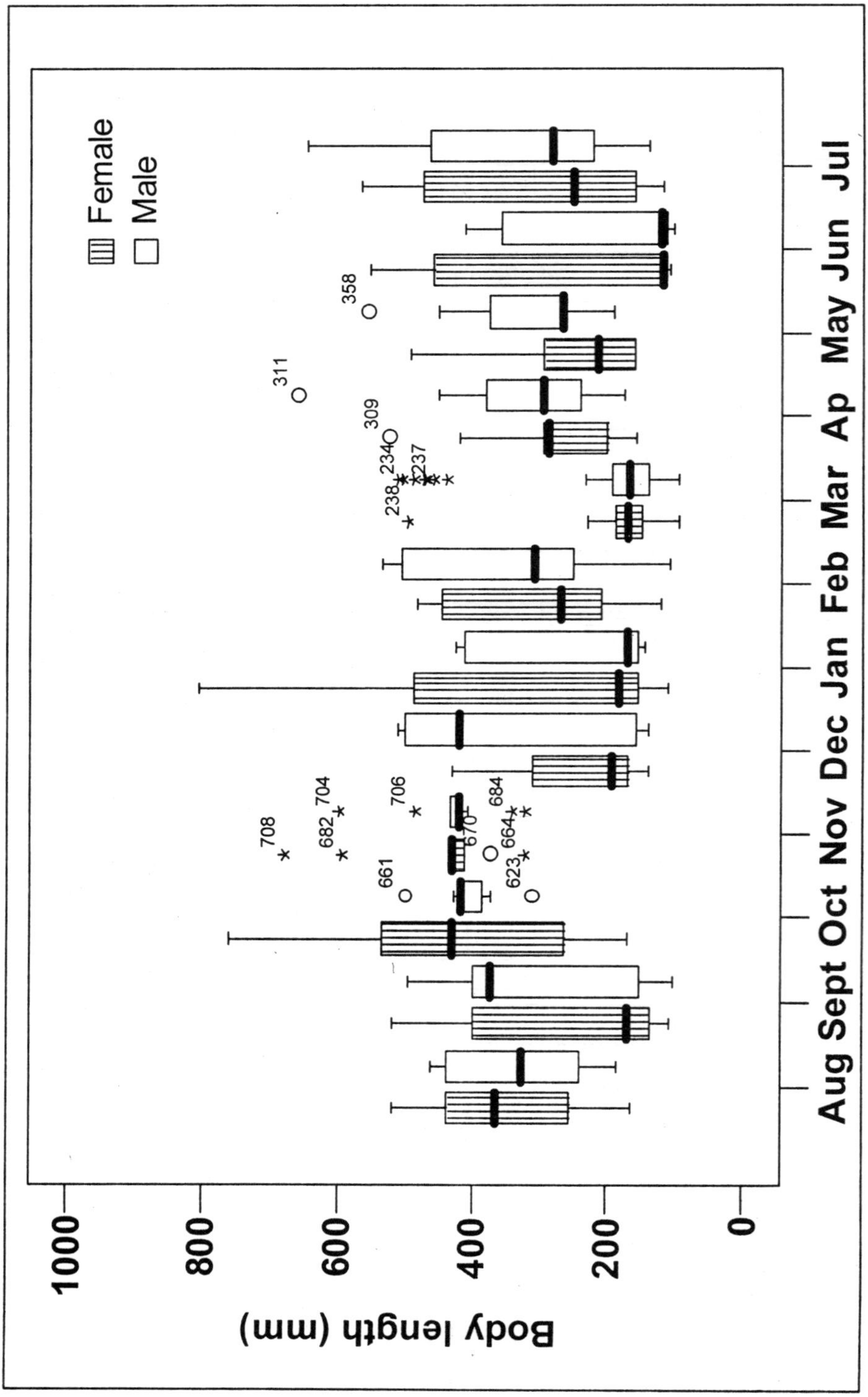

Figure 1.2: Monthly Variation in Amplitude of Total Body Length of Males and Females of *S. brasiliensis* during September 2005 to August 2006

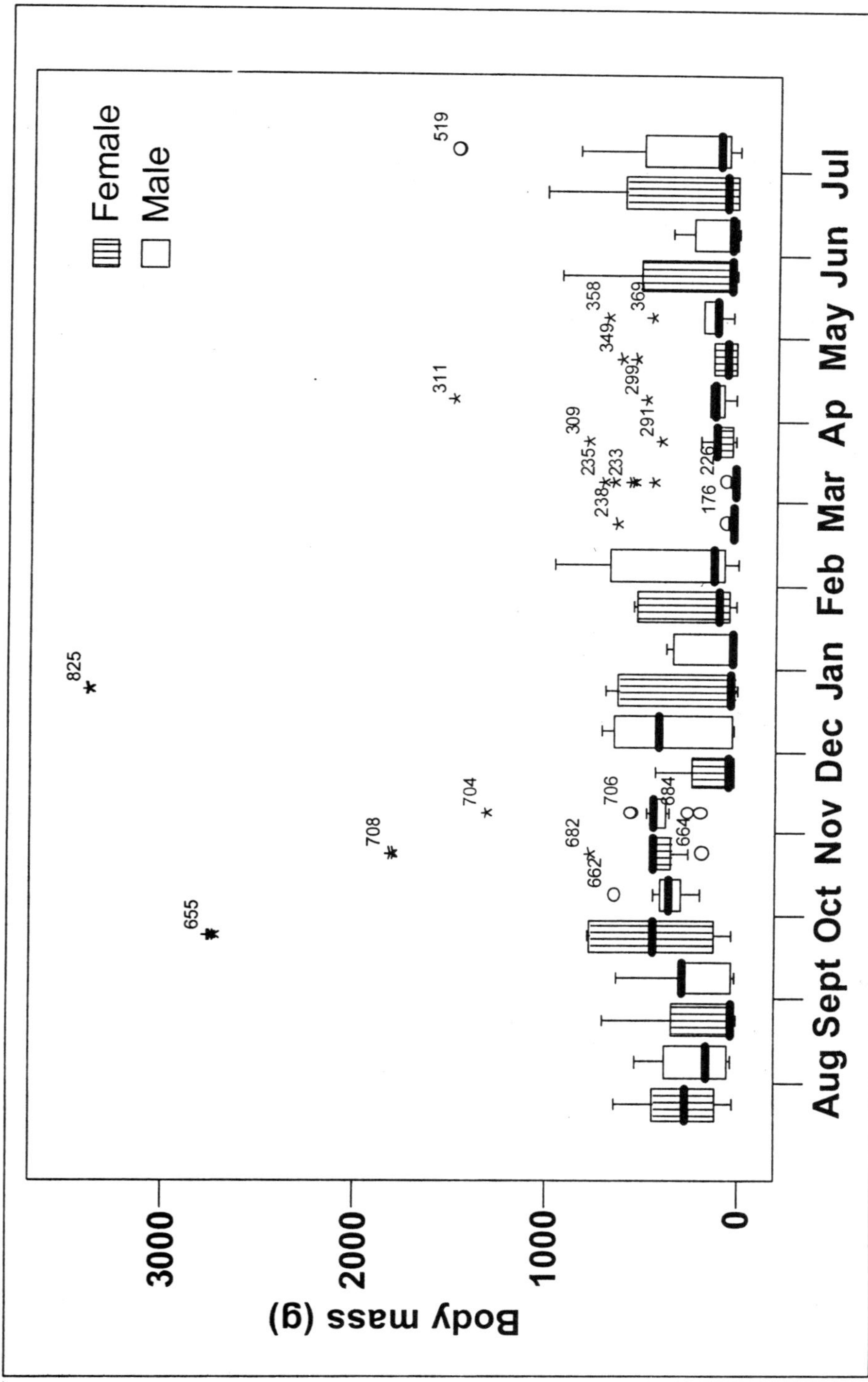

Figure 1.3: Monthly Variation in Amplitude of Total Body Mass of Males and Females of *S. brasiliensis* during September 2005 to August 2006

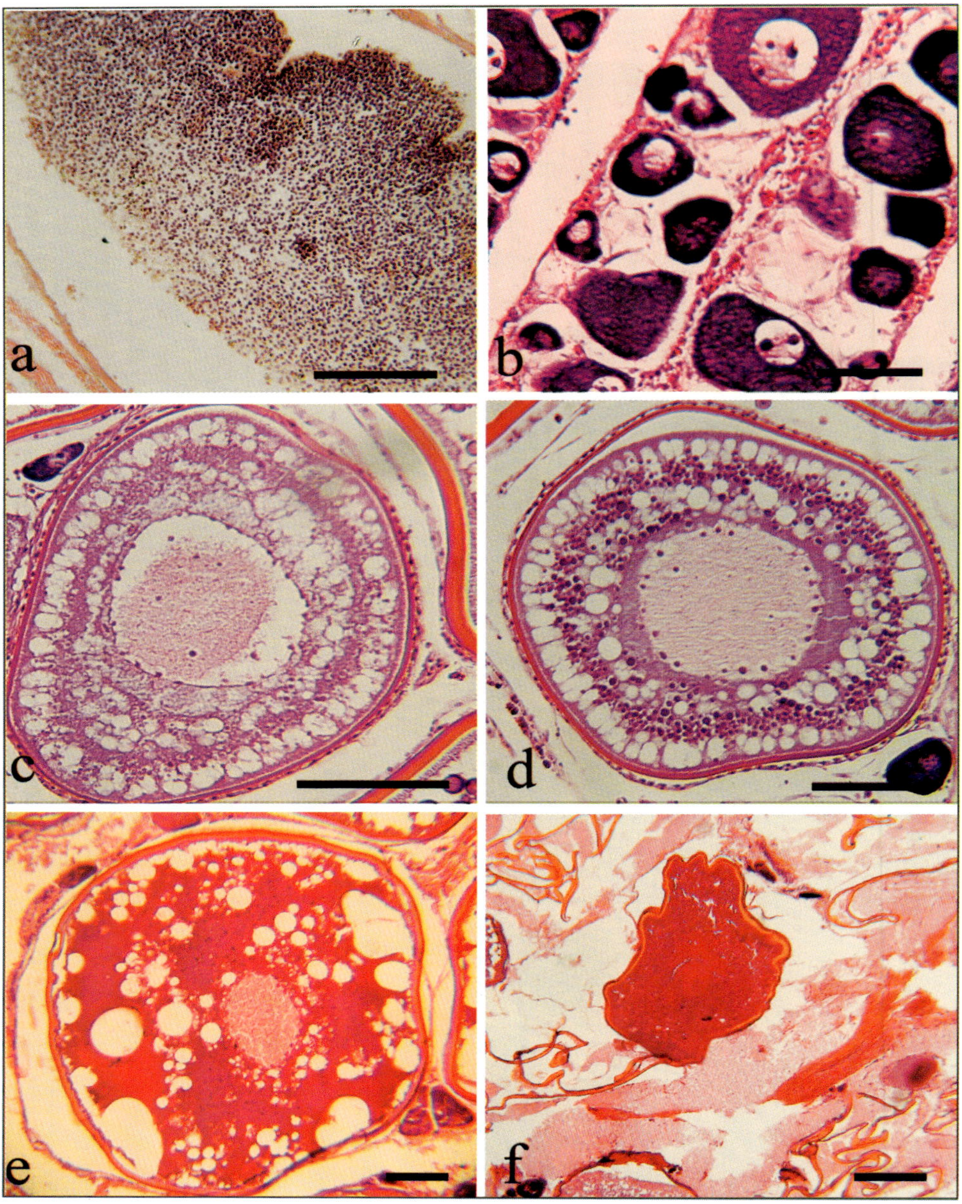

Figure 1.5: Histological Aspects of Oocyte Stages in Ovarian Development of *S. brasiliensis*

(a) Nest of oogonia; (b) Chromatin nucleolus stage and early perinucleolar stage oocytes; (c) Oocyte in yolk vesicle stage; (d) Oocyte with yolk granules and oil vesicles stage; (e) Mature oocyte; (f) Oocyte in the process of atresia in a spent ovary (Scale bar =100 μm)

Table 1.2: Macroscopic and Histological Classification and Descriptions of the Testicular Maturity Stages of *S. brasiliensis*

Stage	*Macroscopic Description*	*Histological Description*
Immature	Testes small, extremely thin and translucent (Figure 1.4a).	Groups of germinative cells (spermatogonia) inside the testes, with basophylic nucleus and reduced cytoplasm.
Developing	Testes are lobed, whitish with blood vessels appearing in the periphery (Figure 1.4b).	Cysts of spermatocytes and spermatids in the tubes and central canal of the testes.
Mature	Testes large and turgid, whitish with presence of blood vessels (Figure 1.4c).	Testes with plenty of spermatozoa in the tubes and central canal.
Spent and Resting	Testes flacid, light brown with blood vessels (Figure 1.4d).	Testes with residual spermatozoa in the central canal.

Gonadosomatic Index (GSI)

Mean GSI for grouped sex was 0.25 per cent, which varied from 0.01 per cent to 7.14 per cent (± 0.66). GSI of males varied from 0.01 to 2.2 (0.16 ± 0.25) and females from 0.02 to 7.14 (0.34 ± 0.91). Monthly values of GSI of developing, mature and spent females showed a period of peak reproductive activity during March-June. In April the adult females (n =4) had GSI varying from 0.14 to 3.24 (1.69 ± 1.78) and in June (n = 16) from 0.08 to 4.14 (1.10 ± 1.81). GSI of adult males in February (n =24) varied from 0.11 to 1.18 (0.51 ± 0.40) and in July (n =15) from 0.10 to 1.16 (0.46 ± 0.44) (Figure 1.6). Reproductive activity occurred during rainy season which lasted from February to August. During the rainy period GSI of males varied from 0.01 to 2.2 (0.16 ± 0.25) and that of females from 0.02 to 4.14 (0.27 ± 0.67). During drought GSI of males varied from 0.01 to 1.36 (0.16 ± 0.18), and that of females from 0.02 to 7.14 (0.46 ± 1.19). The highest GSI of females registered was in November (7.14) due to the presence of only one mature individual with high GSI. The differences in the median values of GSI did not show significant difference (P >0,05).

Fecundity and Breeding

Mature females had a total body length of 712.7 mm (± 117.3) and weighed 2476.9 g (± 1043.82), with ovaries on an average weighing 71.3 g (± 36.8). Total fecundity was 871,523 mature eggs and relative fecundity was 952 mature eggs g^{-1}. Mature ovaries showed the reserve stock of perinucleolar stage oocytes (phase II) with diameter size <120 µm and mature oocytes with diameters ranging from 600 µm to 750 µm. Relative frequency distributions of the oocyte diameter sizes indicate that *S. brasiliensis* is a total spawner. Evidence of spawning was found in the histologically processed ovaries. The period of peak reproductive activity occurred during March–June, coinciding with the rainy season which lasted from February to August.

Rainfall

During the study period rainfall varied from 1.2 mm in November, 2005 to 427.9 mm in April, 2006 (138.9 ± 135.9). During this period northeastern Brazil experienced

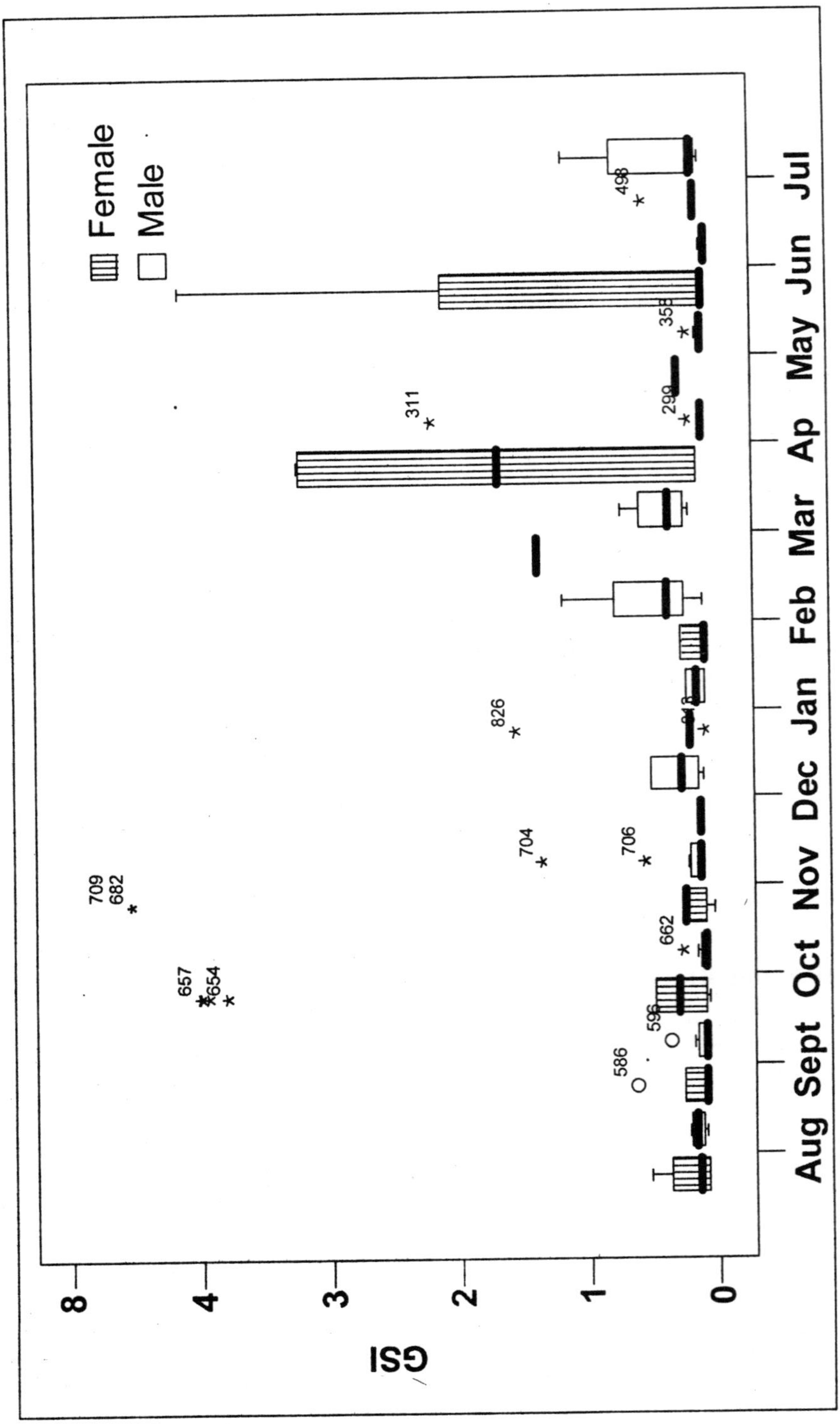

Figure 1.6: Monthly Variation in GSI in Females and Males of *S. brasiliensis* during September 2005 to August 2006

5 months of drought, from September to January with rainfall ranging from 1.2 mm to 43.9 mm (21.3 ± 17.4), and 7 months of rain, from February to August ranging from 87.2 mm to 427.9 mm (206.7 ± 128.1).

Discussion

The present study documents the sex ratio, onset of sexual maturity, changes in the reproductive activity, fecundity and breeding in *S. brasiliensis* occurring in the coastal waters of Northeastern Brazil.

Sex ratio and size structure constitute basic information in assessing reproductive potential and estimating stock size in fish populations. A balanced (1 male to 1 female) sex ratio was observed for *S. brasiliensis* with no significant temporal differences. Regional differences in fishing gear can affect catch and probably more females can be caught due to the type of fishing gear used, thus resulting in biased sex ratios as have been observed by several workers. For *S. maculatus* captured in Venezuela, Franco (1992) observed a sex composition of 54.9 per cent of males and 44.1 per cent of females. In the Maranhão coast of Brazil, a significant difference in sex ratio for *S. brasiliensis* was recorded (Silva *et al.*, 2005), where females were predominant and outnumbered the males (1M:2F).

Fish population size structure is very important to assess the stock. This study establishes that females of *S. brasiliensis* are slightly bigger and heavier than males, in accordance with previous works carried out on this species (Fonteles-Filho, 1988; Franco, 1992). Growth of females most probably reflect the reproductive cycle, and as the ovaries mature they increase in weight and hence in total weight of fish. Differences in growth patterns of males and females can thus provide a mechanism for an adaptive phenotypic response to changes in tropical coastal environments (Araújo and Chellappa, 2002).

The onset of sexual maturity represents a critical transition in the life history, since resource allocation is related mainly to growth before and to reproduction after the sexual maturity (Potts and Wootton, 1984). Thus, size at maturity of males and females is an important reproductive characteristic of fishes (Luksenburg and Pedersen, 2002; Chellappa *et al.*, 2005; Murua and Motos, 2006). In the current study, the onset of sexual maturity in males *S. brasiliensis* was earlier than in females, as development of testes occurred at a smaller body size than for ovaries. The males matured earlier than the females both during the rainy and drought periods. All males were mature at 440 and females at 520 mm and all fish below 200 mm were immature, in accordance with works conducted with on this species (Gesteira and Mesquita, 1976; Franco, 1992). Fish can achieve sexual maturity at sizes which are small in comparison with those found in other populations of the same species due to phenotypic plasticity, which allows them to respond adaptively to the environmental change. Detailed information on size at sexual and maturity gonad development of fish permits the calculation of minimum size at capture, in order not to deplete the breeding stock.

The present study establishes four macroscopic stages for males and females and five stages of ovarian follicle development for females of S. *brasiliensis*. The macroscopic classification and histological analyses of the ovaries and testes suggest that there is a regular pattern of gonadal development for each maturity stage, the immature, developing, mature and spent stages. Dias *et al.* (1998) discussed the possible omissions of the various gonadal developmental stages due to macroscopic analyses and high-lighted the importance of histological studies involving microscopic observations in order to correct the same. Macroscopic analyses of gonads enable only gross information and may involve errors and omissions, whereas, histological studies have revealed various ovarian developmental stages for fishes, such as, in blue warehou, *Seriolella brama* (Günther) (Knuckey and Sivakumaran, 2001), flying fish, *Hirundichthys affinis* (Günther) (Araújo and Chellappa, 2002) and Argentine hake, *Merluccius hubbsi* Marini (Honji *et al.*, 2006).

Fecundity plays a key role as both a critical parameter of stock assessment based on egg production and as a fundamental aspect of fish population dynamics. Fecundity in *S. brasiliensis* show a decreasing tendency since 1972, from a total fecundity of 2,047,000 eggs and relative fecundity of 1,892 eggs female g^{-1} (Gesteira, 1972) to a total fecundity of 871,523 mature eggs and relative fecundity of 952 eggs female g^{-1} in the present study. The following reasons can possibly be contributing to this situation. First, the indiscriminate harvesting of *S. brasiliensis* populations in the coastal waters of northeastern Brazil is possibly leading to a situation where the number of fish that reach maturity is being progressively reduced, thereby lowering their reproductive capacity. Second, the decrease in fecundity over the period 1972-2006 in *S. brasiliensis* can indicate a decrease in reproductive effort among spawning females during the latter period. This also suggests that the decrease in fecundity was primarily due to a decrease in the number of eggs g^{-1} in the ovaries. *S. brasiliensis* is a total spawner releasing all of mature oocytes at the same time, on the other hand multiple spawners are characterised by the temporal pattern of ovarian stages, with the release of mature oocytes in batches as in *Cichla monoculus* Spix and Agassiz (Chellappa *et al.*, 2003).

The reproductive process, such as gonad maturation, in tropical fishes is influenced by various environmental changes induced by the onset of rains. Drought is a natural climatic situation which is characteristic of northeastern Brazil, with irregular distribution of rain in the region (Serhid, 2006). The spawning characteristics of males and females of *S. brasiliensis* showed that the peak period of reproductive activity was during March-June coinciding with the rainy season and very few spawning fish were found outside these months. The macroscopic and histological analysis of gonads equally confirmed that breeding season commenced with the onset of rains.

It is interesting to observe that the peak breeding season of *S. brasiliensis* in northeastern Brazil takes place in a sequential manner, March-June in the coastal areas of Natal, September in the littoral of Ceará and October in Maranhão coastal waters (Gesteira and Mesquita, 1976; Batista and Fabré (2001). A similar trend was observed in Australia for Spanish mackerel *S. commerson* Lacepède (Mackie *et al.*, 2005). These probably reflect their migration for feeding and reproductive purposes. In tropical regions the rainfall plays an important role in determining the reproductive cycles of fishes and collective reproduction occurs during the time when environmental conditions are favorable for the survival of juvenile forms and when adequate food is available, besides protection from predators. Monthly variations of GSI of *S. brasiliensis* showed that peak breeding coincides with rainy season and possibly better environmental conditions.

Conservation of fish stocks in their natural habitat are usually endangered by abusive fishing of immature fishes which have not yet completed their reproductive cycle, as recruitment via reproduction is the means by which the resource is renewed (Lucena *et al.*, 2004). The traditional fishing communities depend on small scale artisanal fishery, which reflects their way of making a living and sustain their lifestyle. Though it is important to preserve this, it is also vital to programme the sustainability and conservation of the fisheries resource. In the predatory fishing technique of beach seine nets with small mesh size are used, in order to catch shrimps, which accounts for a large by-catch of small sized immature *S. brasiliensis*. Measures should be taken to regulate this fishery in order to conserve this valuable fishery resource.

Acknowledgments

The authors wish to thank the National Council for Scientific and Technological Development of Brazil (CNPq) for the financial support awarded during the study period (J. T. A. Ximenes de Lima, Grant n°. 141651/2005-9) and for the Research grants awarded (S. Chellappa, Grant n°. 307497/2006-2, A. Araújo, Grant n°. 302012/2006-0 and N.T. Chellappa, Proc. No. 306274/2003-5).

References

Araújo, A.S. and Chellappa, S., 2002. Estratégia reprodutiva do peixe voador, *Hirundichthys affinis* Günther (Osteichthyes, Exocoetidae). *Revista Brasileira de Zoologia*, 19(3): 691–703.

Batista, V.S. and Fabré, N.N., 2001. Temporal and spatial patterns on serra, *Scomberomorus brasiliensis* (Teleostei, Scombridae) catches from the fisheries on the Maranhão coast. Brazil. *Brazilian Journal of Biology*, 61(4): 541–546.

Carpenter, K.E., 2002. The living marine resources of the Western Central Atlantic. Volume 2: Bony fishes. Part 1 (Acipenseridae to Grammatidae). *FAO Species Identification Guide for Fishery Purposes* and *American Society of Ichthyologists and Herpetologists Special Publication*, FAO, Rome, 5: 601–1374.

Chellappa, S., Câmara, M.R., Chellappa, N.T., Beveridge, M.C.M. and Huntingford, F.A., 2003. Reproductive ecology of a neotropical cichlid fish, *Cichla monoculus* (Osteichthyes, Cichlidae). *Brazilian Journal of Biology*, 63(1): 17–26.

Chellappa, S., Câmara, M.R. and Verani, J.R., 2005. Ovarian development in the Amazonian red discus, *Symphysodon discus* Heckel (Osteichthyes, Cichlidae). *Brazilian Journal of Biology*, 65(4): 609–616.

Collette, B.B., Russo, J.L. and Zavala-Camin, L.A., 1978. *Scomberomorus brasiliensis*, a new species of Spanish mackerel from the western Atlantic. *Fishery Bulletin*, 76(1): 273–280.

De Martini, E.E. and Fountain, R., 1981. Ovarian cycling frequency and batch fecundity in the queenfish, *Seriphus politus:* Attributes representative of serial spawning fishes. *Fisheries Bulletin*, 79(3): 547–559.

Dias, J.F., Peres-Rios, E., Chaues, P.T.C. and Rossi-Wongtschowski, C.L.D.B., 1998. Análise macroscópica dos ovários de teleósteos: problemas de classificação e recomendações de procedimentos. *Revista Brasileira de Biologia, Brazil*, 58(1): 55–69.

Fonteles-Filho, A.A., 1988. Sinopse de informações sobre a Cavala, *Scomberomorus cavalla* (Cuvier) e a serra, *Scomberomorus brasiliensis* Collette, Russo and Zavala-Camin (Pisces: Scombridae), no Estado do Ceará, Brasil. *Arquivos de Ciências do Mar, Brazil*, 27: 21–48.

Fonteles-Filho, A.A., 1989. *Recursos pesqueiros: biologia e dinâmica da população*. Imprensa Oficial do Ceará: Fortaleza, Brazil, 296 p.

Franco, L., 1992. Maduración sexual y fecundidad Del Carite (*Scomberomorus maculatus*) de las costas Del Estado Falcón, Venezuela. *Zootecnia Tropical*, 10 (2): 157–169.

Gesteira, T.C.V., 1972. Sobre a reprodução e fecundidade da serra, *Scomberomorus maculatus* (Mitchill), no Estado do Ceará. *Arquivos de Ciências do Mar*. 12(2): 117–122.

Gesteira, T.C.V. and Mesquita, A.L.L., 1976. Época de reprodução, tamanho e idade na primeira desova da cavala e da serra, na costa do Estado do Ceará (Brasil). *Arquivos de Ciências do Mar, Brazil*, 16(2): 83–86.

Honji, R.M., Vaz-Dos-Santos, A.M. and Rossi-Wongtschowski, C.L.D.B., 2006. Identification of the stages of ovarian maturation of the Argentine hake *Merluccius hubbsi* Marini, 1993 (Teleostei: Merlucciidae): advantages and disadvantages of the use of the macroscopic and microscopic scales. *Neotropical Icthyology*, 4(3): 329–337.

Knuckey, I.A. and Sivakumaran, K.P., 2001. Reproductive characteristics and per–recruit analyses of blue warehou (*Seriolella brama*): Implications for the South East fishery of Australia. *Marine and Freshwater Research, Australia*, 52: 575–587.

Lima, J.T.A.X., Chellappa, S. and Thatcher, V.E., 2005. *Livoneca redmanni* Leach (Isopoda, Cymothoidae) and *Rocinela signata* Schioedte and Meinert (Isopoda, Aegidae), ectoparasites of *Scomberomorus brasiliensis* Collette, Russo and Zavala–Camin (Ostheichthyes, Scombridae) in Rio Grande do Norte, Brazil. *Revista Brasileira de Zoologia*, 22(4): 1104–1108.

Lucena, F., Lessa, R., Kobayashi, R. and Quiorato, A.L., 2004. Aspectos biológico-peaqueiros da serra, *Scomberomorus brasiliensis*, capturada com rede-de-espera no nordeste do Brasil. *Arquivos de Ciências do Mar, Brazil*, 37: 99–104.

Luksenburg, J.A. and Pedersen, T., 2002. Sexual and geographical variation in life history parameters of the shorthorn sculpin. *Jounal of Fish Biology*, 61: 1453–1464.

Mackie, M.C. and Lewis, P.D., 2001. Assessment of gonad staging systems and other methods used in the study of the reproductive biology of the narrow-barred Spanish mackerel, *Scomberomorus commerson*, in Western Australia. *Fisheries Research Report*, 136: 48.

Mackie, M.C., Lewis, P.D., Gaughan, D.J. and Newman, S.J., 2005. Variability in spawning frequency and reproductive development of the narrow-barred Spanish mackerel (*Scomberomorus commerson*) along the west coast of Australia. *Fishery Bulletin, Australia* 103: 344–354.

Murua, H. and Motos, L., 2006. Reproductive strategy and spawning activity of the European hake *Merluccius merluccius* (L.) in the Bay of Biscay. *Jounal of Fish Biology,* 69: 1288–1303.

Palmer, E.E., Sorensen, P.W. and Adelman, I.R., 1995. A histological study of seasonal ovarian development in freshwater drum in the Red Lakes, Minnesota. *Jounal of Fish Biology,* 47: 199–210.

Potts, G.W. and Wootton, R.J., 1984. *Fish Reproduction: Strategies and Tactics.* Academic Press, London, 410 p.

Serhid, R.N., 2006. *Coleção águas potiguares.* (Secretaria de Recursos Hídricos, Águas Potiguares, Rio Grande do Norte) Natal, Brazil, 89 p.

Silva, G.C., Castro, A.C.L. and Gubiani, É.A., 2005. Estrutura populacional e indicadores reprodutivos de *Scomberomorus brasiliensis* Collette, Russo e Zavala-Camin, 1978 (Perciformes: Scombridae) no litoral ocidental maranhense. *Acta Scientiarum,* Maringá, 27(4): 383–389.

Vazzoler, A.E.A.M., 1996. *Biologia de reprodução de peixes teleósteos: teoria e prática.* EDUEM, Maringá, Brazil, 169 p.

West, G., 1990. Methods of assessing ovarian development in fishes: A review. *Australian Journal of Marine and Freshwater Research,* 41: 199–222.

Wootton, R.J., Evans, G.W. and Mills, L.A., 1978. Annual cycle in female three-spined sticklebacks (*Gasterosteus aculeatus* L.) from an upland and lowland population. *Jounal of Fish Biology,* 12: 331–343.

Yoshida, H.O., 1964. Skipjack tuna spawning in the Marquesas Islands and Tuamotu Archipelago. *Fisheries Bulletin,* 65(2): 479–488.

Chapter 2

Anomalies in *Cirrhinus mrigala* (Ham. Buch.), *Catla catla* (Ham. Buch.) and *Labeo rohita* (Ham. Buch.) Inhabiting Freshwater Environments of Jammu (J&K)

☆ *Subash C. Gupta and Touseef A. Zargar*

Abstract

Three deformed specimen of *Cirrhinus mrigala* (Ham. Buch.), showing anomalies like truncated post dorsal body with ventral bulging in anal region; absence of caudal region, and attenuate post dorsal body with a hump in the dorsal fin region; one specimen of *Catla catla* (Ham. Buch.) having truncated post dorsal region, with short scales in the anal region and one specimen of *Labeo rohita* (Ham. Buch.) showing truncated body, with post dorsal depression and spiral like vertebral column, were seen in the samples collected from fish ponds at Birpur, Jammu have been described.

Introduction

Fish anomalies under natural conditions are not uncommon and have been well documented by various workers from India and abroad. Some earlier reports on fish teratology from Jammu and Kashmir are those of Dutta and Malhotra (1984, 86), Dutta *et al.* (1994, 1995a,b, 1997, 1999, 2005), Ara (2000), and Gupta *et al.* (1998, 2000, 2002, 2004). Except for this scanty information on fish teratology, there is no record of any anomaly in fish from lotic and lentic waters of Jammu region of J&K. So the present communication is an addition to the existing knowledge of fish abnormalities from this part of the country and forms an important aspect of study of fish biology.

Materials and Methods

Three abnormal specimens of *Cirrhinus mrigala*, one specimen each of *Catla catla* and *Labeo rohita* were seen in the collections made from Birpur ponds, Jammu. The abnormal fish specimens were measured for their length and weight and examined for various morphological deformities. The fishes were then preserved in 10 per cent formaldehyde solution and were brought to the laboratory for further detailed studies. For the study of vertebral deformities, the fishes were exposed to X-ray radiations.

Observations

Deformities as observed in fishes like *Cirrhinus mrigala, Catla catla* and *Labeo rohita* noticed in the collections made from Birpur ponds, Jammu, include:

Cirrhinus mrigala (Ham. Buch.)

Truncated Post-dorsal Body with Ventral Bulging in the Anal Region

A single aberrant specimen of *Cirrhinus mrigala*, measuring 23.6 cm and weighing 175 gms, was recognized from its truncated post-dorsal body and a ventral bulging in the anal region.

Body of the normal fish (Figure 2.1) is streamlined, dorsal fin origin is more towards the snout than the caudal region. Longest pectoral ray falls short of pelvic fin, longest pelvic ray falls short of anal fin and longest anal fin ray falls short of bilobed caudal fin. In the aberrant specimen, post-dorsal body is truncated. Like the normal fish, dorsal fin origin in the abnormal fish (Figure 2.2) is more towards the snout than the caudal region. The longest pectoral ray falls short of pelvic fin, longest pelvic ray falls short of anal fin, and the longest anal ray reaches upto caudal peduncle due to shortening of post dorsal body. In the aberrant specimen, vertebral column forms a sort of spiral structure.

Comparison of morphological features of normal specimen and abnormal specimen of *Cirrhinus mrigala* has shown certain variations. The number of fin rays in dorsal, pectoral, pelvic, anal and caudal fin in the normal fish are 6, 15, 9, 8, 15 whereas in the abnormal fish, the number of fin rays are 6, 14, 8, 8 and 15 respectively.

An X-ray analysis has revealed the presence of 38 vertebrae in the vertebral column of normal specimen (Figure 2.3) and 33 vertebrae in the aberrant specimen (Figure 2.4). Various aberrations in the vertebral column of this deformed fish are:

1. 1st to 7th vertebrae are loose with normal intervertebral spaces.
2. 8th to 10th vertebrae are tightly fused together.
3. 11th to 14th vertebrae are tightly fused and bent upwards forming a sort of depression in the pelvic region.
4. 15th to 17th vertebrae are again compact and fused together, have deformed or no haemal spines and forms a sort of diffused spiral from 8th to 17th vertebrae with diffused shapes and structures.

5. The vertebral column from 18th vertebrae shows slight downward bend upto 25th vertebrae.
6. 26th to 38th vertebrae form a sort of spiral shaped structure ending into the urostyle. This results in the reduction in the post anal length of the body.

The haemal and neural spines are normal throughout the vertebral column.

The various body ratios like: head length in total body length, head length in standard body length, eye diameter in head length, head height in head length, pre-ocular length in head length, body height in total body length, body height in standard body length, pre-dorsal length in total body length, pre-dorsal length in standard body length, post-dorsal length in total body length, post-dorsal length in standard body length, pre-anal length in total body length, pre-anal length in standard body length, post-anal length in total body length, post-anal length in standard body length, in normal and abnormal fish are (Table 2.1):

5.53, 4.47, 5.7, 1.27, 3.8, 4.63, 3.75, 2.62, 2.12, 1.61, 1.31, 1.58, 1.28, 2.71, 2.19 and 5.0, 4.0, 5.2, 1.1, 3.6, 3.8, 3.0, 2.5, 2.0, 3.8, 3.0, 1.6, 1.3, 4.3, 3.4 respectively.

Table 2.1: Comparison of Morphological Features of Normal and Abnormal Specimen of *Cirrhinus mrigala* (Ham. Buch.)

Sl.No.	Body Ratios	Normal Fish	Deformed Fish
1.	Head length in total body length	5.53	5.0
2.	Head length in standard body length	4.47	4.0
3.	Eye diameter in head length	5.7	5.2
4.	Head height in head length	1.27	1.1
5.	Pre-ocular length in head length	3.8	3.6
6.	Body height in total body length	4.63	3.8
7.	Body height in standard body length	3.75	3.0
8.	Pre-dorsal length in total body length	2.62	2.5
9.	Pre-dorsal length in standard body length	2.12	2.0
10.	Post-dorsal length in total body length	1.61	3.8
11.	Post-dorsal length in standard body length	1.31	3.0
12.	Pre-anal length in total body length	1.58	1.6
13.	Pre-anal length in standard body length	1.28	1.3
14.	Post-anal length in total body length	2.71	4.3
15.	Post-anal length in standard body length	2.19	3.4

Absence of Caudal Region

This single aberrant specimen of *Cirrhinus mrigala* measuring 28.4 cm and weighing 825 gms was identified from the absence of its caudal region including caudal fin. In the aberrant specimen (Figure 2.5), post-dorsal body is truncated and there is complete absence of post-anal region including caudal peduncle.

Figure 2.1: Normal Specimen of *Cirrhinus mrigala* (Ham. Buch.)

Figure 2.2: Abnormal Specimen of *Cirrhinus mrigala* (Ham. Buch.) showing Truncated Post Dorsal Body with Ventral Bulging in Anal Region

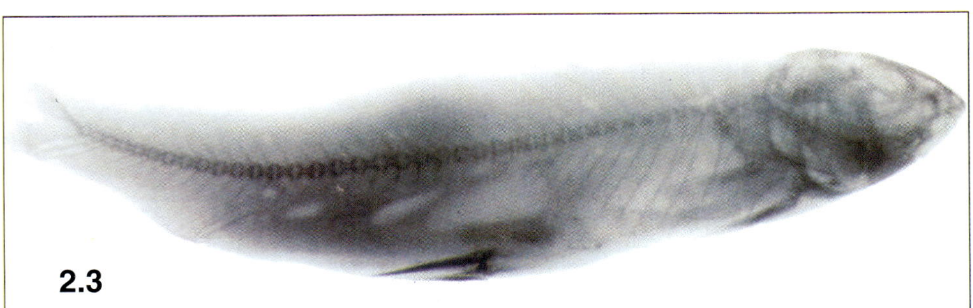

Figure 2.3: X-Ray Photograph of Normal Specimen of *Cirrhinus mrigala* (Ham. Buch.)

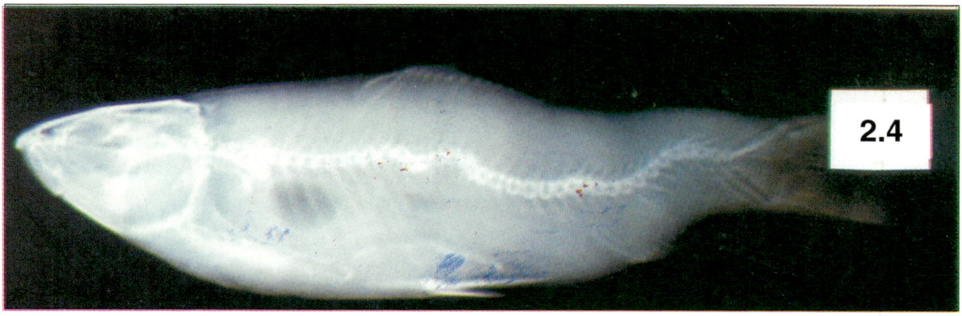

Figure 2.4: X-Ray Photograph of Abnormal Specimen of *Cirrhinus mrigala* (Ham. Buch.) showing Truncated Post Dorsal Body with Ventral Bulging in Anal Region

Figure 2.5: Abnormal Specimen of *Cirrhinus mrigala* (Ham. Buch.) showing Absence of Caudal Region

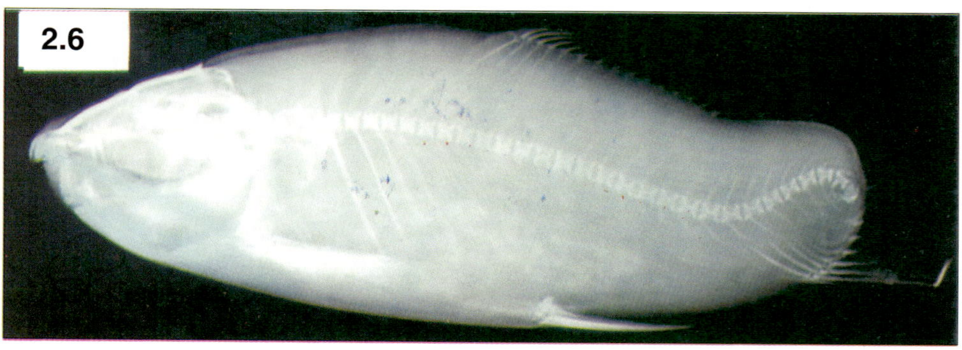

Figure 2.6: X-Ray Photograph of Abnormal Specimen of *Cirrhinus mrigala* (Ham. Buch.)

Dorsal fin is situated more towards the posterior end with longest dorsal ray extending upto the posterior end of body. Longest pectoral ray, like normal fish, falls short of pelvic fin, longest pelvic ray extends beyond the middle of pelvic fin and anal fin whereas anal fin is situated at the posterior extremity of the body. Lateral line runs in the middle of body upto post-dorsal region from where it is slightly curved upwards.

An X-ray study of the aberrant fish (Figure 2.6) has shown the presence of 28 amphicoelous vertebrae in the vertebral column. Vertebral column from 21st vertebrae onwards goes upwards forming a sort of depression ending up at the last vertebra and thus resulting into slight reduction in the post-dorsal length of the body. Though neural and haemal spines are normal but they are slightly shorter in size along the anal fin region.

The various body ratios like: head length in total body length, head length in standard body length, eye diameter in head length, head height in head length, pre-ocular length in head length, body height in total body length, body height in standard body length, pre-dorsal length in total body length, pre-dorsal length in standard body length, post-dorsal length in total body length, post-dorsal length in standard body length, pre-anal length in total body length, pre-anal length in standard body length, post-anal length in total body length, post-anal length in standard body length, in the normal and abnormal fish are (Table 2.2):

5.53, 4.47, 5.7, 1.27, 3.8, 4.63, 3.75, 2.62, 2.12, 1.61, 1.31, 1.58, 1.28, 2.71, 2.19 and 5.1, 4.0, 5.5, 1.0, 3.6, 4, 3.1, 2.3, 1.8, 4.3, 3.4, 1.8, 1.4, 4.6, 3.6 respectively.

Table 2.2: Comparison of Morphological Features of Normal and Abnormal Specimen of *Cirrhinus mrigala* (Ham. Buch.)

Sl.No.	*Body Ratios*	*Normal Fish*	*Deformed Fish*
1.	Head length in total body length	5.53	5.1
2.	Head length in standard body length	4.47	4.0
3.	Eye diameter in head length	5.7	5.5
4.	Head height in head length	1.27	1.0
5.	Pre-ocular length in head length	3.8	3.6
6.	Body height in total body length	4.63	4.0
7.	Body height in standard body length	3.75	3.1
8.	Pre-dorsal length in total body length	2.62	2.3
9.	Pre-dorsal length in standard body length	2.12	1.8
10.	Post-dorsal length in total body length	1.61	4.3
11.	Post-dorsal length in standard body length	1.31	3.4
12.	Pre-anal length in total body length	1.58	1.8
13.	Pre-anal length in standard body length	1.28	1.4
14.	Post-anal length in total body length	2.71	4.6
15.	Post-anal length in standard body length	2.19	3.6

Truncated Body with a Hump in the Dorsal Fin Region and Attenuate Post-dorsal Region

This single aberrant specimen of *Cirrhinus mrigala* measuring 25 cm and weighing 198 gms was recognized from a hump in the dorsal fin region and attenuate post-dorsal body (Figure 2.7). The dorsal fin originates in the middle of body, like normal fish. Again, like normal fish, the longest pectoral ray falls short of pelvic fin, the longest pelvic ray falls short of anal fin whereas the longest anal ray reaches upto the caudal peduncle.

An X-ray analysis of the aberrant specimen (Figure 2.8) has shown the presence of 38 vertebrae in the vertebral column. Various aberrations as studied are: 1st to 15th vertebrae are normal with normal intervertebral spaces and thickness. Neural spines though normal from 10th to vertebrae but haemal spines are not differentiated, 16th to 20th vertebrae are fused and form a compact and undifferentiated mass, 20th to 25th vertebrae bent downwards and then go straight forming a sort of arc ending up in the urostyle.

The various body ratios like: head length in total body length, head length in standard body length, eye diameter in head length, head height in head length, pre-ocular length in head length, body height in total body length, body height in standard body length, pre-dorsal length in total body length, pre-dorsal length in standard body length, post-dorsal length in total body length, post-dorsal length in standard body length, pre-anal length in total body length, pre-anal length in standard body length, post-anal length in total body length, post-anal length in standard body length in the normal and abnormal fish are (Table 2.3):

Table 2.3: Comparison of Morphological Features of Normal and Abnormal Specimen of *Cirrhinus mrigala* (Ham. Buch.)

Sl.No.	Body Ratios	Normal Fish	Deformed Fish
1.	Head length in total body length	5.53	5.2
2.	Head length in standard body length	4.47	4.1
3.	Eye diameter in head length	5.7	4.8
4.	Head height in head length	1.27	1.1
5.	Pre-ocular length in head length	3.8	3.0
6.	Body height in total body length	4.63	4.1
7.	Body height in standard body length	3.75	3.3
8.	Pre-dorsal length in total body length	2.62	2.6
9.	Pre-dorsal length in standard body length	2.12	2.1
10.	Post-dorsal length in total body length	1.61	3.3
11.	Post-dorsal length in standard body length	1.31	2.68
12.	Pre-anal length in total body length	1.58	1.6
13.	Pre-anal length in standard body length	1.28	1.3
14.	Post-anal length in total body length	2.71	3.9
15.	Post-anal length in standard body length	2.19	3.1

5.53, 4.47, 5.7, 1.27, 3.8, 4.63, 3.75, 2.62, 2.12, 1.61, 1.31, 1.58, 1.28, 2.71, 2.19 and 5.2, 4.1, 4.8, 1.1, 3.0, 4.1, 3.3, 2.6, 2.1, 3.3, 2.68, 1.6, 1.3, 3.9, 3.1 respectively.

Catla catla (Ham. Buch.)

Truncated Post Dorsal Body with Short Scales in the Anal Region

A single aberrant specimen of *Catla catla* measuring 25.5 cm and weighing 455 gms was recognized from its truncated post-dorsal body and short scales in the anal region.

Body in the normal fish (Figure 2.9) is streamlined. Dorsal fin origin is more towards the snout than the tip of the caudal fin, longest pectoral ray falls short of pelvic fin, longest pelvic ray falls short of anal fin and the longest anal ray falls short of caudal fin. Caudal fin is bilobed. In the abnormal specimen on the other hand (Figure 2.10), post-dorsally, the body is truncated. Unlike the normal fish, dorsal fin originates in the middle of the body, longest pectoral ray extends beyond the middle of the pectoral fin. Like normal fish, the longest pelvic ray falls short of anal fin and the longest anal ray reaches upto caudal fin base.

Table 2.4: Comparison of Morphological Features of Normal and Abnormal Specimen of *Catla catla* (Ham. Buch.)

Sl.No.	*Body Ratios*	*Normal Fish*	*Deformed Fish*
1.	Head length in total body length	5.30	3.1
2.	Head length in standard body length	4.42	2.5
3.	Eye diameter in head length	5.36	7.2
4.	Head height in head length	1.47	1.1
5.	Pre-ocular length in head length	3.10	3.4
6.	Body height in total body length	4.53	2.4
7.	Body height in standard body length	3.78	1.9
8.	Pre-dorsal length in total body length	2.52	2.3
9.	Pre-dorsal length in standard body length	2.10	1.8
10.	Post-dorsal length in total body length	2.18	4.6
11.	Post-dorsal length in standard body length	1.82	3.6
12.	Pre-anal length in total body length	1.63	1.7
13.	Pre-anal length in standard body length	1.35	1.3
14.	Post-anal length in total body length	2.58	2.4
15.	Post-anal length in standard body length	2.15	1.9

X-ray examination has revealed the presence of 37 vertebrae in the vertebral column of the normal specimen (Figure 2.11), and 34 vertebrae in the aberrant specimen (Figure 2.12). Various aberrations in the vertebral column of this deformed fish are:

1. 1st and 2nd vertebrae are fused together with diffused vertebral thickness and intervertebral spaces.

2. 5th to 11th, 13 to 15th, vertebrae are fused together forming a compact mass.
3. 16th to 22nd vertebrae are small with narrow intervertebral spaces and reduced vertebral thickness.
4. 23rd to 26th vertebrae again fused together forming a compact mass.

All these deformities form slight depressions in the pelvic and anal regions, thereby, contributing towards the formation of a spiral like vertebral column.

The various body ratios like: head length in total body length, head length in standard body length, eye diameter in head length, head height in head length, pre-ocular length in head length, body height in total body length, body height in standard body length, pre-dorsal length in total body length, pre-dorsal length in standard body length, post-dorsal length in total body length, post-dorsal length in standard body length, pre-anal length in total body length, pre-anal length in standard body length, post-anal length in total body length, post-anal length in standard body length, in the normal and abnormal fish are (Table 2.4):

5.30, 4.42, 5.36, 1.47, 3.10, 4.53, 3.78, 2.52, 2.10, 2.15, 1.82, 1.63, 1.35, 2.58, 2.15 and 3.1, 2.5, 7.2, 1.1, 3.4, 2.4, 1.9, 2.3, 1.8, 4.6, 3.6, 1.7, 1.3, 2.4, 1.9 respectively.

Labeo rohita (Ham. Buch.)

Truncated Body with a Post-dorsal Depression and Spiral Like Vertebral Column

A single aberrant specimen of *Labeo rohita* measuring 28 cm and weighing 400 gms was recognized from its truncated body, with a post-dorsal depression and reduced scales in anal region.

Body in the normal fish (Figure 2.13) is streamlined. Dorsal fin origin is more towards the anterior side of body. Longest pectoral ray falls short of pelvic fin, longest pelvic ray falls short of anal fin, anal fin originates posterior to the last dorsal ray and falls short of bilobed caudal fin. In the aberrant specimen on the other hand (Figure 2.14), body is truncated post-dorsally. Unlike the normal fish, dorsal fin originates in the middle of the body. Like the normal fish, the longest pectoral ray falls short of pelvic fin and the longest pelvic ray falls short of anal fin whereas the longest anal ray reaches upto caudal peduncle.

X-ray analysis has revealed the presence of 36 vertebrae in the vertebral column of normal specimen (Figure 2.15) and 32 vertebrae in the aberrant specimen (Figure 2.16). Various aberrations in the vertebral column of this deformed fish include:

1. 11th and 12th vertebrae are tightly placed with reduced intervertebral spaces and thickness.
2. 13-16th vertebrae turns upward forming a compact and undifferentiated mass.
3. 19 and 20th vertebrae are fused.

From 21st vertebrae onwards, the vertebral column bent downwards and then again turns upward forming a trough in the anal region, thus giving an aberrant shape to the fish.

Figure 2.7: Abnormal Specimen of *Cirrhinus mrigala* (Ham. Buch.) showing Attenuate Post Dorsal Body with a Hump in Dorsal Fin Region

Figure 2.8: X-Ray Photograph of Abnormal Specimen of *Cirrhinus mrigala* (Ham. Buch.)

Figure 2.9: Normal Specimen of *Catla catla* (Ham. Buch.)

Figure 2.10: Abnormal Specimen of *Catla catla* (Ham. Buch.) showing Truncated Post Dorsal Body with Short Scales in the Anal Region

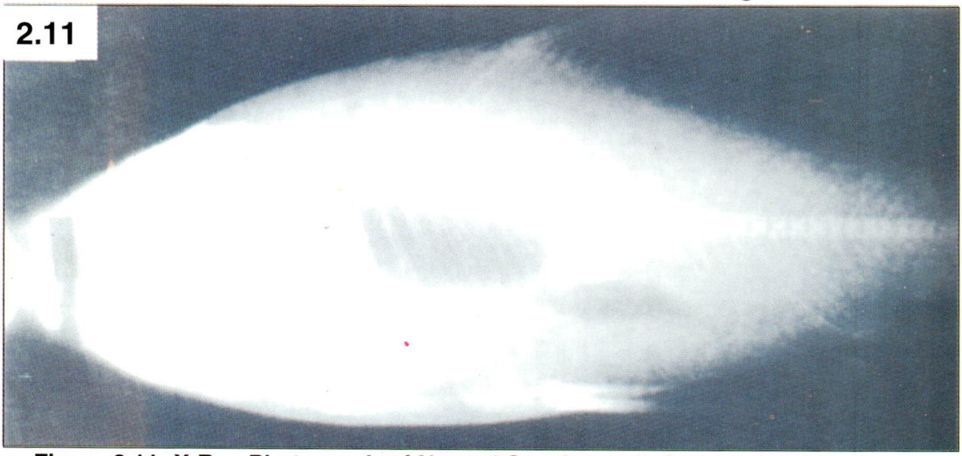

Figure 2.11: X-Ray Photograph of Normal Specimen of *Catla catla* (Ham. Buch.)

Figure 2.12: X-Ray Photograph of Abnormal Specimen of *Catla catla* (Ham. Buch.) showing Truncated Post Dorsal Body with Short Scale in the Anal Region

Figure 2.13: Normal Specimen of *Labeo rohita* (Ham. Buch.)

Figure 2.14: Abnormal Specimen of *Labeo rohita* (Ham. Buch.) showing Truncated Body with a Post Dorsal Depression and Spiral like Vertebral Column

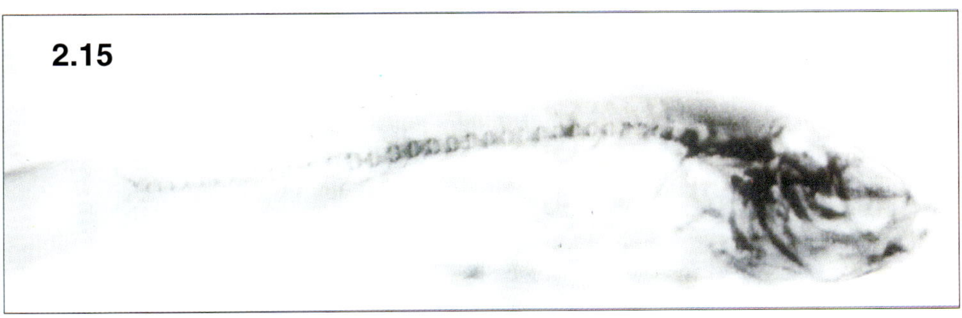

Figure 2.15: X-Ray Photograph of Normal Specimen of *Labeo rohita* (Ham. Buch.)

Figure 2.16: X-Ray Photograph of Abnormal Specimen of *Labeo rohita* (Ham. Buch.) showing Truncated Body with a Post Dorsal Depression and Spiral like Vertebral Column

The various body ratios like: head length in total body length, head length in standard body length, eye diameter in head length, head height in head length, pre-ocular length in head length, body height in total body length, body height in standard body length, pre-dorsal length in total body length, pre-dorsal length in standard body length, post-dorsal length in total body length, post-dorsal length in standard body length, pre-anal length in total body length, pre-anal length in standard body length, post-anal length in total body length, post-anal length in standard body length in the normal and abnormal fish are (Table 2.5):

4.5, 3.64, 5.4, 1.3, 2.9, 4.5, 3.54, 2.6, 2.1, 1.6, 1.3, 1.53, 1.22, 2.9, 2.3 and 4.56, 3.6, 6.1, 1.1, 2.6, 3.1, 2.5, 2.4, 1.9, 3.6, 2.9, 3.5, 2.8, 5.6, 4.5 respectively.

Table 2.5: Comparison of Morphological Features of Normal and Abnormal Specimen of *Labeo rohita* (Ham. Buch.)

Sl.No.	*Body Ratios*	*Normal Fish*	*Deformed Fish*
1.	Head length in total body length	4.5	4.56
2.	Head length in standard body length	3.64	3.6
3.	Eye diameter in head length	5.4	6.1
4.	Head height in head length	1.3	1.1
5.	Pre-ocular length in head length	2.9	2.6
6.	Body height in total body length	4.5	3.1
7.	Body height in standard body length	3.54	2.5
8.	Pre-dorsal length in total body length	2.6	2.4
9.	Pre-dorsal length in standard body length	2.1	1.9
10.	Post-dorsal length in total body length	1.6	3.6
11.	Post-dorsal length in standard body length	1.3	2.9
12.	Pre-anal length in total body length	1.53	3.5
13.	Pre-anal length in standard body length	1.22	2.8
14.	Post-anal length in total body length	2.9	5.6
15.	Post-anal length in standard body length	2.3	4.5

Discussion

Anomalies in fishes have been attributed to various factors like salinity fluctuations, parasitic infections, low concentration of dissolved O_2, radiations, U.V. radiations, dietary vitamin deficiency, hereditary factor, defective embryonic development, pollution etc.

The fish abnormalities as given above cannot be attributed to any of these factors, as the present collection sites are free from any pollution and various physico-chemical characteristics of water fall within the safe limits of fish survival. The fish abnormality as studied in one specimen of *Cirrhinus mrigala i.e.* absence of caudal region, appears to be caused by an injury or predatory effect. Fish, anomalies due to injury or predation, have also been reported by Devadoss (1983), Srivastava (1983–84), Singh, Kohli and Goswami (1986–87), Dutta and Kumar (1991) and Khan (2001), whereas in other specimens, anomalies are caused due to some developmental errors during the course of development of fish. Fish teratology due to developmental error has been described earlier by Dutta and Tilak (1962), Saxena and Tyagi (1978), Dutta (1989–90), Dutta and Malhotra (1986), Dutta *et al.* (1995 and 2005), Shekhar and Dutta (1993) and Gupta *et al.* (2000).

Thus from the above discussion, it is concluded that fish anomalies are very complex and cannot be assigned to any single factor but a complex of factors operating in the water body.

Acknowledgements

The authors are grateful to the Department of Environmental Sciences, University of Jammu, Jammu (J&K) for providing necessary facilities to undertake the present investigations.

References

Ara, R., 2000. Anomalies in fry of rainbow trout, *Salmo gairdneri gairdneri* (Richardson) from Kokernag Trout Fish Farm, Kashmir (J&K). *M.Sc. Dissertation*, University of Jammu.

Datta, Gupta, A.K. and Tilak, R., 1962. Caudal fin anomaly in *Heteropneustes fossilis. Zoological Polliniae*, 12(3): 305–308.

Devadoss, P., 1983. On some specimens of abnormal elasmobranches. *Matsya*. 9 and 10: 186–188.

Dutta, S.P.S., 1989–90. Vertebral deformity in *Barilius vagra* (Ham.) from Jammu. *Matsya*, 15–16: 166–168.

Dutta, S.P.S., Gupta, K., Gupta, S.C. and Verma, M., 1997. A report on some abnormal fishes inhabiting the aquatic environments of Jammu (J&K state). *Himalayan Journal of Environment and Zoology*, 11: 87–92.

Dutta, S.P.S., Jan, N.A. and Bali, J.P.S., 1995a. Multiple deformities in *Salmo gairdneri gairdneri* (Richardson) from Kokernag Trout Fish Farm, Kashmir (J&K). *Journal of Freshwater Biology*, 7(3): 183–186.

Dutta, S.P.S., Jan, N.A., Bali, J.P.S., Gupta, S.C. and Mahajan, A., 1999. A report on the abnormalities of fry of rainbow trout *Salmo gairdneri gairdneri* (Richardson) from Kokernag Trout Fish Farm, Kashmir (J&K). *Oriental Science*, 6(1): 9–19.

Dutta, S.P.S., Gupta, Subash C., Rathore, V. and Sharma, A., 2005. Fish fauna of some tributaries of River Ravi, District Kathua, J&K State. In: *Trends in Biodiversity and Limnology*, (Eds.) Ashwani Wanganoo and R.K. Langer. Daya Publishing House, Delhi, pp. 443–452.

Dutta, S.P.S. and Kumar, S., 1991. Deformity in dorsal fin in *Puntius conchonius* (Ham.) from Jammu. *Geobios New Reports*, 2: 173–174.

Dutta, S.P.S. and Malhotra, Y.R., 1984. An upto date checklist and a Key to identification of fishes of Jammu. *Jammu Univ. Review*, 2: 65–92.

Dutta, S.P.S. and Malhotra, Y.R., 1986. Absence of pelvic fins in *Ompok bimaculatus* (Bloch). *Geobios New Reports*, 2: 73–74.

Gupta, S.C., Dutta, S.P.S. and Verma, M., 1998. A report on abnormal specimen of *Puntius sarana* (Ham.) from river Basantar, Samba, Jammu (J&K). *Journal of Freshwater Biology*, 10(3–4): 137–140.

Gupta, S.C., Dutta, S.P.S. and Sharma, N., 2000. A report on some morphological deformities in silver carp *Hypophthalmicthyes molitrix* (Vallenciennes) inhabiting aquatic environments of Jammu (J&K state). *Himalayan Journal of Environment and Zoology*, 14: 25–30.

Gupta, S.C., Dutta, S.P.S., Sharma, N. and Bala, N., 2002. Morphological deformities in *Cirrhinus mrigala* (Ham.) inhabiting lentic environments of Jammu. *Aquacult.*, 3(2): 149–154.

Gupta, S.C., Dutta, S.P.S., Sharma, N. and Bala, N., 2004. A report on teratology of *Labeo rohita* (Hamilton): An important freshwater food fish inhabiting the lentic and lotic environments of Jammu (J&K state). In: *Advances in Fish and Wildlife Ecology and Biology*, 3: 51–48.

Khan, F.A., 2001. Teratology in some fishes *viz., Puntius sarana* (Ham. Buch.), *Puntius chola* (Ham.), *Cirrhinus mrigala* (Ham. Buch.), *Wallago attu* (Bloch and Schn.), *Channa orientalis* (Bloch and Schn.) and *Labeo dyocheilus* (Mc.Cll) inhabiting freshwater environments of Jammu. *M.Sc. Dissertation*, University of Jammu.

Shekhar, C. and Dutta, S.P.S., 1993. An abnormal specimen of *Schizothorax richardsonii* (Gray and Hard) with vertebral deformity. *Himalayan Journal of Environment and Zoology*, 7: 101–102.

Singh Kohli, M.P. and Goswami, U.C. (1986–1987). Teratological manifestation in air breathing cat fishes, *Heteropneustes fossilis*. *Matsya*, 12 and 13: 188–191.

Srivastava, S.K., 1983–1984. Caudal fin deformity in freshwater spiny eel, *Macrognathus aculeatus*. *Matsya*, 9 and 10: 189–190.

Chapter 3

Histopathological Alterations in Interrenal and Chromaffin Cells of *Channa punctatus* (Bloch) Exposed to Sublethal Concentration of Carbaryl and Cartap

☆ *D.K. Mishra, K. Bohidar and A.K. Pandey*

Abstract

In order to record the histopathological changes in interrenal and chromaffin cells, *Channa punctatus* were exposed to sublethal concentration (30 per cent of LC_{50} for 96 hours) of Carbaryl (5.20 mg l^{-1}) and Cartap (0.18 mg l^{-1}) for 24, 48, 72 and 96 hours. Interrenal as well as chromaffin cells of both the treated groups responded initially (by 24 hours) with the loss of staining affinity and hypertrophy of the endocrine cells. At 48 hours of exposure, there were significant increase in size of interrenal and chromaffin cells (hypertrophy) and increased vascularization. Hyperplasia was noticed in these cells of both the treated groups at 72 hours. Though hypertrophy and hyperplasia were prominent by 96 hours, increased vascularization and vacuolization as well as pyknotic nuclei were frequently observed in the interrenal and chromaffin cells of the murrels treated with both the pesticides.

Keywords: *Carbaryl, Cartap, Interrenal cells, Chromaffin cells, Hypertrophy, Hyperplasia, Channa punctatus.*

Introduction

Though the impacts of pesticides on various aspects of toxicology have been documented during the past, studies pertaining to the effects of carbamates on endocrine glands of the teleosts attracted attention during the recent years (Cairns *et al.*, 1984; Pandey *et al.*, 1999; Sinha *et al.*, 1991; Brown *et al.*, 2005). Carbaryl (1-naphthyl, N-methyl carbamate) and Cartap ((S,S'-[2-dimethylamino)-1, 3-propanediyl] dicarbamethioate) are the widely used carbamate pesticides in agriculture on a variety of crops for pest control (Kwak *et al.*, 2000; Bondarenko *et al.*, 2004; Anon, 2005a, b). There are instances that pollutants serve as stressors in fish and affect the hypothalamo-pituitary-interrenal (HPI) axis (Brown, 1993; Sumpter, 1996; Harvey, 1996; Hontella 1997; Ontella, 1998; Norris, 2000). Interrenal and chromaffin cells of fish correspond to the adrenal cortex and medulla of the higher vertebrates and secrete adrenocorticosteroids as well as epinephrine and norepinephrine (Chester Jone and Phillips, 1986). As nothing is known about the effects of Carbaryl and Cartap on the interrenal and chromaffin cells of teleosts, an attempt has been made to record the effects of sublethal exposure of both the pesticides on interrenal and chromaffin cells of the freshwater air-breathing teleost, *Channa punctatus*.

Materials and Methods

Healthy *Channa punctatus* (measuring 14.28±1.03 cm; weighing 41.9±8.75 g) were collected from Bhubaneswar (Orissa). They were acclimatized to the laboratory conditions for two weeks before initiation of the experiment. Fishes were fed with minced goat liver and earthworm *ad libitum* during entire period of the experiment to avoid the effect of starvation. Technical grade Carbaryl (Sevin: Bayer Crop Science, India; Batch No. SEVSS 3072) and Cartap 50 SP (Caldan: Dhanuka Pesticides Ltd., Gurgaon: Batch No. DCS/09087) were procured and dissolved in water. Static bioassays were conducted according to the method recommended by APHA (1991). From their mortality, the LC_{50} values for 24, 48, 72 and 96 hours were determined (Litchfield and Wilcoxon, 1949) which were 30, 25, 21 and 17 mg l^{-1} for Carbaryl and 2.4, 1.8, 1.6 and 0.6 mg l^{-1} for Cartap (Mishra and Bohidar, 2005). 30 per cent of LC_{50} values for 96 h of Carbaryl (5.20 mg l^{-1}) and Cartap (0.18 mg l^{-1}) were taken as sublethal concentration for the present study.

Sixty five fishes were randomly selected and divided into three equal groups. In group 1, fishes were maintained in tap water and served as control whereas in group 2 and 3, they were exposed to sublethal concentration of Carbaryl and Cartap. Fishes of all the groups were maintained in glass aquaria containing 20 litres of water. The test solution was changed everyday to give constant effect of the pesticides. Fishes from all the groups were killed after 0, 24, 48, 72 and 96 hours. Head kidney were taken out and fixed immediately in freshly prepared Bouin's solution. After 24 hours, the tissues were washed thoroughly in running tap water, dehydrated in ascending series of alcohol, cleared in xylene and embedded in paraffin wax at 60°C. Serial sections were cut at 6 µm and stained in haematoxylin and eosin (H&E).

Results and Discussion

The head kidney of *Channa punctatus* consisted mainly of haemopoietic tissue with only few renal tubules interspersed in it. The interrenal cells were arranged in

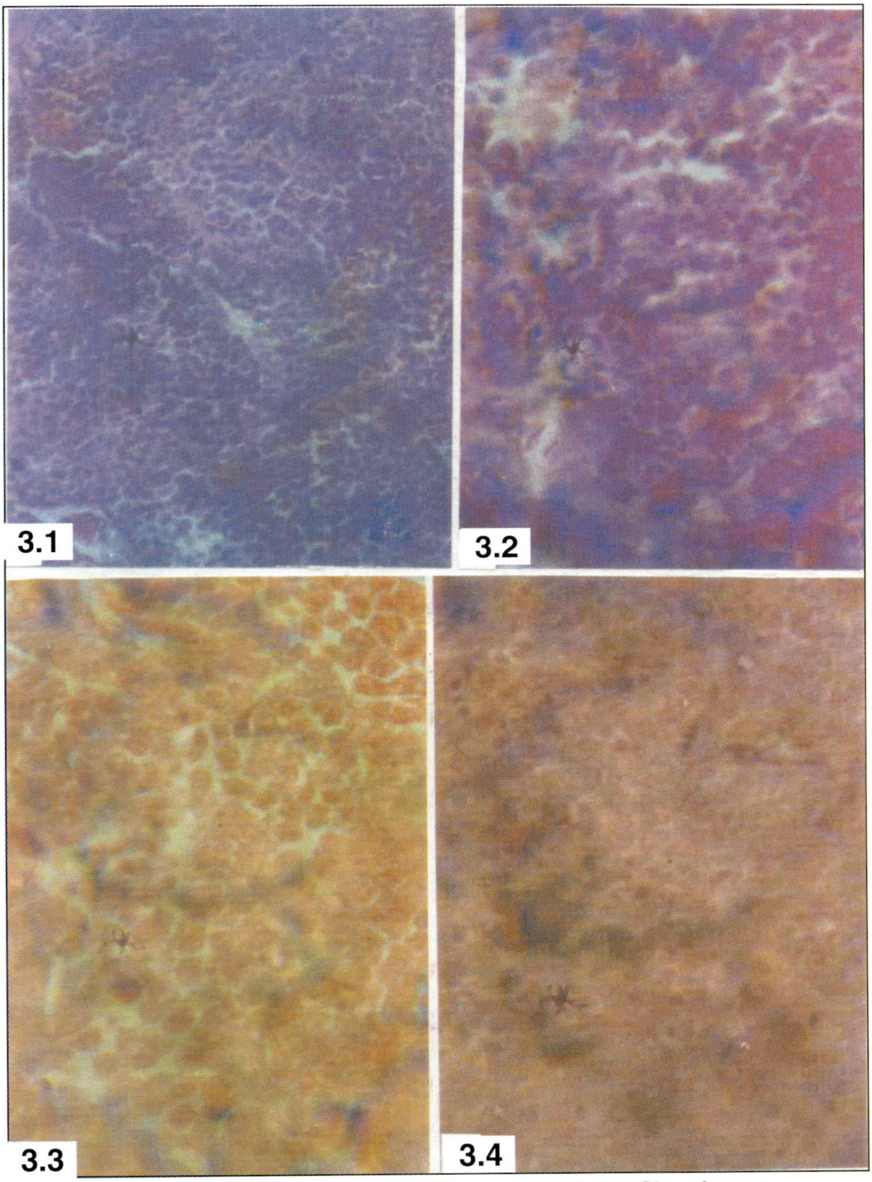

Figure 3.1: Head Kidney of *Channa punctatus* Showing Distribution of Interrenal Cells. x 200.

Figure 3.2: Head Kidney of *Channa punctatus* Exhibiting Eosinophilic Interrenal Cells Arranged in Cords/Strands. x 400.

Figure 3.3: Head Kidney of *Channa punctatus* Exposed to Carbaryl for 24 Hours Depicting Partial Loss of Staining Affinity of the Interrenal Cells. x 400.

Figure 3.4: Head Kidney of *Channa punctatus* Exposed to Cartap for 24 Hours Showing Partial Loss of Staining Affinity of the Interrenal Cells. x 400.

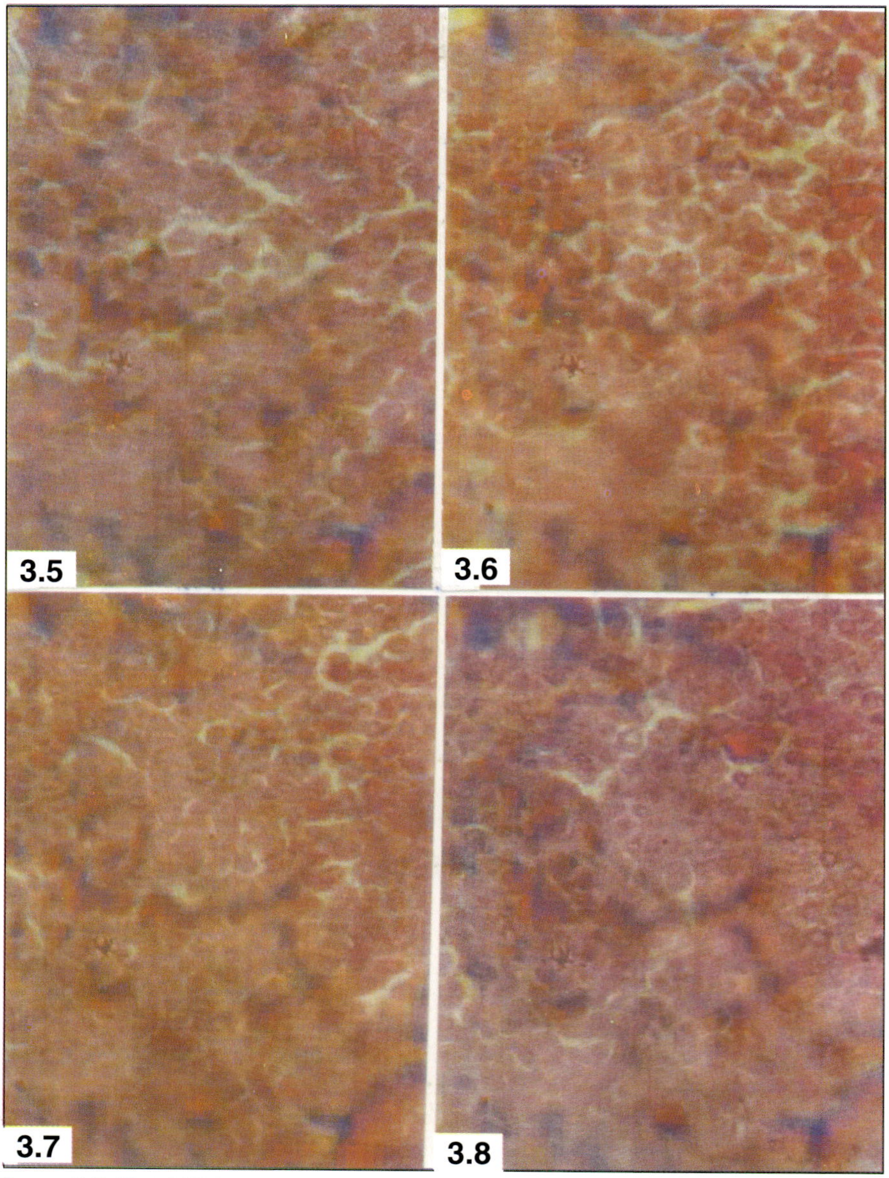

Figure 3.5: Head Kidney of *Channa punctatus* Exposed to Carbaryl for 48 Hours Exhibiting Hypertrophy of the Interrenal Cells. x 400.

Figure 3.6: Head Kidney of *Channa punctatus* Exposed to Cartap for 48 Hours Depicting Hypertrophy of the Interrenal Cells. x 400

Figure 3.7: Head Kidney of *Channa punctatus* Exposed to Carbaryl for 72 Hours Showing Hyperplasia of the Interrenal Cells. x 400.

Figure 3.8: Head Kidney of *Channa punctatus* Exposed to Cartap for 72 Hours Exhibiting Hyperplasia of the Interrenal Cells. x 400.

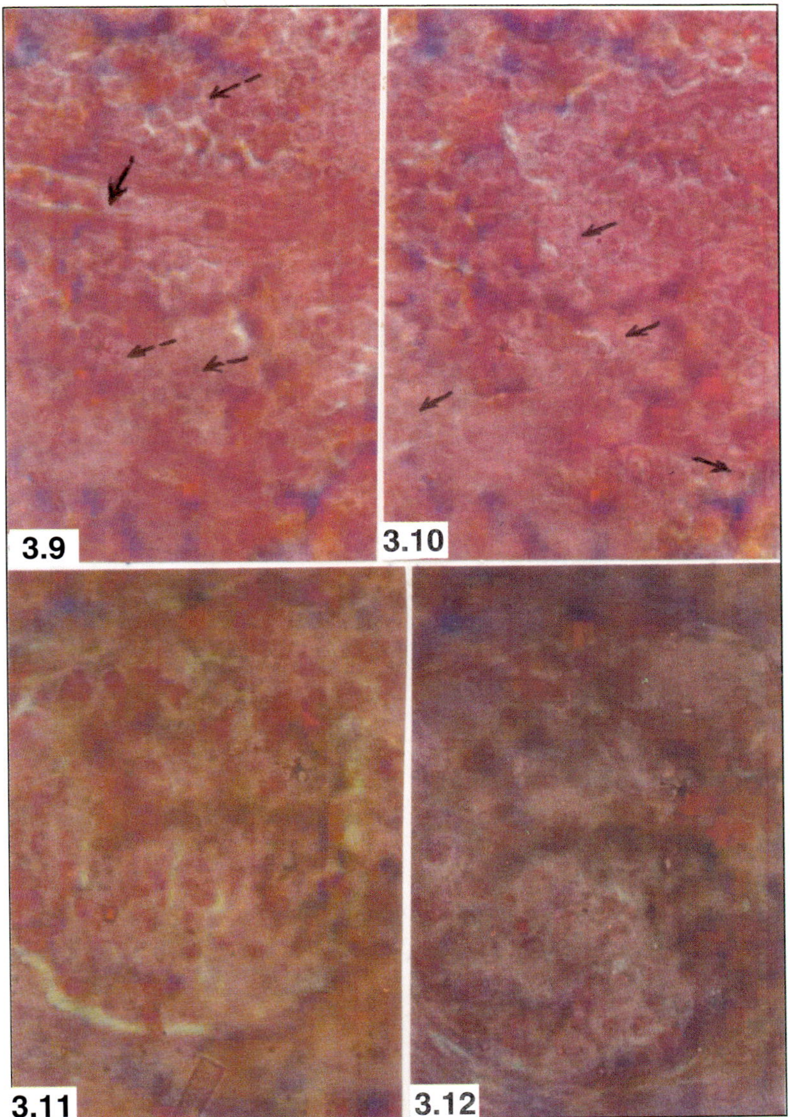

Figure 3.9: Head Kidney of *Channa punctatus* Exposed to Carbaryl for 96 Hours Depicting Hyperplasia of Interrenal Cells. Mark the blood vessel (arrow) and cells with pyknotic nuclei (broken arrow). x 400.

Figure 3.10: Head Kidney of *Channa punctatus* Exposed to Cartap for 96 Hours Showing Hyperplasia of the Interrenal Cells and Dilated Sinusoids. Few cells with pyknotic nuclei are also seen (arrow). x 400.

Figure 3.11: Head Kidney of Control *Channa punctatus* Showing Distribution of Chromaffin Cells in Cords. H&E. x 400.

Figure 3.12: Chromaffin Cells of *Channa punctatus* Exposed to Carbaryl for 24 Hours Depicting Partial Loss of Staining Affinity. H&E. x 400.

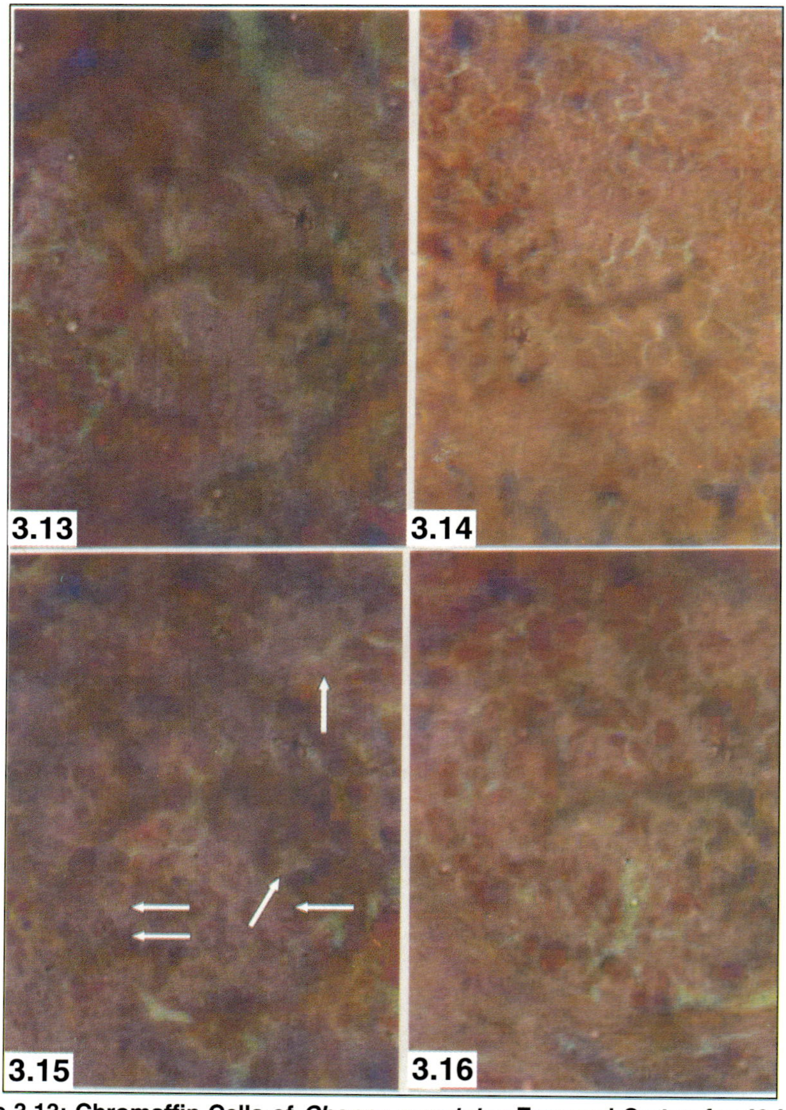

Figure 3.13: Chromaffin Cells of *Channa punctatus* Exposed Cartap for 48 Hours Depicting Hypertrophy and Increased Vascularization. H&E. x 400

Figure 3.14: Chromaffin Cells of *Channa punctatus* Exposed Carbaryl for 72 Hours Showing Hypertrophy and Hyperplasia. Few cells with pyknotic nuclei are also seen. H&E. x 400.

Figure 3.15: Chromaffin Cells of *Channa punctatus* Exposed Cartap for 72 Hours Exhibiting Hypertrophy and Hyperplasia. Mark the pyknotic nuclei (arrow). H&E. x 400.

Figure 3.16: Chromaffin Cells of *Channa punctatus* Exposed Carbaryl for 96 Hours Depicting Hypertrophy and Hyperplasia. Mark a few cells with pyknotic nuclei. H&E. x 400.

cords or strands in head kidney near the cardinal vein. They were polygonal, columnar or cuboidal with large round nuclei and more eosinophilic cytoplasm as compared to the renal tubular cells (Figures 3.1 and 3.2). The interrenal cells of Carbaryl as well as Cartap treated *Channa punctatus* responded initially with the loss of staining affinity and mild hypertrophy at 24 hours (Figures 3.3 and 3.4). After 48 hours of exposure to both the pesticides, there were significant increase in size of interrenal cells (hypertrophy) and increased vascularization (Figures 3.5 and 3.6). Hyperplasia was noticed in the interrenal cells of both the treated groups at 72 hours (Figures 3.7 and 3.8). Though hyperplasia persisted by 96 hours, increased vascularization of the gland and marked cytoplasmic vacuolization as well as pyknotic nuclei were frequently encountered in the interrenal cells in response to the treatment of carbamate pesticides (Figures 3.9 and 3.10).

Chromaffin tissue of *Channa punctatus* was scattered throughout the interrenal cells of the head kidney but in some cases observed in cords or strands too. These cells were ovoid or polygonal, possessed nuclei with sparse chromatin and eccentrically located nucleolus (Figure 3.11). The chromaffin cells of Carbaryl as well as Cartap treated murrels responded initially with the loss of staining affinity with a few hypertrophied cells at 24 hours (Figure 3.12). After 48 hours of exposure to both the pesticides, there were significant increase in size of chromaffin cells (hypertrophy) and increased vascularization (Figure 3.13). Hyperplasia was noticed in the these cells of both the treated groups at 72 hours (Figures 3.14 and 3.15). Though hypertrophy and hyperplasia were prominent by 96 hours, marked cellular vacuolization as well as pyknotic nuclei were frequently encountered in the chromaffin cells of the fish treated with sublethal concentrations of Carbaryl and Cartap (Figure 3.16).

The distribution and micromorphology of the interrenal and chromoffin cells of *Channa punctatus* resemble to those reported for other freshwater teleostean species (Belsare, 1974; Belurkar and Belsare, 1976; Kulkarni and Sathyanesan, 1979; Joshi and Sathyanesan, 1979, 1980; Chester Jones and Phillips, 1986; Dhande and Patil, 1998; Roberts, 2001). Chromaffin cells are derived from neighbouring paraganglion cells of the neural crest complex and then migrate to lie adjacent to the adrenocortical (interrenal) cell groups (Chester Jones and Phillips, 1986). There are growing evidences that contaminants do affect normal physiology of fish by interfering with the endocrine systems (Cairns *et al.*, 1984; Adams, 1990; Atterwill and Flack, 1992; Heath, 1995; Pandey *et al.*, 1993, 1995; Hontella *et al.*, 1996; Brown *et al.*, 2005) and the hypothalamo-pituitary-interrenal (HPI) axis plays pivotal role in adjustment of the teleosts to the altered environmental conditions (Brown, 1993; Sumpter, 1996; Harvey, 1996; Hontella 1997; Ontella, 1998; Norris, 2000). In the present study, the interrenal as well as chromaffin cells of *Channa punctatus* gets stimulated by sublethal exposures to Carbaryl and Cartap as evident by the occurrence of hypertrophy, hyperplasia and increased vascularization in the glands of the treated fish. Ram and Singh (1988) observed hypertrophy as well as hyperplasia in the interrenal cells of *Channa punctaus* under long-term exposure to ammonium sulfate fertilizer. Dhande and Patil (1998) also recored enhanced activity in the interrenal as well as chromaffin cells of *Channa punctatus* exposed to sublethal concentrations of mercuric chloride, cadmium chloride and cupric chloride. The observed vacuolization and pyknotic nuclei in both the endocrine tissues of the murrels at 96 hours of exposure might be depicting the signs of exhaution due to perpetual pesticidal stress.

Acknowledgements

One of us (DKM) is grateful to the Principal, Deogarh College, Deogarh and the Director, Higher Education, Government of Orissa, Bhubaneswar for granting Study Leave. Thanks are due to the Head, Department of Zoology, Utkal University and the Director, CIFA, Bhubaneswar for providing laboratory facilities.

References

Adams, S.M., 1990. Biological indicators of stress in fish. *Am. Fish. Soc. Symp. Ser.*, 8: 1–191.

Anon, 2005a. *Report of the Pesticide Information Project of Cooperative Extension Office.* Cornell University, Ithaca, New York and Michigan State University, Michigan.

Anon, 2005b. *Pesticides Standardized in Punjab with Dose(s), Crop(s) and Target Pest(s).* National Pak. Commission Pakistan Agriculture Online. Agriculture Department, Government of Punjab, Pakistan.

APHA, 1991. *Standard Methods for the Examination of Water and Wastewater.* American Water Works Association and Water Pollution Control Federation, American Public Health Association, Washington.

Atterwill, C.K. and Flack, J.D., 1992. *Endocrine Toxicology.* Cambridge University Press, Cambridge.

Belsare, D. K., 1974. Histophysiological response of the adrenal tissue homologues to adrenergic drugs in *Heteropneustes fossilis* (Bloch). *Acta Biol. Sci. Hung.*, 25: 247–253.

Belurkar, B.R. and Belsare, D.K., 1976. Effect of hydrocortisone acetate on interrenal cells in the normal and hypophysectomized catfish, *Heteropneustes fossilis* (Bloch). *Anat. Anz.*, 139: 363–368.

Bondarenko, S., Gen, J., Haver, D.L. and Kabashima, J.N., 2004. Persistence of selected organophosphate and carbamate insecticides in water from a coastal watershed. *Environ. Toxicol. Chem.*, 23: 2649–2654.

Brown, J.A., 1993. Endocrine responses to environmental pollutants. In: *Fish Ecophysiology: Fish and Fisheries Series* (Eds.) Rankin, J.C. and Jensen, F.B. Chapman and Hall, London, p. 276–296.

Brown, S.B., Adams, B.A., Cyr, D.G. and Eales, J.E., 2005. Contaminant effects on the teleost fish thyroid. *Environ. Toxicol. Chem.*, 23: 1680–1701.

Cairns, V.W., Hodson, P.V. and Nraigu, J.O., 1984. *Contaminants Effects on Fisheries.* John Wiley and Sons, New York.

Chester Jones, I. and Phillips, J.G., 1986. The adrenal and interregnal glands. In: *Vertebrate Endocrinology: Fundamentals and Biomedical Implications. Morphological Considerations* (Eds.) Pang, P.K.T. and Schreibman, M.P. Academic Press, San Diego, p. 319–350.

Dhande, R.R. and Patil, G.P., 1998. Effect of heavy metal pollutants on the endocrine kidney of fish, *Channa punctatus* (Bloch). *Proc. Acad. Environ. Biol.*, 7: 149–154.

Harvey, P.W., 1996. *The Adrenals inToxicology*. Francis and Taylor, London.

Heath, A.G., 1995. *Water Pollutants and Fish Physiology. 2nd Edn.* CRC Press, Boca Ratan, Florida.

Hontella, A., 1997. Endocrine and physiological responses of fish to xenobiotics. Role of glucocorticoide hormone. *Rev. Toxicol.*, 1: 159–206.

Hontella, A., Daniel, C. and Ricard, A.C., 1996. Effects of acute and subacute exposures to cadmium on the interrenal and thyroid function in rainbow trout (*Oncorhynchus mykiss*). *Aquat. Toxicol.*, 35: 171–182.

Joshi, B.N. and Sathyanesan, A.G., 1979. Adrenal histochemistry and histoenzymology of the teleost, *Channa punctatus* (Bloch). *Monit. Zool. Ital.*, 13: 77–82.

Joshi, B.N. and Sathyanesan, A.G., 1980. A histochemical study on the adrenal components of the teleost, *Cirrhinus mrigala* (Ham.). *Z. mikrosk.–anat. Forsch.*, 94: 327–336.

Kulkarni, R.S. and Sathyanesan, A.G., 1979. Adrenal histochemistry of the freshwater teleost, *Labeo rohita* (Ham.). *Arch. Anat. Ital. Embryol.*, 84: 171–181.

Kwak, H.I., Bae, M.O., Lee, M.H., Sung, H.J., Shin, J.S., Ahn, G.H., Kim, Y.H., Lee, C.Y. and Cho, M.H., 2000. Effects of Cartap on the early life stages of medaka (*Oryzias latipes*). *Bull. Environ. Contam. Toxicol.*, 65: 717–724.

Litchfield, J.T. and Wilocox, P., 1949. A simplified method of evaluating does–effect experiment. *J. Pharma. Exp. Ther.*, 96: 99–133.

Mishra, D.K. and Bohidar, K., 2005. Toxicity of pesticides, Carbaryl and Cartap hydrochloride on a freshwater teleost, *Channa punctatus*. *J. Adv. Zool.*, 26: 20–23.

Norris, D.W., 2000. Endocrine disruptors of the stress axis in natural populations: How can we tell?. *Am. Zool.*, 40: 393–401.

Ontella, A.H., 1998. Interrenal dysfunction in fish from contaminated sites: *in vivo* and *in vitro* assessment. *Environ. Toxicol. Chem.*, 17: 44–48.

Pandey, A.C., Pandey, A.K. and Das, P., 1999. Fish and fisheries in relation to aquatic pollution. In: *Environmental Issues and Resource Management* (Eds.) Das, P., Verma, S.R. and Gupta, A. K. Nature Conservators, Mizaffaranagar, p. 76–112.

Pandey, A.K., George, K.C. and Mohamed, M.P., 1993. Effect of mercuric chloride on thyroid gland of *Liza parsia* (Hamilton–Buchanan). *J. Adv. Zool.*, 14: 15–19.

Pandey, A.K., George, K.C. and Mohamed, M.P., 1995. Effect of DDT on thyroid gland of the mullet, *Liza parsia* (Hamilton–Buchanan). *J. Mar. Biol. Assoc. India*, 37: 287–290.

Ram, R.N. and Singh, S.K., 1988. Long–term effect of ammonia sulfate fertilizer on histophysiology of adrenal in the teleost, *Channa punctatus*. *Bull. Environ. Contam. Toxicol.*, 41: 245–252.

Roberts, R.J., 2001. *Fish Pathology.* 2nd *Edn.* W.B. Saunders, London and Philadelphia.

Sinha, N., Lal, B. and Singh, T.P., 1991. Carbaryl induced thyroid dysfunction in the freshwater catfish, *Clarias bnatrachus. Ecotoxicol. Environ. Saf.*, 21: 240–247.

Sumpter, J.P., 1996. The endocrinology of stress. In: *Fish Stress and Health in Aquaculture* (Eds.) Iwama, G.K., Pickering, A.D., Sumpter, J.P. and Schreck, C.B. Cambridge University Press, Cambridge, p. 95–118.

Chapter 4

Art of Freshwater Prawn, *Macrobrachium rosenbergii*, Culture in Saline Water

☆ P.K. Roy

Abstract

Considering the high export potential, the giant fresh water prawn, *Macrobrachium rosenbergii*, the scampi, enjoys immense potential for culture in India. This research paper contains back ground information on the culture of *Macrobrachium rosenbergii* in saline water with low cost feed and minimum input in management. This is intended to provide back ground information for the extension workers. The world market for scampi is expanding with attractive prices. There is great scope for scampi production and export. Pond was fertilized with raw cattle dung to increase adequate Zooplankton density. Home made low cost supplementary feed was applied. For good survival, juveniles of 30±5 mm length and 50±2 mg weight, were used for stocking. Prawns were hand counted and stocked at 30,000/ha into the pond.

Introduction

Considering the high export potential, the giant fresh water prawn, *Macrobrachium rosenbergii*, enjoys immense potential for culture in India. About 4 million hectare of impounded fresh water bodies in various states of India, offer great potential for fresh water prawn culture. Scampi can be cultivated for export through monoculture in existing as well as new ponds or with compatibles fresh water fishes in existing ponds. Since, the world market for scampi is expanding with attractive prices, there is great scope for scampi production and export.

Macrobrachium rosenbergii (de Man) is the largest among all the freshwater prawn. This is indigenous to south East Asia, parts of oceanic and some specific islands. The giant fresh water prawn *M. rosenbergii* commonly known as scampi. Due to its first growing habit *M. rosenbergii* is the best and most common candidate species selected for culture operation in our water, comparable for monoculture and polyculture and has mixed growth potential, preferring an omnivorous diet. There is good consumer preference and demand both in domestic and international markets, has stimulated the interest in culture of freshwater prawn, in many parts of the world particularly in the Southeast Asian countries.

About a dozen of species of freshwater prawns belonging to the genus *Macrobrachium.* At least, three species, *viz., M. rosenbergii, M. malcolmsonii* and *Birmaricum chopra* attain sufficiently large sizes and economically very important of these, *M. rosenbergii,* which is the largest prawn in the world attaining over 300 mm in length and 400 gram in weight and popularly known as scampi or the giant long legged prawn, is now cultivated on a large scale in Asia also in Latin America.

In India, success in prawn breeding achieved in 1975 (Subrahmanyam) at the prawn breeding center of CIFRI, Kakinada. Though hatcheries are now established in public and private sector in Andhra Pradesh, Kerala, Maharashtra, Orissa, Tamil Nadu, Gujarat and West Bengal. However, juveniles collected from natural resources some of the maritime states are now increasingly being used for aquaculture along with other species or monoculture.

Scampy can cultured as an alternative to *P. monodon* in saline water areas. *P. monodon* was severely affected with bacterial and viral diseases recently in all the maritime states of east and west coast of India; therefore, it is advisable to go for fresh water prawn culture. The scampy can grow well up to 10 ppt. Saline water, but the experiments have also been conducted at higher salinities up to 20 ppt (A.K. Reddy, 2005) where the growth rate was slow and the culture period was prolonged. Thereby, farming may be taken up in low saline water ponds during July to January.

The commercial scientific farming of scampy has started only recently in some parts of our country and several new farms are being developed scientifically for monoculture of scampi or polyculture with other fish species as major carps like rohu, mrigala, and catla. Due to its high price in international markets, most of the states prefer to do the culture of *M. rosenbergii,* especially in the Andhra Pradesh, West Bengal, Kerala, Orissa, Maharashtra, Punjab, Haryana and Gujarat. Apart from developing the new areas, several existing fish farms which were once upon a time used for fish culture and have become less lucrative are now being used either for the mono or polyculture of *M. rosenbergii.*

The giant fresh water prawn, *M. rosenbergii,* is one of the biggest Natantians of the world. Freshwater prawns were earlier collected from nature but the scientific culture systems of the species were developed after 1962. Larval rearing of *M. rosenbergii* was achieved by Ling and Mercian (1961). Ling (1969) for the first time succeeded in tracing the complete life cycle of the giant fresh water prawn.

In present study an effect has been made to study the culture practice of *M. rosenbergii,* in saline water of 5 to 7.5 ppt with low cost feed and minimum input in management. It would go a long way in helping all those who are engaged or intend to take up culture at minimum input.

Materials and Methods

Pond Preparation and Stocking

Two weeks prior to the anticipated stocking date the experimental pond located at Sunderbans was drained and allowed to dry. Less than one week prior to stocking, ponds were filled with water. The water surface area of the pond was 0.06 ha, and average depth of water approximately 4.07 ft. To achieve algal bloom phosphorus at a rate of 9.0 kg/ha was applied quick lime at a rate of 600 kg/ha and raw cattle dung at a rate of 10,000 kg/ha was applied to increase adequate Zooplankton density. Before stocking of seed water inlets were guarded for preventing the entry of predatory fishes and their eggs and also prevent escape of prawns. For good survival, juveniles of 30±5 mm length and 50±2 mg weight were used for stocking and stocked on 02-04-2007. The mean stocking weight was determined from a sample of 100 prawns that were blotted free of surface water and individually weighed. Prawns were hand counted and stocked at 30,000/ha in to the pond.

Samples

A 3.2 mm mesh seine was used to collect a sample of prawn from the pond every month. Each sample was weighed, and then returned to the pond.

Feed and Feeding

In order to enhance the growth of prawns supplementary feed was given. A home made feed was used consisting of 50 per cent cooked broken rice, 25 per cent rice bran, and 25 per cent fish meal. This feed cost was Rs. 19/- per kg to make and was used throughout the rearing period. Feed was normally presented once per day between 1600 h and 1700 h. Prawns were initially fed at a set rate of 25 kg/ha/day until an average individual weight of 5 g was achieved in samples from the pond. For weight greater than 5 g prawns were fed a percentage of body weight by increasing daily allotments of 20 per cent. A daily upper limit was set at 50 kg/ha/day to avoid potential water quality problems. Feeding rates were adjusted weekly, based on an assumed feed conversion rate (FCR) of 2.5 and an assumed survival of 100 per cent.

Water Quality Management

Dissolved oxygen (DO) was monitored daily at 0900h by modified Wrinkler's method (APHA, 1998). Winkler's A ($MnSO_4$) and Winkler's B (KI) were used as fixatives. Finally the fixed samples were titrated by N/40 Sodium thiosulphate using starch as an indicator. Before titration the samples were acidified with 2 ml concentrated sulphuric acid. Dissolved free carbon-dioxide was estimated in the field immediately after collection of sample and titrating by N/44 Sodium hydroxide (N/44 NaOH) using phenolphthalein indicator. pH was estimated by electronic pH meter. Total alkalinity was estimated by titrating the sample with N/50 sulphuric acid by using Phenolphthalein and methyl orange indicators. Water temperature was estimated with the help of mercury thermometer graduated up to 50°C at the sampling spot. Turbidity was measured by a standard Secchi Disc. The disc was lowered on the graduated line into the water and the depth (d1) at which it disappeared

Table 4.1: Water Parameters of the Experimental Pond during Culture Period

Sl.No.	Parameters	April	May	June	July	August	September	October	Average
1.	Depth of water (ft)	3.0	3.5	3.5	4.0	4.5	5.0	5.0	4.07
2.	pH	8.0	8.0	8.5	8.0	7.6	7.4	7.4	7.84
3.	Temp. (°C)	32.0	35.0	35.5	30.0	30.0	30.0	29.5	31.71
4.	Dissolved oxygen (ppm)	7.0	6.5	5.0	6.0	7.0	5.0	7.5	6.28
5.	Dissolved carbon dioxide (ppm)	1.0	2.0	Nil	1.5	1.0	Nil	Nil	0.78
6.	Total alkalinity (ppm)	120.0	130	140	135.0	150.0	150.0	150.0	135.0
7.	Transparency (cm)	20.0	18	19	12.0	14.0	20.0	20.0	16.42
8.	Salinity (ppt)	7.5	7.0	6.5	5.0	5.5	5.0	5.0	5.92

was noted. Then the disc was lifted slowly and the depth (d2) at which the disc reappeared was noted. The reading d1+d2/2 in cm given a measure of light penetration.

Harvest

Prawns were cultured for a period of 180 days. One day prior to harvest, the water level of the pond lowered to approximately 2 ft. at the drain end, and pond was seined three times with a 1.3 cm square mesh seine and then completely drained. Remaining prawns were manually harvested from the pond bottom. Total bulk weight and number of prawns was recorded.

Results

During culture period it was found that all the physico-chemical parameters remain with in the optimum limits (Table 4.1). Length and weight calculated in juveniles stage 30±5 mm and 50±2 mg. During harvesting calculated growth rate (length–weight) was 150±10 mm and 80±15gm respectively. Total 216 kg prawn was harvested and observed 50 per cent survibility during 180 days culture period. Total 2700 prawn were harvested out of 5400 number of juveniles.

Table 4.2: Growth Parameters of *M. rosenbergii* during the Culture Period of 180 Days

Month	*Length (mm)*	*Weight (g)*
April	30±5	0.05±2
May	50±10	1.5±0.8
June	100±15	35±10
July	120±5	50±10
August	130±10	60±15
September	140±10	65±12
October	150±15	80±15

Conclusion

Aquaculture is considered as one of the cheapest means of producing animal protein food that is fish which can play a great role in improving the nutrition of the weaker sections of the population.

In India, the production of *M. rosenbergii* in the year 1989 was to the tune of 150 tonnes, it increased to 198 tonnes and 2000 tonnes in 1990 and 1992, respectively. At present the total area under fresh water prawn farming in the country is 37000 ha with an estimated production of 38700 tonnes.

The culture part of *M. rosenbergii* was studied in a pond of 0.06 ha in which juveniles prawn were stocked in the month of April, juveniles/ha. After a culture period of six months the prawn production was 216 kg, which estimated 7200 kg/ha/year. Keeping palm leaves and bamboos strips inside water increased the food resources. According to Rajalakshmi (1991) the optimal range of physico-chemical parameters should be temperature 26–31°C, dissolved oxygen 6–8 ppm, pH 7.5–8.0, salinity 12±2 ppt. The water quality of the pond was constantly monitored (Table 4.1) and it was found that all the physico-chemical parameters were with in the required ranges.

The culture practice of this species would further get a boost when better hatchery management and grow out pond culture systems would be improved further. Many inland states in the country are trying to take up culture of *M. rosenbergii* on commercial scales. Lack of proper extension services, technical manpower and availability of quality seed are the main reasons for not developing the fresh water prawn farming in the country.

It is concluded from the study that the giant fresh water prawn is highly suitable for culture in saline water and also shows more profit in comparison to the culture of other species. Apart from early export income, the giant freshwater prawn farming also provides rural employment opportunities and augments the socio-economic development in remote villages.

Chapter 5

Studies on the Neuroendocrine Regulation of Egg Maturation in the Giant Freshwater Prawn, *Macrobrachium rosenbergii* (de Man)

☆ *A.K. Pandey and Anjani Kumar*

Abstract

Neurosecretions from the eyestalk, brain and thoracic ganglia play important role in gonadal maturation of crustaceans. An attempt was made to record the histomorphology of the neurosecretory cells in the eyestalk, brain (supraoesophageal) and thoracic ganglia during different stages of ovarian maturation of *Macrobrachium rosenbergii*. There were five types of neurosecretory cells (NSCs) in the eyestalk having size in range of 5-35 µm with or without axons. Their shape varied from round to oval. These cells were distributed in medulla externa, medulla interna and medulla terminalis. A blood sinus-like structure measuring 30–35 µm in size was also observed in the medulla interna region. Axonal terminals of these neurosecretory cells were found to terminate in this structure. Brain and thoracic ganglia possessed five types of neurosecretory cells such as giant neuron (>80 µm), A (60-80 µm), B (40-60 µm), C (20-40 µm) and D (<20 µm). All types of NSCs were recorded in the brain. They were seen arranged in several groups in different parts of brain. In anterior region B, C and D cells were located whereas in posterior region giant neurons and A cells predominated. In lateral regions A, B, C and D cells were recorded. The thoracic ganglionic mass was divided into anterior, middle and posterior regions. The NSCs were distributed in anterior and posterior portions but were lacking in middle portion. A and B cells were present in anterior-most region followed by C and D cells. In posterior-most region, the giant neurons and A cells were present. Histochemical tests demonstrated that the neurosecretory cells of the giant freshwater prawn were strongly positive to acid fuchsin, paraldehyde fuchsin but exhibited feeble reaction to Sudan black B and periodic acid-Schiff's reagent (PAS) too. Unilateral eyestalk ablation for 15 days induced ovarian maturation by increasing GSI and ova diameter as well as enhancing the secretory activities of the giant neurons (GN) and A type of neurosecretory cells of the brain and thoracic ganglia of *M. rosenbergii*.

Keywords: *Neurosecretory cells, Eyestalk, Brain, Thoracic ganglia, Unilateral eyestalk ablation, Ovarian maturation, Macrobrachium rosenbergii.*

Introduction

Freshwater prawn is gaining importance among the aquatic commodities for export because of the setback of brackishwater shrimp culture owing to white spot disease and environmental problem. Giant freshwater prawn, *Macrobrachium rosenbergii,* is an important species for diversifying aquaculture in India due to its attributes like reproduction under captivity, established technique for larval rearing, excellent growth rate and survival, absence of major disease problem, wide consumer acceptance and high market value (Kutty, 2001). Out of 150 freshwater prawn species recorded world over, 40 species are inhabitant of the Indian subcontinent. *M. rosenbergii*, the largest species of the Genus, is distributed throughout the tropical and subtropical zones of the world. It is cultured on commercial scale in many Asian countries including Thailand, China, Taiwan, Philippines, Vietnam, Malaysia, Indonesia, Bangladesh and India. It is mostly found in inland freshwaters including lakes, rivers, swamps, irrigation canals, ditches, ponds and estuaries (Rao, 1991; Achuthankutty and Desai, 2001).

During the past, seed collected from natural environment were used in prawn farming but in recent years, the natural seed resources have been drastically declined due to several anthropogenic interventions. As such the natural seed supply is insufficient to cater to the growing demand of freshwater farming (Upadhyay *et al.*, 2006). Since quality seed is the most critical input in prawn farming, it is imperative to understand the reproductive physiology of the candidate species. The neurosecretory systems play vital roles in reproductive physiology of crustaceans by transducing the environmental stimuli into physiological processes (Adiyodi and Adiyodi, 1970; Quackenbush, 1986; Subramoniam, 1999; Huberman, 2000). The histomorphological changes occurring in the neurosecretory cells of the eyestalk, brain and thoracic ganglia of *M. rosenbergii* in relation to ovarian maturation have been recorded. Preliminary observations related to the effects of unilateral eyestalk ablation on the neurosecretory cells and ovarian recrudescence has also been discussed.

Maturity Stages

The female giant freshwater prawn possessed paired ovaries located dorsally to the stomach and hepatopancrease in cephalothorax region. They gave paired oviducts which opened into gonopores on the basal segment of the third pleopods. Morphologically, the ovaries showed marked variations in relation to maturity. It was commonly observed that ovaries in each stage contained eggs of more than one stage but high proportion of ova were in a single maturity stage. Accordng to occurrence of the maximum percentage of ova, ovary of the freshwater prawn was divided into five maturity stages. However, it is pertinent to remark that after spawning, ovaries returned either to stage 1 or 2 depending upon the type of successive moult. Some females followed somatic growth just after spawning while reproductively active females followed ovarian growth after spawning. During this study, it was observed that even in egg-bearing stage, the individual could attain maturity stage 3 too. Initially the prawns were segregated into different maturity stages on external morphology but they were examined histologically. Based on the size of ova, position as well as size of nucleus, yolk deposition and distribution inside ova, maturity of the female *M. rosenbergii* was divided into the following five stages (Table 5.1).

Immature

In this group, females were recognized by clear, transparent ovary inside the rostrum. It was small in size just arising at the junction between carapace and first abdominal somite. The gonadosomatic index (GSI) of the animal was 0.6250±0.26. Histologically, oocytes were in the first stage of meiotic prophase. They were small, round and irregular cells with large cytoplasm. Maximum size of oocytes was 19 μm and minimum 6 μm with average 11.54±0.25 μm. Generally, weight of the animal in this stage was below 20 gm.

Previtellogenic

In this stage, prawns were recognized by development of colouration and increase in size of the ovary. Pigmentation (light orange or yellow) in the ovary was noticed inside the rostrum. The gonadosomatic index (GSI) of the animal was 2.28±0.45. Development of ovaries towards rostrum was clearly visible. Histological observations showed that maximum and minimim size of oocytes were 70 and 30 μm, respectively with mean value of 55.52±7.8 μm and standard deviation (SD) 13.42 μm. Average weight of the prawn in this stage ranged between 20-25 gm.

Primary Vitellogenic

In this stage, ovaries could be distinguished by deep orange colour. There was marked increase in size and extended anteriorily upto base of the rostrum. The gonadosomatic index (GSI) of the animal was 4.7±1.52. Maximum and minimum sizes of the oocytes were 160 and 100 μm, repectively. Mean ova diameter was 128±2.69 μm with SD 14.99 μm. The cytoplasm was acidophilic at the periphery but was still basophilic around the nucleus. The animals weighed between 25-30 gm.

Vitellogenic

This stage was characterized by considerable increase in the ovarian size as it extended upto first spine of the rostrum. It was clearly visible from outside as deep orange mass. The gonadosomatic index was 6.62±1.08. Maximum and minimum sizes of oocytes were 400 and 250 μm, respectively. Mean value and standard deviation were recorded 347±5.29 and 39.27 μm, respectively. Heavy accumulation of yolk globules was observed in the oocytes. The cytoplasm of oocytes displayed acidophilic character. Weight of prawn during this stage ranged between 30-40 gm.

Table 5.1: Different Maturity Stages in Ovary of *M. rosenbergii* (n=100)

State of Female	*No. of samples*	*GSI*	*Oocyte Measured (N)*	*Mean of Sample* ξ *(µm)*	*Standard Error of Mean* ξ	*Standard Deviation*	*Max. Oocyte Size (µm)*	*Min. Oocyte Size (µm)*	*Population Mean (µm)*
Immature	10	0.63±0.26	88	11.54	0.24	2.30	19	6	11.54±0.24
Previtellogenic	10	2.28±0.45	50	55.52	1.89	13.42	70	30	55.52±1.89
Primary vitellogenic	10	4.70±1.32	50	128.70	2.69	14.99	160	100	128.70±2.69
Vitellogenic	10	6.62±1.08	55	347	5.29	39.27	400	250	347.0±5.29
Mature	10	6.79±1.10	50	544	22.0	155.59	700	180	544.0±22.0

Ripe (Mature)

In this stage, ovary was large, deep orange in colour, extended anteriorly upto base of second and third spine of the rostrum. Ovary could be recognized as bulky mass beneath the rostrum. The gonadosomatic index (GSI) of the animal was 6.79±1.10. Maximum and minimum sizes of the oocytes ranged between 700 and 180 µm, respectively. Average ova diameter was 544±15.29 µm and standard deviation 155.59 µm. The animal weighed above 40 gm during this stage.

Morphology and Distribution of Neurosecretory Cells

Eyestalk X-organ Complex

In *M. rosenbergii,* neurosecretory cells aggregations were recorded in medulla externa (ME), medulla interna (MI) and medulla terminalis (MT) of eyestalk. No such type of cells was noticed in lamina ganglionaris (LG). Based on the morphological (size and shape) and histochemical (staining affinity) characteristics, these cells were divided into five categories (Table 5.2). However in all the three aggregations, more than one type of cells were recorded.

Table 5.2: Different Types of Neurosecretory Cells in Eyestalk of Female *M. rosenbergii*

Cell Type	*Cell Diameter (µm)*	*Nucleus Diameter (µm)*	*Shape*	*Staining Intensity*	*Other Features (in Mallory's Triple)*
A	30-35	10-12	Round/ oval	+++	Largest cells found in the eyestalk in medulla externa region, few in number, stained orange or red.
B	20-25	6-10	Round/ oval	++	Medium sized cells in medulla interna and medulla terminalis, few in number, stained orange or blue in colour.
C	15-20	<5	Round	++	Small in size, abundant, found in all the regions in bunch, stained orange, blue or red in colour.
D	10-15	_	Round	++	Small in size, abundant, mostly in medulla interna and terminalis, stained orange, blue or red in colour.
E	<10	_	Round	++	Smallest in size, nucleus not visible, abondant in number, orange or blue in colour.

(N = 50, each cell type) (Staining intensity: + Low; ++ Medium, +++ High).

First type of cells (A) were mostly oval in shape having cell diameter between 30–35 µm and nuclear size 10-12 µm. These cells were recorded at the junction of medulla externa and medulla interna. Most of these cells were with axons but few cells were also without axons. They displayed homogenous staining response in all stages of prawns and no cyclicity has been observed. Second type of cells (B) were more in number, present along with bigger cells in medulla externa. They were also located in medulla interna and medulla terminalis. Cell and nuclear sizes were 25-30 and 6-10 µm, respectively. Axonal endings of these cells were terminated in the medulla interna region. Morphological and histochemical differences in cell diameter, granulation and staining intensity were noticed in these cells at different stages of maturity.

Third category of cells (C) were more in number as compared to A and B. They were present in medulla interna and medulla terminalis. They were round in shape, with or without axons. Diameter of cell and nucleus were measured between 20-25 and 6-10 µm, respectively. Axonal endings of these cells were also terminated in the medulla interna. Morphological and histochemical differences in respect of cell diameter, degranulation and staining intensity were observed in these cells at different maturity stages. Fourth group of cells (D) was located in medulla interna and medulla terminalis regions. They were round in shape and the diameter ranged between 10-15 µm and nucleus measured less than 6.0 µm. No marked change in mrophological as well as staining affinity was recorded in these cells in relation to ovarian maturity.

The last category of cells (E) were located in medulla interna and medulla terminalis. These cells were round in shape with diameter of less than 10 µm while the nucleus was too small to measure in light microscopy. They did not display marked change in morphological as well as staining response in relation to ovarian maturation.

A blood sinus-like structure, measuring 30-35 µm was observed in the medulla interna. Axonal terminals of the neurosecretory cells were found to terminate in this structure. It looked like a depository of axonal endings. Variations in staining intensity of the content were noticed at different maturity stages.

Based on the changes such as cell and nuclear diameters, granulation, staining intensity and migration of secretory granules towards axonal portion encountered in the neurosecretory cells (NSCs) of eyestalk at different maturity stages, it was found that more than 75 per cent NSCs were active in immature females whereas such cells were below 50 per cent among the mature individuals.

Brain and Thoracic Ganglia

Nervous system of *M. rosenbergii* consisted of a large supra-oesophageal ganglionic mass (brain) and a ventral nerve cord with a pair of ganglia corresponding to each embryonic somite. The ganglia were joined longitudinally by connectives and transversely by commissures. The nerve cord passing through thoracic region is known as thoracic ganglia. In brain and thoracic ganglia of female *M. rosenbergii*, several types of neurosecretory cells of different size, shape and features were encountered at different stages of maturity. The five categories of cells designated as giant neuron (GN), A, B, C and D were observed in brain and thoracic ganglia (Table 5.3).

The first category of cells were giant neuron (GN) having cell and nuclear diameter >80 and 20-40 µm, respectively. They were oval or round in shape with or without axon. These cells though very few in number were mostly confined to the peripheral region but also seen in the middle area. They displayed marked changes in cell size, number of nucleolus, granulation and vacuolation as well as migration of secretory granules towards axonal portion. Cell and nuclear diameters of the second group cells (A) ranged between 60-80 and 15-36 μm, respectively. They were oval or round in shape and located mostly in middle region of the brain. A type of cells are more in number than giant neurons in brain and thoracic ganglia. They showed marked histological changes such as increase in size and number of nucleolus as well as staining intensity at different maturity stages. Migration of secretory granules towards axons has also been noticed in matured specimens.

Table 5.3: Different Types of Neurosecretory Cells in Brain and Thoracic Ganglia of *M. rosenbergii*

Cell Type	*Cell Diameter (μm)*	*Nucleus Diameter (μm)*	*Shape*	*Staining Intensity*	*Other Features*
Giant Neurons	80-120	20-40	Oval or round	++ to ++++	Very few in number, giant cells having bigger nucleus, with or without axons, stain very deeply in Mallory's triple.
A	60-80	15-36	Oval or round	++ to ++++	Large cells with comparatively large nucleus, with or without axons, granulated in secretory phase, more in number, generally red colour in Mallory's stain
B	40-60	15-25	Oval or round	+ to +++	High in number, with or without axons less secretory activity and granulation, red in colour in Mallory's stain
C	20-40	7-15	Oval to round	+ to +++	Very high in number, with or without axons, secretory activity and granulation is less, red in colour in Mallory's stain
D	5-20	2-10	Oval to round	+ to +++	Very numerous, with or without axons cytoplasm content is very less, nucleus prominent, secretory activity is very less, faint red in colour in Mallory's stain.

(N = 50, each cell type) (Staining intensity: + Low; ++ Medium, +++ High).

The third group of cells (B) were more in number in comparison to giant neurons and A cells. They were oval or round in shape with or without axons. Their size ranged between 40-60 μm with bigger nucleus (15-25 μm). Histological changes were noticed in these cells in relation to ovarian maturation. The fourth group of neurosecretory cells (C) were having size between 20-40 μm and nuclear diameter 7-15 μm with scant cytoplasm. Cells were round with or without axon. They exhibited less secretory activity as compared to GN, A and B cells. Last type of cells (D) were more in number with size range between 5-20 μm and nuclear diameter 2-10 μm. They were oval to round in shape and appear to show less secretory activity in relation to ovarian maturation.

All the above types of neurosecretory cells (NSCs) were seen in different parts of brain as several cell groups. In anterior region of the brain B, C and D cells were more in number whereas in posterior region giant neurons (GN) and A cells dominated. In lateral groups, A, B, C and D cell were uniformly distributed. The neurosecretory cells in thoracic ganglia were divided into anterior, lateral (anterolateral, midlateral and posterolateral), posterior and two median groups. B and C cells were noticed in the anterior group. C and D cells were located in lateral groups of thoracic ganglia. In posterior and median groups, only A and B cells were noticed.

Secretory Activity of the NSCs

Based on the histological as well as histochemical profiles, secretory activity of the NSCs has been assigned–quiescent or resting (Q), secretory (S) and vacuolar (V) phases. When the NSCs displayed comparatively less cell diameter, granulation and staining intensity, they were grouped under quiescent (Q) phase. When the cells were full of secretory granules, large number of nucleolus, migration of secretory granules towards periphery, high staining intensity and hypertrophy, they were grouped as active or secretory (S) phase. In vacuolar (V) phase, vacuoles were seen towards periphery due to release of the neurosecretory materials. Interestingly, number of nucleolus were less during quiescent phase of secretory activity found in thoracic ganglia and brain as compared to secretory as well as vacuolar phases.

Changes in activity of different types of neurosecretory cells in brain and thoracic ganglia in relation to ovarian maturation have been summarized in Table 5.4. As maturation advanced in the female freshwater prawn, number of secretory (S) as well as vacuolar (V) cells increased rapidly in brain and thoracic ganglia. Though brain and thoracic ganglia of immature females had majority of cells in quiescent (Q) phase and a very few in secretory (S) phase, mature females had more than 75 per cent cells in secretory phase.

Table 5.4: Percentage of Neurosecretory Cells in Brain and Thoracic Ganglia during Different Maturity Stages of Female *M. rosenbergii*

Maturity , Stage	*Brain*			*Thoracic Ganglia*		
	Quiescent (Q) Phase	*Vacuolar (V) Phase*	*Secretory (Q) Phase*	*Quiescent (V) Phase*	*Vacuolar (Q) Phase*	*Secretory (V) Phase*
Immature	62.1±0.40	30.1±0.44	7.6±0.45	77.8±0.29	12.7±0.15	9.6±0.26
Previtellogenic	60.2±0.35	23.5±0.34	16.4±0.33	42.3±0.39	24.5±0.52	33±0.61
Primary vitellogenic	50.1±0.45	25±0.42	24.9±0.45	39.9±0.27	28.2±0.72	31.6±0.49
Vitellogenic	29±0.57	34.9±0.50	35.3±0.7	27.5±0.68	34.2±0.46	38.3±0.55
Mature	20.2±0.35	30.4±0.26	49.4±0.52	12.8±0.44	31.6±0.61	55.6±0.68

Histochemistry of Neurosecretory Cells

Different histochemical stainings were done to find out the chemical nature of secretory material released by NSCs (Table 5.5). Among different cells, giant neuron (GN), A and B cells exhibited strong affinity in almost all tests as compared to C and

D cells. Poor reactions of the latter may probably be due to less secretory granules and scanty cytoplasm. Though all these cells were positive for Mallory's triple but staining affinity varied. The neurosecretory cells of eyestalk were positive for Mallory's triple and aldehyde fuchsin (AF). Similarly, the giant neurons (GN), A and B cells of brain and thoracic ganglia also stained intensely with acid fuchsin (in Mallory's triple) and aldehyde fuchsin (AF). These results indicate proteinaceous nature of their secretion and presence of cysteine/cystine-containing amino acids. B cells of eyestalk and giant neuron as well as A cells of brain and thoracic ganglia were positive (with low intensity) to Sudan black B. B and C cells of eyestalk, giant neuron (GN), A as well as B cells of brain and thoracic ganglia were positive to periodic acid-Schiff reagent (PAS) indicating presence of carbohydrates/mucopolysaccharides.

Table 5.5: Staining Affinity of the Neurosecretory Cells of Female *M. rosenbergii*

Test	Eyestalk		Brain			Thoracic Ganglion		
	B	C	GN	A	B	GN	A	B
Mallory's triple	++	++	+++	++	++	+++	++	++
Paraldehyde fuchsin	++	+	++	++	+	++	++	+
Sudan Black B	+	–	+	+	–	+	+	–
Periodic acid Schiff (PAS)	++	–	++	++	+	++	++	+

Staining intensity: + Low; ++ Medium, +++ High.

Effects of Unilateral Eyestalk Ablation

Effects of unilateral eyestalk ablation on the ovarian maturation as well as neurosecretory cells of brain and thoracic ganglia *M. rosenbergii* were recorded. Gonadosomatic index (GSI) of the control prawn ranged between 0.62-0.69 during the experimental period of 15 days. Unilateral eyestalk ablation induced ovarian maturation as GSI increased from 0.68 to 3.20 whereas in unilateral ablated+eyestalk extract injected prawns, it increased from 0.70 to 1.10. Ova diameter also exhibited similar trend as in control it registered increased from 46±6.8 to 80±6.80 µm, in unilateral ablated+eyestalk extract administered prawns from 50±2.20 to 110±4.80 µm and in unilateral eyestalk ablated specimens from 54±3.80 to 330±10.20 µm (Table 5.6).

Unilateral eyestalk ablation induced changes in activity of the neurosecretory cells of brain and thoracic ganglia of *M. rosenbergii*. In control female prawn, secretory activity of giant neurons (GN) and A type cells were comparable to the previtellogenic females as percentage of the secretory cells (secretory+vacuolar phases) were 35 in brain and 54 in thoracic ganglia. After eyestalk ablation, population of the secretory cells increased to 59.6 per cent in brain and 69.9 per cent in thoracic ganglia. In unilateral ablated+eyestalk extract adminstered prawns, percentage of secretory cells were 44.2 and 61.75 in brain and thoracic ganglia, respectively (Table 5.7).

Hypertrophy was observed in the giant neurons (GN) and A type cells of brain and thoracic ganglion of female *M. rosenbergii* after eyestalk ablation. In brain of the control prawns, cell diameter of giant neurons on day 0 and day 15 were

Table 5.6: Effects of Eyestalk Ablation on Gonadosomatic Index (GSI) and Ova Diameter of *M. rosenbergii*

Group	*Mean Total Length (mm) (n=10)*	*Mean Total Weight (g)*	*G.S.I.*		*Ova Diameter (μm) (n=50)*	
			Initial	*Final*	*Initial*	*Final*
1 (Saline-injected)	148.2±6.5	30.2±0.60	0.62±0.12	0.69±0.21	46.3±6.8	80.3±9.6
2 (Eyestalk-ablated)	152.5±6.2	32.4±0.28	0.68±0.12	3.20±0.28	54.4±3.8	330.8±10.2
3 (Eyestalk ablated + eyestalk extract-injected)	151.3±7.4	32.3±0.46	0.70±0.11	1.10±0.21	50.2±2.2	110.2±4.8

Values are mean±S.E. of 5 specimens.

81.2±0.64 µm and 83.70± 0.90 µm, respectively. However, in unilateral ablated prawns, average cell diameters of giant neurons during the corresponding periods were 80.2±0.58 and 92.62±1.48 µm, respectively. Similar trends were also observed in A cells of brain and giant neurons as well as A cells of thoracic ganglia (Table 5.8). In general, there was cellular hypertrophy in NSCs of brain and thoracic ganglia after eyestalk ablation in the giant freshwater prawn. Staining intensity of the neurosecretory cells of brain and thoracic ganglia of the unilateral eyestalk ablated prawns also enhanced from ++ to +++. Interestingly, 60 per cent of the unilateral eyestalk ablated females moulted by day 12th indicating relation of moulting with eyestalk ablation.

Table 5.7: Percentage Occurrence of Giant Neurons (GN) and A Cells in Brain and Thoracic Ganglia of Female *M. rosenbergii* After Unilateral Eyestalk Ablation

Group	*Brain*			*Thoracic Ganglia*		
	Quiescent (Q) Phase	*Vacuolar (V) Phase*	*Secretory (Q) Phase*	*Quiescent (V) Phase*	*Vacuolar (Q) Phase*	*Secretory (V) Phase*
1	64.2±0.29	20.5±0.35	15.0±0.26	45.8±0.32	25.0±0.48	29.2±0.22
2	40.4±0.32	30.5±0.29	29.1±0.18	30.1±0.16	32.8±0.29	37.1±0.26
3	55.8±0.32	25.5±0.29	18.7±0.33	39.25±0.22	30.0±0.29	31.75±0.26

Values are mean±S.E. of 5 specimens.

Table 5.8: Changes in Cell Diameter (µm) of Giant Neuron (GN) and A Cells (n=50) in the Brain and Thoracic Ganglia of Female *M. rosenbergii* After Eyestalk Ablation

Group	*Brain*				*Thoracic ganglia*			
	GN		*A Cells*		*GN*		*A cells*	
	Initial	*Final*	*Initial*	*Final*	*Initial*	*Final*	*Initial*	*Final*
1	84.0±0.9	86.6±1.2[a]	64.1±0.7	65.2±0.7[a]	84.2±0.8	86.4±0.8[a]	63.1±0.5	63.8±0.5[a]
2	91.3±0.8	103.8±1.4[b]	61.6±2.4	73.7±0.5[b]	93.5±1.0	103.1±1.0[b]	61.4±2.3	73.8±0.5[b]
3	90.8±.0.7	92.0±0.8[a]	63.6±1.4	64.4±0.4[a]	91.3±.0.9	93.8±0.9[a]	64.3±0.4	65.8±0.5[a]

Values are mean ± S.E. of 5 specimens. Significance: [a] P<0.05, [b] P<0.001.

Discussion

Ovarian development and maturity stages in *M. rosenbergii*, has been studied by few workers in the past (Rajyalakshmi, 1961; Ling, 1969; Rao, 1991; Damrongphol *et al.*, 1991). Rajyalakshmi (1961) and Rao (1991) have described four maturity stages in the prawn but when the virgin female were classified separately, five scale classification pattern was followed by O'Donovan and Cohen (1984), Ra'anan *et al.* (1991), Mohamed and Diwan (1994) and Nandkumar (2001). In *M. rosenbergii*, five-scale classification was followed as females were divided into immature, previtellogenic, primary vitellogenic, vitellogenic and mature stages.

Morphology of the ovary of the giant freshwater prawn changes during maturation. Colouration of the ovary varies from colourless to deep orange as maturity advanced. In immature stage, size of the ovary was tiny and confined to the posteriormost region of carapace while in mature ones, it occupied the entire carapace cavity. Similar observations have also been recorded by other workers in this species (Rajyalakshmi, 1961; Ling, 1969; Rao, 1991; Damrongphol *et al.*, 1991). When detailed histological observations of ovary of the giant freshwater prawn was recorded, the ova diameter varied from 11.54±0.25 to 544±22 μm in immature and mature specimens, respectively. O'Donovan and Cohen (1984) have recorded ova diameter in immature *M. rosenbergii* as 8.5 μm and in mature individuals 530 μm. In the present study, GSI value was found to be 0.62±0.26 in immature and 6.7± 1.10 in mature prawns. Our observations corroborate the findings of O'Donovan and Cohen (1984) and Damrongphol *et al.* (1991) in the same species. Though *M. rosenbergii* breeds throughout the year, the peak may occur in particular month (Rao, 1991). Though oocyte development shows multimodal peak in the present study too, there is a clear indication that a particular batch of ova are more in number representing state of maturation of the particular individual.

The basic cytoarchitecture of eyestalk of *M. rosenbergii* is quite similar to those reported for *Macrobrachium kistnensis* (Mirajkar *et al.*, 1984). The adult possessed well-developed lamina ganglionaris, medulla externa ganglionic-X-organ (m.e.t.x.) and medulla interna ganglionic-X-organ (m.i.g.x.) and medulla terminalis-X-organ (m.t.g.x.). Similar to other prawn species, the giant neurons (GN) were not observed in eyestalk of the giant freshwater prawn. There were five types of neurosecretory cells (A, B, C, D and E) in eyestalk of *M. rosenbergii* (size ranging between 10-35 μm). A and B cells of eyestalk were smaller as compared to their counterparts in brain and thoracic ganglia. Secretory activity of B and C cells of the giant freshwater prawn exhibited changes with the ovarian maturation. In immature prawns more than 70 per cent of these cells were in active phase whereas activity declined in mature individuals as percentage of such cells were below 50. Mirajkar *et al.* (1983-*M. kistnensis*), Deecaraman and Subramonium (1983a-*Squilla holoschista*), Mohamed and Diwan (1991-*Penaeus indicus*), Pandey *et al.* (1999a-*Parapenaeopsis stylifera*), Pandey *et al.* (1999b-*Penaeus indicus*) and Pandey *et al.* (2000- *Metapenaeus dobsonii*) have also recorded similar changes in the eyestalk of prawns in relation to ovarian maturity. It may be likely that B and C cells exert inhibitory influence on ovarian maturation by elaborating maturation-inhibiting hormone (MIH) (Adiyodi and Adiyodi, 1970; Anilkumar and Adiyodi, 1985).

Sinus gland measured 30-35 μm, took bright orange stain and appeared swollen among immature female *M. rosenbergii* while it was shrunken, devoid of granules and staining intensity declined among mature females. In *Squilla holoschista*, Deecaraman and Subramoniam (1983b) have also recorded higher granulation in immature crabs whereas the gland was devoid of granules in mature individuals.

Histochemical observations have shown that neurosecretory cells of the eyestalk of *M. rosenbergii* were positive for acid fuchsin (in Mallory's triple) and aldehyde fuchsin (AF) indicating proteinaceous nature (rich in cysteine/cystine) of the secretion. These observations support the findings of Rao *et al.* (1985-*M. affinis*), Mohamed and

Diwan (1991-*P. indicus*), Pandey *et al.* (1999a-*P. stylifera*), Pandey *et al.* (1999b-*P. indicus*) and Pandey *et al.* (2000-*M. dobsonii*). Similar to *M. affinis* (Rao *et al.*, 1985), *P. stylifera* (Nagabhusanam *et al.*, 1986), only B type cells in the giant freshwater prawn exhibited weak positive response to periodic acid-Schiff reagent (PAS) and Sudan black B.

Attempts have been made in the past to classify different types of neurosecretory cells (NSCs) in brain and thoracic ganglia of crustasceans. Nagabhushanam *et al.* (1986) identified eight types of NSCs in *P. stylifera*. Rao *et al.* (1985) reported five types of cells in *M. affinis* while Mirajkar *et al.* (1984) found four types in *M. kistnensis*. Deecaraman and Subramoniam (1983b) observed only three types of cells in stomatopod crustacean (*S. holoschista*). However, five types of neurosecretory cells were identified in *M. rosenbergii*. They are giant neuron (>80 μm), A (60-80 μm), B (40-60 μm), C (20-40 μm) and D (5-20 μm). Shape of these cells varied from oval to round. In brain of giant freshwater prawn, NSCs were distributed in all portions while the cells were observed only in upper and lower portion of the thoracic ganglion similar to those reported by Nagabhushanam *et al.* (1986) and Rao *et al.* (1988). Based on size, shape and staining affinity, A cells of *M. affinis* could be compared with giant neurons (GN) of *M. rosenbergi*, B with A, C with B, D with C and E with D (Rao *et al.*, 1988). Joshi and Khanna (1985) and Joshi (1989) also observed four types of cells almost of same size in the crab, *Potamon koolooense*. In *P. indicus*, Mohamed and Diwan (1991b) observed giant neurons, A and B cell types in different neuroendocrine centres of the brain and thoracic ganglia. Recently, Jadhav *et al.* (2001) reported three types of cells in brain and four types in thoracic ganglion of the male crab (*Ucalatea annulipes*).

Marked changes like increase in cell diameter (hypertrophy), enhanced granulation of cytoplasm, increase in number of nucleolus, enhanced staining intensity and migration of secretory granules towards axonal portion were observed in giant neuron (GN), A and B cell types of brain and thoracic ganglia of *M. rosenbergii*. Mirajkar *et al.* (1983), Deecaraman and Subramoniam (1983b), Joshi (1989), Mohamed and Diwan (1991) and Jadhav *et al.* (2001) have also recorded similar observations in different crustaceans. Interestingly, in brain of immature prawns 62.1±0.40 per cent cells (giant neurons, A, B types) were in quiescent (Q), 30.1±0.44 per cent in secretory (S) and 7.6±0.45 per cent in vacuolar (V) phase while in mature individuals 20.2±0.35 per cent cells were in quiescent (Q), 30.4±0.26 per cent in secretory (S) and 49.4±0.52 per cent in vacuolar (V) phase. In thoracic ganglia of immature prawns 77.8±0.29 per cent were in quiescent (Q), 12.7±0.15 per cent in secretory (S) and 9.6±0.26 per cent in vacuolar (V) phase as against 12.8±0.44 per cent in quiescent (Q), 31.6±0.61 per cent in secretory (S) and 55.6±0.68 per cent in vacuolar (V) phase of the matured specimens. Similar observations have also been recorded by Mohamed and Diwan (1991) during ovarian maturation of *P. indicus*. There exist reports on the presence of gonad stimulating hormones (GSH) in brain and thoracic ganglia of the crustaceans (Mirajkar *et al.*, 1983; Deecaraman and Subramoniam, 1983a, b; Rao *et al.*, 1988; Jayalakshmi *et al.*, 1989; Nagabhushanam *et al.*, 1989; Mohamed and Diwan, 1991; Yano, 1992; Yano and Wyban, 1992; Jadhav *et al.*, 2001). Our observations suggest the involvement of giant neuron (GN), A and B types of neurosecretory cells in ovarian maturation of the giant freshwater prawn.

Histochemical tests indicated that giant neuron, A and B cells of brain and thoracic ganglia of *M. rosenbergii* were aldehyde fuchsin (AF) and acid fuchsin (in Mallory' triple stain) positive indicating the secretion to be rich in cysteine/cystine amino acids. Mirajkar *et al.* (1983), Victor and Sarojini (1985), Rao *et al.* (1985) and Mohamed and Diwan (1991) have also observed similar staining responses of these cells in decapod crustaceans. Positive response of giant neuron, A and B cells to periodic acid-Schiff reagent (PAS) indicates the presence of carbohydrate-like compounds/mucopolysaccharides in the cells. Besides, giant neurons (GN) and A type cells also exhibited mild positive response to Sudan black B suggesting lipoidal nature of the secretion. The present observations corroborate the findings of Mirajkar *et al.* (1983-*M. kistnensis*), Nagabhushanam *et al.* (1986-*P. stylifera*) Rao *et al.* (1988-*M. affinis*) and Mohamed and Diwan (1991-*P. indicus*).

Unilateral eyestalk ablation in *M. rosenbergii* induced ovarian maturation as evident by the increase in gonadosomatic index (GSI) as well as ova diameter of the female prawns (Table 5.6). Enhanced ovarian development in response to unilateral eyestalk ablation has also been reported by several workers in a number of crustaceans (Caillout, 1972; Arnstein and Beard, 1975; Muthu and Laxminarayana, 1977; Santiago, 1977; Primavera, 1978; Beard and Wickins, 1980; Kelemec and Smith, 1980; Dietz, 1983; Kelemec and Smith, 1984; Browdy and Samocha, 1985; Browdy *et al.*, 1986; Makinouchi and Primavera, 1987; Bray *et al.*, 1990; Har, 1991; Tan-Fermin, 1991; Mohamed and Diwan, 1991; Wilder *et al.*, 1994; Cavalli *et al.*, 1997; Sagi *et al.*,1997; Browdy, 1998; Ragunathan and Arivazhagan, 1999; Fox, 2000; Stella *et al.*, 2000).

Though studies on the role of neurosecretory cells of brain and thoracic ganglia in reproduction of crustaceans attracted wide attention but the observations are equivocal (Otsu, 1963; Rao *et al.*, 1981; Joshi, 1989; Nagabhushanam *et al.*, 1989, 1992; Har, 1991; Mohamed and Diwan, 1991; Yano, 1992; Yano and Wyban, 1992). Though oocytes of *M. kistnensis* reared in thoracic ganglia or brain extracts exhibited larger size as compared to the controls, the brain extract was found to be more effective than the extract from thoracic ganglia (Nagabhushanam *et al.*, 1989). Rao *et al.* (1981) reported increased cytological activities in the neurosecretory A, B and C cells of thoracic ganglia of *M. lanchesteri* during ovarian maturation. Joshi (1989) observed marked annual cyclic changes in synthesis and release of cytoplasmic materials from brain and thoracic ganglia of the female crab, *Potamon koolooense* in relation to ovarian cycle. Nagabhushanam *et al.* (1992) recorded increased secretory activities in neurosecretory A and B cells of brain and thoracic ganglia of *M. affinis* during reproductive peaks suggesting their role in reproduction. Our observation revealed that unilateral eyestalk ablation enhances secretory activity in the neurosecretory GN and A cells of brain and thoracic ganglia of *M. rosenbergii* suggesting the inhibitory influence of eyestalk on synthesis and secretion of hormone(s) from these neuroendocrine centres. Mohamed and Diwan (1991) have also reported enhanced activity in the GN and A cells of the unilaterally–as well as bilaterally eyestalk ablated female *Penaeus indicus*. The observed hypertrophy in GN and A cells of brain and thoracic ganglia of female *M. rosenbergii* after eyestalk ablation supports the findings of Joshi (1989) in female *Potamon koolooense*. We observed enhanced staining intensity from ++ (mild positive) to +++ (strong positive) of the neurosecretory cells in brain and thoracic ganglia of the eyestalk ablated female *M. rosenbergii*. Babu *et al.* (1979–*Menippe rumphii*) and Joshi (1989–*Potamon koolooense*) also recorded a similar response to eyestalk ablation and ovarian maturation of crabs, respectively. There exist reports that eyestalk ablation induces moulting in a number of crustacean species (Adiyodi and Adiyodi, 1970; Babu *et al.*, 1979; Chang, 1985; Kumari and Pandian, 1987; Stella *et al.*, 2000). In the present study also, about 60 per cent of the unilateral eyestalk ablated females moulted on day 12 indicating relation of moulting with eyestalk ablation in *M. rosenbergii*.

Acknowledgements

We are grateful to Dr. K. Janaki Ram, former Director, Central Institute of Freshwatewr Aquaculture, Bhubaneswar for providing the necessary facilities to carry out the work. Financial assistance provided by the Indian Council of Agricultural Research, New Delhi to one of the authors (AK) as Junior Research Fellow of Central Institute of Fisheries Education, Mumbai is thankfully acknowledged.

References

Achuthankutty, C.T. and Desai, U.M., 2001. Captive shrimp broodstock production: relevance of eyestalk ablation and regeneration in gonad maturation. In: *International Symposium on Fish for Nutritional Security in the 21st Century (December 4–6, 2001)*. Central Institute of Fisheries Education, Mumbai, p. 144–153.

Adiyodi, K.G. and Adiyodi, R.G., 1970. Endocrine control of reproduction in decapod crustacean, *Biol. Rev.*, 45: 121–165.

Anilkumar, G. and Adiyodi, K.G., 1985. The role of eyestalk hormones in vitellogenesis during the breeding season in the crab, *Paratelphusa hydrodromous* (Herbst) *Biol. Bull.*, 169: 689–695.

Arnstein, D.R. and Beard, T.W., 1975. Induced maturation of prawn, *Penaeus orientalis* Kishinouye in the laboratory by means of eyestalk removal. *Aquaculture*, 20: 79–89.

Babu, D.E., Shyamasundari, K. and Rao, K.H., 1979. Effects of bilateral ablation of eyestalks on the brain and moulting in the crab, *Menippe rumphii* Fabricius. *Indian J. Exp. Biol.*, 17: 1394–1396.

Beard, T.W. and Wickins, J.F., 1980. Breeding of *Penaeus monodon* Fabricius in laboratory recirculation system. *Aquaculture*, 20: 79–89.

Bray, W.A., Lawrence, A.L. and Lester, L.J., 1990. Reproduction of eyestalk ablated *Penaeus stylirostris* fed various levels of total dietary lipids. *J. World Aqua. Soc.*, 21: 41–52.

Browdy, C.L., 1998. Recent developments in penaeid broodstock and seed production technologies improving the outlook for superior captive stocks. *Aquaculture*, 164: 3–21.

Browdy, C. L and Samocha, T. M.,1985. The effects of eyestalk ablation on spawning, molting and mating of *Penaeus semisulcatus* de Haan. *Aquaculture*, 49: 19–29.

Browdy, C.L., Hadani, A., Samocha, T.M. and Loyola, Y., 1986. The reproductive potential of wild and pond reared *Penaeus semisucatus* de Haan. *Aquaculture*, 59: 251–258.

Caillout, C.W., 1972. Ovarian maturation induced by eyestalk ablation in pink shrimp, *Penaeus duorarum* (Burkenroad). *Proc. World Aqua. Soc.*, 3: 205–225.

Cavalli. R.O., Scaruda, P. and Wasielesky, W., 1997. Reproductive performance of different sized wild and pond–reared *Penaeus paulensis* females. *J. World Aqua. Soc.*, 28: 260–267.

Chang, E.S., 1985. Hormonal control of molting in decamped Crustacea. *Am. Zool.*, 25: 179–185.

Damrongphol, P., Eangchuan, N. and Poolsanguan, 1991. Spawning cycle and oocyte maturation in laboratory–maintained giant freshwater prawns, *Macrobrachium rosenbergii*. *Aquaculture*, 95: 347–357.

Deecaraman, M. and Subramoniam, T., 1983a. Endocrine regulation of ovarian maturation and cement glands activity in a stomatopod crustacean, *Squilla holoschista*. *Proc. Indian Acad. Sci. (Anim. Sci.)*, 92: 399–408.

Deecaraman, M. and Subramoniam, T., 1983b. Histochemistry of the neurosecretory systems in a stomatopod crustacean, *Squilla holoschista*. *Proc. Indian Acad. Sci. (Anim. Sci.)*, 92: 387–398.

Dietz, R.A., 1983. Eyestalk histology and the effects of eyestalk ablation on the gonads of the shrimp, *Macrobrachium rosenbergii* (de Man). *Ph. D. Thesis*. Texas A. and M. University, Austin.

Fox, J., 2000. Eyestalk ablation. In: *Encyclopedia of Aquaculture* (R.R Stickney, ed.). John Wiley and Sons, New York, p. 329–330.

Har, C.H., 1991. Effects of eyestalk ablation, thoracic ganglion extract and gonad extract from 'spent spawners' on ovarian maturation in pond–reared shrimps, *Penaeus monodon* Fabricius. *J. Aquacult. Fish. Mgmt.*, 22: 463–471.

Huberman, A., 2000. Shrimp endocrinology: a review. *Aquaculture*, 191: 191–208.

Jadav, S., Raghunathan, M.G. and Deecaraman, M., 2001. Changes in the neurosecretory cells of the brain and thoracic ganglion of male crab, *Uca* (*Celuca*) *lactea annulipes* with respect to season. *J. Environ. Biol.*, 22: 311–314.

Jayalakshmi, K., Sarojini, R., Rao, S.S. and Nagbhushanam, R., 1989. An ovary stimulating factor in the freshwater prawn, *Macrobrachium lamerri*. *Uttar Pradesh J. Zool.*, 9: 56–60.

Joshi, P.C., 1989. Neurosecretion of brain and thoracic ganglion and its relation to reproduction in the female crab, *Potamon koolooense* (Rathbun). *Proc. Indian Natn. Sci. Acad.*, 98: 41–49.

Joshi, P.C. and Khanna, S.S., 1985. Studies on the relationship between the neurosecretion of the brain and testicular cycle in a crab, *Potamon koolooense* (Rathbun) (Crustacea, Decapoda). *Zool. Anz.*, 214: 320–332.

Kelemec, J.A. and Smith, I.R., 1980. Induced ovarian development and spawning of *Penaeus plebejus* in a recirculatory laboratory tank after unilateral eyestalk enucleation. *Aquaculture*, 21: 55–62.

Kelemec, J. A. and Smith, I. R., 1984. Effects of low temperture storage and eyestalk enucleation of gravid eastern king prawns, *Penaeus plebejus*, on spawning, egg fertilization and hatching. *Aquaculture*, 40: 67–76.

Kumari, S.S. and Pandian, T.J., 1987. Effects of unilateral eyestalk ablation on moulting, growth, reproduction and energy budget of *Macrobrachium nobilii*. *Asian Fish. Soc.*, 1: 1–17.

Kutty, M.N., 2001. Diversification of aquaculture. In: *Sustainable Indian Fisheries* (T.J. Pandian, ed.). National Academy of Agricultural Sciences, New Delhi, p. 189–212.

Ling, S.W., 1969. The general biology and development of *Macrobrachium rosenbergii*. *FAO Fish. Rep.* No. 57: 589–606.

Makinouchi, M. and Primavera, J. H., 1987. Maturation and spawning of *Penaeus indicus* using different ablation methods. *Aquaculture*, 62: 73–81.

Matsumato, K., 1954. Neurosecretion of the thoracic ganglion of the crab, *Eriocheir japonicus Biol. Bull.*, 106: 60–68.

Mirajkar, M.S., Sarojini, R. and Nagabhushanam, R., 1983. The neurosecretory control of the annual reproductive cycle in the freshwater prawn, *Macrobrachium kistnensis*. *Curr. Sci.*, 52: 967–970.

Mirajkar, M.S., Sarojini, R. and Nagabhushanam, R., 1984. Histochemistry of the neurosecretory cells in the freshwater prawn, *Macrobrachium kistnensis* (Palaemonidae). *Proc. Nat. Acad. Sci. India*, 54B: 189–193.

Mohamed, K. S. and Diwan, A. D., 1991. Neuroendocrine regulation of ovarian maturation in the Indian white prawn *Penaeus indicus* H. Milne Edwards. *Aquaculture*, 98: 381–393.

Mohamed, K.S. and Diwan, A.D., 1994. Vitellogenasis in the Indian white prawn, *Penaeus indicus* (Crustacea: Decapoda) *J. Aqua. Trop.*, 9: 157–171.

Muthu, M.S. and Laxminarayana, A., 1977. Induced maturation and spawning of Indian penaeid prawns. *Indian J. Fish.*, 24: 172–180.

Nagabhushanam, R. and Sarojini, R., 1969. Neurosecretion in the central nervous system of the hermit crab, *Diogenes bicristimanus*. *Proc. Indian Acad. Sci.*, 69B: 20–28.

Nagabhushanam, R., Sarojini, R. and Joshi, P.K., 1986. Observation on the neurosecretory cells on the marine penaeid prawn, *Parapenaeosis stylifera J. Adv. Zool.*, 7: 63–70.

Nagabhushanam, R., Machale, P.R. and Sarojini, R., 1989. Pheromonal and hormonal control of reproduction in the freshwater prawn, *Macrobrachium kistnensis*. *Proc. Indian Natn. Sci. Acad.*, 98: 95–98.

Nagabhushanam, R., Sarojini, R. and Rao, S.S., 1992. Reproductive endocrinology of the Indian marine commercial shrimp, *Metapenaeus affinis*. In: *Aquaculture Research Needs for 2000 A.D* (J. K. Wang and P. V. Dehadrai, eds.). Oxford and IBH Pub. Co., New Delhi, p. 163–179.

Nandkumar, G., 2001. Reproductive biology of the speckled shrimp, *Metapenaeus monoceros* (Fabricius). *Indian J. Fish.*, 48: 1–8.

O'Donovan, P. and Cohen, D., 1984. The ovarian cycle during the intermoult in ovigerous *Macrobrachium rosenbergii. Aquaculture*, 36: 347–358.

Otsu, T., 1963. Biohormonal control of sexual cycle in the freshwater crab, *Potamon dobzani, Embryologica*, 8: 1–20.

Pandey, A.K., Sridar, N. and Mohamed, M.P., 1999a. Changes in the neurosecretory cells of eyestalk of female *Parapenaeopsis stylifera* in relation to gonadal maturation. In: *The Indian Agricultural Scientists and Farmers Congress (February 20–21, 1999)*. Bioved Research Society, Allahabad. p. 53.

Pandey, A.K., Sridar, N. and Mohamed, M.P., 1999b. Histomorphology of the neurosecretory cells of eyestalk of female prawn, *Panaeus indicus*, in relation to gonadal maturation. In: *The 20th Annual Session of the Academy of Environmental Biology (December 2–4, 1999)*. Department of Zoology, Andhra University, Visakhapatnam, p. 36.

Pandey, A.K., Sridar, N. and Mohamed, M.P. (2000). Histomorphology of the neurosecretory cells of eyestalk of female *Metapenaeus dobsonii* (Miers) in relation to gonadal maturation. In: *Eight Annual Session of Nature Conservators and National Symposium on Sustainable Development and Conservation of Fish Genetic Resources of Coldwaters (June 7–8, 2000)*. Department of Fisheries, Himachal Pradesh Krishi Visvavidhyalaya, Palampur, p. 29.

Primavera, J. H., 1985. Induced maturation and spawning in five month old *Penaeus monodon* Fabricius by eyestalk ablation. *Aquaculture*, 13: 355–359.

Quackenbush, L.S., 1986. Crustacean endocrinology: a review. *Can. J. Fish. Aquat. Sci.*, 43: 2271–2282.

Ra'anan, Z., Sagi, A., Wax, Y., Karplus, I., Hulata, G. and Kuris, A., 1991. Growth, size rank and maturation of the freshwater prawn, *Macrobrachium rosenbergii*: anylysis of marked prawns in an experimental population. *Biol. Bull.*, 181: 379–386.

Raghunathan, M.G. and Arivazhagan, A., 1999. Influence of eyestalk ablation and 5–hydroxytryptamine on gonadal development of a female crab, *Paratelphusa hydrodomous* (Herbst). *Curr. Sci.*, 76: 583–587.

Rajyalakshmi, T., 1961. Studies on maturation and breeding in some estuarine palaemonid prawns. *Proc. Natl. Inst. Sci.*, 27B: 179–188.

Rao, K.J., 1991. Reproductive biology of the giant freshwater prawn, *Macrobrachium rosenbergii* (de Man) from Lake Kolleru (A.P.). *Indian J. Anim. Sci.*, 61: 780–787.

Rao, N.Ch., Shakuntala, K. and Reddy, S.R., 1981. Studies on the neurosecretion of the thoracic ganglion in relation to reproduction of female *Macrobrachium lanchesteri. Proc. Indian Acad. Sci. (Anim. Sci.)*, 90: 503–511.

Rao, S.S., Sarojini, R. and Nagabhushanam, R., 1988. Histomorphology and histochemistry of the neurosecretory system in a penaeid prawn, *Metapenaeus affinis*. *J. Adv. Zool.*, 9: 113–122.

Sagi, A., Shoukrun, R., Levy, T., Barki, A., Hulata, G. and Karplus, I., 1997. Reproduction and molt in previously spawned and first–time spawning red–claw crayfish *Cherax quadricarinatus* females following eyestalk ablation during the winter reproductive–arrest period. *Aquaculture*, 156: 101–111.

Santiago, A. C. Jr., 1977. Successful spawning of cultured *Penaeus monodon* Fabricius after eyestalk ablation. *Aquaculture*, 11: 185–196.

Stella, S.V., Greco, L.S.L. and Rodriguez, E.M., 2000. Effects of eyestalk ablation at different times of the year on molting and reproduction of the estuarine grapsid crab, *Chasmagnathus granulata* (Decapoda, Brachyura) *J. Crus. Biol.*, 20: 239–244.

Subramoniam, T., 1999. Endocrine regulation of egg production in economically important crustaceans. *Curr. Sci.*, 76: 350–360.

Tan-Fermin, 1991. Effect of unilateral eyestalk ablation on ovarian maturation: Histology and oocyte size frequency of wild and pond-reared *Penaeus monodon* (Fabricius) broodstock. *Aquaculture*, 93: 77–86.

Upadhyay, A.S., Kulkarni, B.G. and Pandey, A.K., 2006. Recent advances in giant freshwater prawn culture in India. In: *Proceedings of Recent Advances in Applied Zoology* (H.S. Singh, A.K. Chaubey and S.K. Bhardwaj, eds.). Ch. Charan Singh University, Meerut, p. 116–132.

Victor, B. and Sarojini, R., 1985. Histomorphology of cerebral neurosecretory system in the caridean prawn, *Caridina rajadhari*. *Curr. Sci.*, 54: 817–818.

Wilder, M.N., Okumura, T., Suzuki, Y., Fusetani, N. and Aida, K., 1994. Vitellogenin production induced by eyestalk ablation in juvenile giant freshwater prawn, *Macrobrachium rosenbergii* and trial methyl farnesoate administration. *Zool. Sci.*, 11: 45–53.

Yano, I., 1992. Effect of thoracic ganglion on vitellogenin secretion in kuruma prawn, *Penaeus japonicus*. *Bull. Natl. Res. Inst. Aquaculture*, 21: 9–14.

Yano, I. and Wyban, J.A., 1992. Induced ovarian maturation of *Penaeus vannamei* by injection of lobster brain extract. *Bull. Natl. Res. Inst. Aquaculture*, 21: 1–7.

Chapter 6

Freshwater Fish Diversity of Tunga and Bhadra Rivers, Western Ghats, Karnataka

☆ *A. Shahnawaz and M. Venkateshwarlu*

Abstract

The fish diversity of aquatic system plays a major role in our national economy. The present study was undertaken with the purpose of assessing fish diversity of two important rivers of Karnataka namely Tunga and Bhadra. The study has shown that both the rivers support a rich fish diversity including 77 fish species. According to conservation status (IUCN-1994), 20 fish species are categorized into LR-nt, 33 NA, 10 UV, 7 EN, 2 DD, 4 and 1 as CR and LR-lc respectively. For the proper management and utilization of this fish wealth, it is mandatory to take up the sustainable steps to monitor the rivers constantly for future generations.

Introduction

Fishes are cold-blooded vertebrates which breathe by means of pharyngeal gills, propelling and balancing themselves with the help of fins. They are completely adapted to aquatic life than any other animals. They inhibit different kinds of habitats ranging from deep water to open spaces and from torrential streams to stagnant pools. They are closely bound to the landmasses and are inescapably confined to their own drainage systems. Their dispersal from one to another system can only be effected through hydrographic changes caused by the geological and climatic factors, with the exception of human interference. The fish species are the most numerous among all vertebrates, of which 41.2 per cent inhibit the freshwater and remaining 58.2 per cent are marine of the world (Figure 6.1 and Table 6.1). The fish species diversity, which is currently, recognized worldwide show the number of 25,000 species, of which 10,000 species are found in the freshwater ecosystems. The fish specialists estimate that at least 5,000 fish species await discovery (Jenkins, 2005). Among the faunal community, more attention has been paid towards fishes, because the whole human community depends on fishery resources for a variety of economic importance. They serve an excellent source of protein and food for increasing human population in developing countries. Fishes occupy significant status in socio-economic sector of South-Asian countries by providing the population not only the nutritious food but also an employment opportunity.

India is endowed with a vast expanse of inland waters in the form of rivers, canals, estuaries, natural and manmade lakes, backwaters, brackish water impoundments and mangrove wetlands. From the point of potential, the inland fish resources of are the richest in the world. In terms of production also, India ranks second in the world after China. Indian fisheries constitute an important sector of national economy for various regions. According to the national income statistics, Indian fisheries constitute about 2 per cent of the gross domestic product and 3.3 per cent of the income (Central statistics organization).

India has a vast potential for development of Inland fisheries. According to available resources and statistical data, there are about 1.6 million hectares of tanks, ponds and 3 million hectares of reservoirs and lakes, besides 29,000 km length of river which could be suitably developed for both capture and culture fisheries. The inland fisheries production will have to increase seven folds if 50 per cent of the minimum expected demand is to be satisfied from the inland sector.

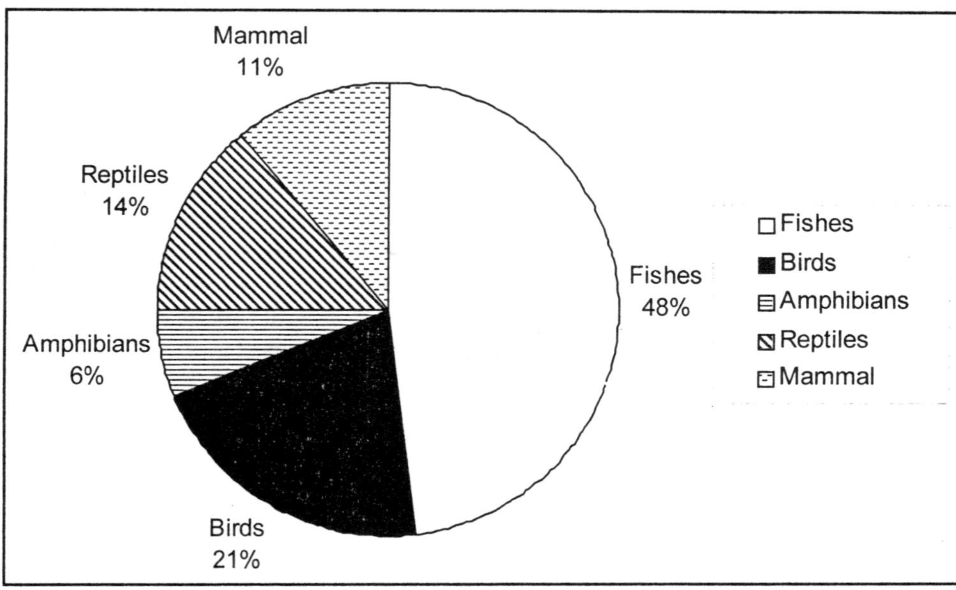

Figure 6.1: Percentage Proportion of Fishes by Groups of the some 41,600 Species of Recent Vertebrates

Table 6.1: Percentage (Per cent) of Fresh and Marine Water Fishes of the World

Sl.No.	Habitat Type	Percentage (Per cent)
1.	Freshwater fish species	41.2 per cent
2.	Marine fish species	
	a. Diodromous fishes	0.6 per cent
	b. Marine fishes (shore and continental up to 200 m) in warm water.	39.6 per cent
	c. Marine fishes (shore and continental up to 200 m) in cold water.	5.6 per cent
	d. Deep sea, benthic below 200m	6.4 per cent
	e. Epipelagic above 200m	1.3 per cent
	f. Deep sea pelagic below 200m	5.0 per cent

India is fortunate to be gifted with a bounty of natural habitats, including snow covered Himalayas, the Indo-Gangetic plains, the Deccan Plateu, Western Ghats, coastal regions and seas. Such a vast areas support a broad extent of water resources, harboring abundant fish genetic resources. The "Western Ghats" constitute one of the most important biodiversity "hotspots" of the world, which harbor a unique fish wealth and have a high degree of endemicity. In spite of constant and adverse human impact, it still supports a good number of endemic flora and fauna including fish diversity. The Western Ghats is so called because of its occupying position in Peninsular India between 8°–40°N and 7°–77°E from the Tapti Valley in Gujarat to Kanyakumari in Tamil Nadu. These series of hill ranges run north south along the west coast traveling the states of Gujarat, Maharashtra, Goa, Karnataka, Kerala and Tamil Nadu. Karnataka has enough number of water bodies including rivers, lakes and ponds. They are rich in fish diversity and play a vital role in state economy.

Tunga and Bhadra Riverine System

The Tunga and Bhadra rivers are the major sub-basins of the east flowing largest peninsular river, the Krishna and two west flowing rivers that are Seetanadi and Netaravati. The course of Tunga and Bhadra rivers are obstructed at several places for human use like hydroelectric projects and the main purpose of this is irrigation. Tunga river is dammed for irrigation and fishery purposes at Gajanur place. Bhadra river has a big reservoir project and the main purse of this project is power generation, irrigation and fishery.

Status of Fish Diversity in Tunga and Bhadra Rivers

The investigation of the fish diversity in the Bhadra and Tunga riverine system indicates that both these rivers are high in fish diversity and could be comparable to other parts of the Western Ghats. The study reports 77 fish species belonging to various families and genera. The family Cyprinidae dominated the other groups in the fish diversity. Though, the Cyprinoids are found to be more specious than Siluroids but the latter is found to be more diversified. Among the fish species recorded from the Bhadra and Tunga rivers, the following are considered as economically important and cultivable fishes including *Notopterus notopterus, Cyprinus carpio, Catla catla, Gonoproktopterus kolus, Glossogobius guiris, Oreochromis mosambica, Labeo rohita, Cirrhinus mrigala, Sperata seenghala, Sperata oar, Channa marulius* and *Channa punctatus.* These rivers also inhabit the ornamental fishes like *Puntius arulius, P. conchonius, P. amphibious, P. chola, P. sophore, P. filamentosus, P. ticto, P. sahyadreinsis, Rohtee ogilbii, P. sarana sabnastus* and *Etroplus maculatus.* River Tunga still inhabits probably the best freshwater fish of India that is Mahseer. They are recognized as well known sport and food fishes reaches weigh of 20 kg like *Tor khudree* and *Tor mussullah.* They are also considered as National heritage. But due to constant human and anthropogenic activities, Mahseers have come under the threat of threatened species. River Tunga also supports a very important fish species that is like *Danio rerio,* which is being extensively used for genetic research as a model species for neuro-science. The Table 6.2 shows the fish species diversity, habitat and its global biodiversity status of Tunga and Bhadra rivers.

Major Threats and Conservation Measures

Freshwater fishes being a major source of protein to many communities including human beings. They have got important role in recreational, educational and scientific disciplines and are considered very crucial ecologically as they are used as indicators for the health of aquatic ecosystems. Freshwater fishes of the rivers specially face a number of environmental threats resulting in decline in fish diversity or leads to total extinction in several places. Although speciation and extinction are natural phenomenon but human stress and intervention has drastically deteriorated the ichthyofaunal wealth of our nation.

This kind of selfish exploitation of this healthy ecosystem has led to unnatural rate of extinction and added to the number of threatened species. It is reported that about 10 per cent of fish species in the Western Ghats are under the category of extinction. In view of the above, it is strongly recommended to assess the conservation status of fish diversity especially in areas where they are still holding ground. Implementing various programs like ecological importance of fishes and their role in human health can ensure their long-term survival.

Conservation scenario of fish diversity is a difficult task, as most of our aquatic bodies are facing the threat of pollution. In addition to human intervention, there are several factors to be listed responsible for declining the fish species including the introduction of exotic species, deforestation, siltation, use of pesticides, heavy metal toxicity, dynamiting, poisoning, construction of dams and canals and unplanned irrigation. In regard to take up conservation measures for ichthyofauna, a holistic approach and public awareness are of vital importance. Provisions for creation and management of wildlife sanctuaries for viable conservation of threatened species of fishes and maintenance of genetic resources and biomass. This can be done within the habitat (*in situ*) and outside their natural habitat (*ex situ*). The *ex situ* conservation includes cryopreservation techniques (*i.e.* the long term live preservation sperms and eggs of fishes) in liquid nitrogen. Other conservation approaches include a baseline data of identifying areas of threat and finding out species living over there. Then documentation and identification of threatened species should be done carefully.

Table 6.2: Biodiversity Status, Abundance, Habitat and Size Range of Fishes of Bhadra and Tunga Rivers, Karnataka

Sl.No.	Species	Vernacular/ Local Name	Biodiversity Status IUCN-1990	Abundance	Habitat	Size (TL)
1.	*Labeo calbasu*	Karae-kolasa	LR nt	A-2	Rivers and ponds	90.0 cm
2.	*Labeo rohita**	Rohu	LR nt	A-2	Rivers and culture ponds	91.0 cm
3.	*Labeo angra*	–	NA	A-1	Rivers, lakes and ponds	22.0 cm
4.	*Labeo potail*	Hoobali	NA	A-1	Rivers and streams	30.0 cm
5.	*Labeo porcellus*	Kali-dindu	NA	A-1	Rivers and large streams	30.0 cm
6.	*Osteobrama neilli*	Koona	EN	A-2	Fast flowing rivers and streams	12.0 cm
7.	*Osteobrama cotio peninsularis*	Parake	NA	A-2	Rivers and streams	15.0 cm
8.	*Puntius sophore*	Gudda-pakke	LR nt	A-(3-4)	Rivers, streams and ponds	18.0 cm
9.	*Puntius chola*	Dodda-karsae	VU	A-(3-4)	Rivers, streams and tanks	12.0 cm
10.	*Puntius arulius*	Kempu-puthri	EN	A-2	Rivers, streams and lakes	9.5 cm
11.	*Puntius jerdoni*	Saymeen	EN	A-2	Rivers and reservoirs	46 cm
12.	*Tor khudree*	Bili-meenu	VU	A-0	Rivers, rapid streams and mountain lakes	46.0 cm
13.	*Glossogobius guiris*	Bhangi-sidda	LR-nt	A-(3-4)	Rivers, ponds and lakes	30 cm
14.	*Puntius sarana sabnastus*	Gende	NA	A-2	Rivers and tanks	31 cm
15.	*Puntius amphibius*	Pakke-meenu	NA	A-2	Streams and ponds in the plains	9.0 cm
16.	*Salmostoma boopis*	Malli	NA	A-2	Upper reaches of rivers	12.0 cm
17.	*Salmostoma sardinella*	–	NA	A-2	Lower reaches of rivers	15.0 cm
18.	*Barilius gatensis*	Agasa-gatti	NA	A-1	Hill streams with gravelly or rocky bed	15.0 cm
19.	*Barilius bendelisis*	Bilechi	LR-nt	A-2	Hill streams and rivers pebbly and rocky bottom	15.5 cm

Contd...

Table 6.2–Contd...

Sl.No.	Species	Vernacular/ Local Name	Biodiversity Status IUCN-1990	Abundance	Habitat	Size (TL)
20.	*Barilius canarensis*	–	DD	A-2	Clear streams and rivers with sandy and rocky bottom	8.0 cm
21.	*Danio aequipinnatus*	Bidrele-saslu	LR-nt	A-(3-4)	Clear streams and rivulets	8.5 cm
22	*Danio malabaricus*	–	NA	A-2	Hill streams	3.8 cm
23.	*Garra bicornuta*	Mukurti	CR	A-1	Rivers and mountain streams	10.0 cm
24.	*Garra mullya*	Kallu-korava	LR-nt	A-2	Rivers a mountain streams	16.8 cm
25.	*Cirrhinus fulungee*	Arja	LR-nt	A- (3-4)	Rivers and tanks	30.0 cm
26.	*Cirrhinus reba*	Arja	VU	A- (3-4)	Large streams, rivers and tanks	30.0 cm
27.	*Cirrhinus mrigala*	Mrigal	LR-nt	A-2	Large rivers and ponds	99.0 cm
28.	*Rohtee ogilbii*	Sipri	NA	A-2	Fast flowing rivers and streams	50.0 cm
29.	*Catla catla**	Catla	VU	A-2	Rivers, lakes and culture ponds	182.0 cm
30.	*Cyprinus carpio*	Gowri	LR-IC	A-2	Dams and culture ponds	28 cm
31.	*Hypsilobarbus kolus*	Kolasa	EN	A-2	Rivers with sandy bottom	30 cm
32.	*Osteochilichthys nashii*	Kantaka	CR	A-2	Hill streams and rivers	18 cm
33.	*Osteochilichthys thomassii*	Kantaka	CR	A-2	Hill streams and streams	32 cm
34.	*Botia straita*	Handi	NA	A-l	Rivers and mountain streams	7.8 cm
35.	*Sperata oar*	Kappu-surasi	NA	A-2	Large rivers, lakes and reservoirs	1.8 m
36.	*Sperata seenghala*	Bili-suragi	NA	A-2	Large rivers, canals, and inundated fields	1.5 m
37.	*Mystus cavasius*	Girlu	LR-nt	A- (3-4)	Rivers and lakes	46.0 cm
38.	*Mystus armatus*	Girlu	NA	A-1	Fresh and brackish waters	14.5 cm
39.	*Mystus krishnensis*	Haddu	NA	A-1	Rivers and reservoirs	116.0 cm

Contd...

Table 6.2–Contd...

Sl.No.	Species	Vernacular/ Local Name	Biodiversity Status IUCN-1990	Abundance	Habitat	Size (TL)
40.	*Rita gogra*	Karkandaka	LR-nt	A-2	Rivers	150 cm
41.	*Proeutropicthys taakree*	Halathi	NA	A-2	Freshwater and tidal rivers	40.0 cm
42.	*Ompok bimaculatus*	Godalae	EN	A-2	Shallow rivers and ponds with muddy or sandy bottom	45.0 cm
43.	*Ompok pabo*	Godalae	NA	A-2	Rivers, ponds and lakes	16.8 cm
44.	*Wallago attu*	Balae	LR-nt	A-1	Large rivers, tanks and reservoirs with muddy bottom	2 cm
45.	*Clarias batracus*	Murugodu	VU	A-2	Lower reaches of rivers	46.0 cm
46.	*Heteropneustes fossilis*	Chaelu	VU	A- (3-4)	Muddy rivers, ponds and lakes	30.0 cm
47.	*Nangra itchkea*	Urigan	NA	A-1	Rivers with rapid flow	5.0 cm
48.	*Chanda nama*	Bachanikemeenu	NA	A-2	Fresh and brackish with standing or running waters	11.0 cm
49.	*Nemachilichthys rueppelli*	Murrangi	NA	A-1	Rivers	7.4 cm
50.	*Etroplus maculatus*	Matak	NA	A-1	Rivers	8 cm
51.	*Oreochromis mossambica**	Jilebi	NA	A- (3-4)	Lower reaches of rivers, ponds and lakes	29 cm
52.	*Channa marulius*	Avalu	LR-nt	A-2	Rivers, lakes and deep pools in streams	1.2 m
53.	*Channa punctatus*	Korava	LR-nt	A-2	Rivers and ponds	25.0 cm
54.	*Rita pavimentatus*	Keehalu	NA	A-2	Large rivers	26.0 cm
55	*Notopterus notopterus*	Chappali	LR-nt	A- (3-4)	Rivers, lakes, swamps and canals	61 cm
56.	*Xenontodon cancilla*	Konti	VU	A-1	Slow moving streams, rivers and ponds	40.0 cm

Contd...

Table 6.2–Contd...

Sl.No.	*Species*	*Vernacular/ Local Name*	*Biodiversity Status IUCN-1990*	*Abundance*	*Habitat*	*Size (TL)*
57.	*Mastacembelus armatus*	Haavumeenu	LR-nt	A-(3-4)	Rivers, and lakes	61 cm
58.	*Parambassis thomassi*	Mulla-jabba	VU	A-2	Streams, dams and lakes	12 cm
59.	*Mystus malabaricus*	Girlu	EN	A-1	Rivers	15 cm
60.	*Puntius conconius*	Chikkakarsae	VU	A-2	Fast flowing hill streams	12.5 cm
61.	*Labeo* spp.	–	NA	–	–	–
62.	*Psilorhynchus tenura*	–	NA	A-1	Upper streams of rivers	7 cm
63.	*Danio rerio*	Patte-meenu	LR-nt	A-(3-4)	Slow moving streams, ponds and rice fields	3.8 cm
64.	*Lepidocephalus thermalis*	Hunase	NA	A-2	Hill streams	8 cm
65.	*Puntius filamentosus*	Kijan	NA	A-2	Clear streams, Donds and lakes	18.0 cm
66.	*Aplocheilus lineatus*	Moogu-malli	NA	A-2	Hill streams, rivers, tanks and reservoirs	7 cm
67.	*Schistura semiarmata*	–	VU	A-2	Upstreams of rivers	8 cm
68.	*Amplypharyngodon mola*	Enapu-pakke	NA	A-1	Rivers and hill streams	20 cm
69.	*Neotropius khavalchor*	Illi-meenu	DD	A-2	Upper reaches of rivers	15 cm
70.	*Hypselobarbus thomassi*	–	NA	A-1	Upper reaches of rivers	18 cm
71.	*Hypselobarbus lithopidos*	–	EN	A-1	Upper reaches of rivers	30 cm
72.	*Tor mussullah*	Bili-meenu	CR	A-0	Rivers	150 cm
73.	*Puntius sayhadriensis*		NA	A-2	Clear streams	57.0 cm
74.	*Balitora mysorensis*	Kallu-korava	NA	A-2	Torrential streams	5.0 cm

Contd...

Table 6.2–Contd...

Sl.No.	*Species*	*Vernacular/ Local Name*	*Biodiversity Status IUCN-1990*	*Abundance*	*Habitat*	*Size (TL)*
75.	*Puntius ticto*	Naya-paisa	LR-nt	A-1	Rivers, streams and ponds with clear waters	20.0 cm
76.	*Glyptothorax lonah*	Kanta	LR-nt	A-1	Upper streams of rivers	15.0 cm
77.	*Rasbora daniconius*	Golai	LR-nt	A- (3-4)	Forest streams, ponds to mountain streams	7.0 cm

Abundance: A-0: Very rare; A-1: Rare, A-2: Common, A-(3-4): Very common; CR: Critically endangered; EN: Endangered; DD: Data deficient; LR-lc: Lower risk least concern; LR-nt: Lower risk-near threatened; VU: Vulnerable; NA: Not assessed; TL: Total length

* Introduced species.

The threatened freshwater fishes should be included under the Wildlife Protection Act, so that greater attention and protection can be given to them.

Only two Mahseer species like *Tor Khudree* (the Deccan Mahseer) and *Tor musullah* (the Humpback mahseer) so far known from streams and rivers of the Western Ghats, Karnataka are under the great threat of extinction. Though the Karnataka has about 10 Mahseer sanctuaries situated at various rivers and various agencies are involved in the conservation of genetic diversity of Mahseers. Some of them to be listed like the Karnataka State Department of Fisheries (DOF), College of Fisheries (COF), Manglore, Temple Trusts and Non-Governmental Organizations (NGO), but the threats from poachers are still remaining. Identification and documentation of the areas where Mahseers are dwindling are needed and construction of fish ladders along the dams are required in order to ensure the breeding migration of Mahseers. Therefore, by strong implementation of the above said conservation measures we can save the ichthyofaunal diversity of these riverine system from being extincted.

References

Anon., 1998. *Report of the Workshop on Conservation Assessment and Management Plan (CAMP) for Freshwater Fishes of India.* Zoo Outreach Organization and NBFGR, Lucknow, 22–26 September 1997, pp. 156.

IUCN, 1994. *Red List of Threatened Animals.* IUCN, Gland.

Jayaram, K.C., 1999. *The Freshwater Fishes of the Indian Region.* Narendra Publishing House, Delhi, pp. 551.

Jenkins, A.P., 2005. The unique and unheralded freshwater fish diversity of the Pacific Island region. *Wetlands,* 12: 5.

Lagler, K.F., Bardach, J.E., Miller, R.R. and Passino, D.R.M., 2007. *Ichthyology.* John Wiley and Sons, New York.

Maitland, P.S., 1995. The conservation of freshwater fish: Past and present experience. *Biol. Conserve.,* 72: 259–270.

Menon, A.G.K., 1999. *Checklist–Freshwater Fishes of India.* Zoological Survey of India, Calcutta, Occasional Paper, 175: 1–366.

Mittermeir, R.A., Myers, N.P.R. and Mittermeir, C.G., 2000. Hotspots, earths biological riches and most endangered terrestrial eco-region. *CEMEX and Conservation International,* pp. 430.

Nelson, J.S., 1994. *Fishes of the World,* 3rd edn. John Wiley and Sons Inc., New York, USA, XX+600 pp.

Pandey, B.N. and Kulkarni, G.K., 2005. *Fisheries and Fish Biotechnology.* APH Publishing Corporation, New Delhi.

Ponnaiah, A.G. and Gopalakrishnan, A., 2000. *Endemic Fish Diversity of Western Ghats.* NBFGR–NATP Publication–I. National Bureau of Fish Genetic Resources, Lucknow, U.P. India, pp. 347.

Ranga, M.M and Shammi, Q.J., 2005. *Fish Biotechnology.* Agrobios Publications, New Delhi.

Shammi, Q.J. and Bhatnagar, S., 2002. *Applied Fisheries*, Agrobios Publications, New Delhi, pp. 281.

Srivastava, U.K. and Vasthala, S., 1984. *Inland Fishery Resources in India.* Concept Publishing Company, New Delhi.

Talwar, P.K. and Jhingran, A., 1991. *Inland Fishes of India and Adjacent Countries.* Oxford and IBH Publishing Co. Pvt. Ltd., New Delhi, 2(19): 1158.

Chapter 7

Population Growth and Demography of *Moinodaphnia macleayi* (King, 1853) (Crustacea : Cladocera) in Relation to Algal (*Chlorella vulgaris* or *Scenedesmus acutus*) Food Density

☆ *S. Nandini and S.S.S. Sarma*

Abstract

We evaluated the patterns of population growth and the life history characteristics of a common tropical cladocera *Moinodaphnia macleayi* offered two algal food species (*Chlorella vulgaris* and *Scenedesmus acutus*) at three different concentrations (dry weight, 5.8, 116 and 23.2 µg /ml). *M. macleayi* showed increased abundances with increasing availability of algal food in the medium. The cladocerans reached higher abundances on *Scenedesmus* as compared to those on *Chlorella*. The rate of population increase (*r*) derived from the growth data varied from 0.13 to 0. 30 per day, depending on the food type. Age-specific survivorship curves of *M. macleayi* showed better survival of younger age groups on *Scenedesmus*, although the maximal duration of lifespan was higher in *Chlorella*, especially at low food density. Fecundity was lower but spread over a longer period on *Chlorella*, while it was higher and restricted to two peaks on *Scenedesmus*. Within the algal types, regardless food concentration, net reproductive rates showed significant differences depending *Chlorella* or *Scenedesmus* but for the rest of the variables (average lifespan, life expectancy at birth, gross reproductive rate, generation time and the rate of population increase) algal type had no significant influence. These results were discussed in relation to the pertinent studies available in literature.

Keywords: *Life history, Cladocera, Population growth, Reproductive rate, Lifespan, Alga.*

Introduction

Cladocerans in terms biomass are the dominant group in most freshwater ecosystems. Though there are about 95 genera of cladocerans, much of the ecological, aquacultural and ecotoxicological works are concentrated on a few genera such as *Daphnia, Moina* and *Ceriodaphnia* (Forró *et al.*, 2008). There are also other genera such as *Diaphanosoma* and *Moinodaphnia* which are also common in many freshwater bodies, particularly those in tropical waters (Dodson and Frey, 2001). Unlike typically planktonic cladocerans such as *Daphnia* and *Moina* (Dutta *et al.*, 2007), *Moinodaphnia*, which shares some of the morphological characters of these genera, is non-planktonic and monotypic (Elmoor-Loureiro, 2007). Recent ecotoxicological studies show that *M. macleayi* is sensitive to stress from toxicants (Barata *et al.*, 2002). However, in relatively non-contaminated waterbodies, the factors that largely affect cladocerans including *M. macleayi* are possibly the temperature and food densities (Dodson and Frey, 2001).

In general, zooplankton species show similar responses with increase in temperature, *i.e.* accelerate metabolism resulting in reduced lifespan and enhance growth rates (Vijverberg, 1980). However, with reference to food density, the response of cladocerans varies considerably. Though it is known that increase in food density results in higher offspring production for zooplankton (Nandini and Sarma, 2000), within a certain narrow range of high food density, there is no significant difference in the growth rates but above this level, cladocerans actually die of starvation because their appendages and filtration apparatus get clogged with high particulate density (Porter *et al.*, 1982). Therefore, in waterbodies that seasonally show high oscillations in the algal production, the density of cladocerans may show the positive correlation only until a certain food density, and at lower Secchi levels, the corrections are rather feeble.

Most cladocerans including *Moinodaphnia* feed edible phytoplankton such as green algae in natural waterbodies (Dodson and Frey, 2001). Therefore, the use of algae as the natural diet for laboratory cultures of cladocerans is generally recommended (Hyne *et al.*, 1993). In addition, since algal densities in nature vary seasonally, this is considered as an import variable for understanding the life history characteristics of different zooplankton species (DeMott, 1989). Population growth and demography studies are considered complementary for understanding the life history strategies of zooplankton species (Nandini *et al.*, 2000). Under the laboratory test conditions, in population growth studies, there is an intra-specific competition among the individuals of the same species since the offered food is generally limiting. On the other hand, in life table demographic studies the neonates produced for each clutch are removed at frequent intervals (*e.g.*, every 24h) and therefore intraspecific competition, in any, is less intense. Consequently, the limiting food effects are minimum and therefore, a given zooplankton may have a chance to realize the full reproductive potential. Thus, when one can evaluate the food effects on the life history characteristics if both life table demography and population growth experiments are conducted simultaneously.

The aim of this study was therefore to evaluate the population growth and the life history characteristics of *M. macleayi* fed two algal species (*Chlorella vulgaris* and *Scenedesmus acutus*) under three different food concentrations (5.8, 11.6 and 23.2μg dry weight/ml).

Material and Methods

Moinodaphnia macleayi (King, 1853) was obtained collected from a small eutrophic pond in the Veracruz City (Mexico). Clonal population of *M. macleayi* was established starting from a single parthenogenetic individual on moderately hardwater (EPA medium). For the stock *M. macleayi* cultures as well as for the experiments, we maintained temperature at 25±0.5°C, photoperiod: 12L:12D, pH of the medium: 7.0-7.5 and dissolved oxygen: 7-8 mg l^{-1}. The cladocerans were fed one of the two algal species (*Chlorella vulgaris* (strain CL-V-3 CICESE, Ensenada Baja California, Mexico) or *Scenedesmus acutus* (f. Alternans Hortobagy No. 72, University of Texas, USA). The EPA medium was prepared by dissolving 96 mg $NaHCO_3$, 60 mg $CaSO_4$, 60 mg $MgSO_4$ and 4 mg KCl in 1 L of distilled water (Weber, 1993).

Mass cultures of the two algal species were separately established using Bold's basal medium, supplemented with 0.5 g of sodium bicarbonate every alternate day as carbon source (Borowitzka and Borowitzka, 1988). The algae in exponential phase of growth were harvested, centrifuged at 3000 rpm for 5 min., rinsed and resuspended in distilled water. Algal density was estimated using haemocytometer. Since the cell size of two algal species is different, we considered the offered food in equivalents of dry weight and we used 3 food densities *viz.*, 5.8 (low), 11.6 (medium) and 23.2 (high) μg/ml) for each algal species (Alva-Martínez *et al.*, 2004).

Population Growth Experiments

For population growth experiments we used a total of 24 test jars (= 2 food types X 3 concentrations X 4 replicates) of 100 ml capacity, each containing 50 ml EPA medium with one of the algal diet at a chosen concentration. The initial density of (mixed age groups, but predominantly juveniles) cladocerans in each jar was 10 individuals (= 0.2 ind./ml), introduced individually under a stereo microscope at 20X. Following the initiation of the growth experiment, daily we estimated the density of *M. macleayi* in each jar and transferred to fresh jars containing appropriate algal food of a specified density. The experiments were terminated after 15 days by which time cladocerans in most replicates began to decline.

Based on the data collected, we derived the rate of population increase (r) using the exponential phase of the population growth (Krebs, 1985): $r = (\ln N_t - \ln N_0)/t$, where N_0 is the initial population density, N_t is the density after the time t (days).

The differences in the rate of population increase of *M. macleayi* fed either *Chorella* or *Scenedesmus* at different food levels were statistically evaluated using analysis of variance (ANOVA) and Tukey's test (Sokal and Rohlf, 2000) using Statistica version 5 (StatSoft Inc., USA).

Life Table Demography Experiments

For demographic study, we used the same experimental design described above. However, here we used only neonates (*i.e.*, cohort) (<24 h following birth) in each test jar. Daily we quantified the surviving individuals of the original cohort and transferred to fresh jars containing appropriate algal food type and concentration. Neonates produced and dead adults, if any, were also quantified but removed. The life table demography experiments were terminated when every individual of the originals cohort had died.

Based on the data collected, we derived selected life history variables (average lifespan, life expectancy at birth, gross reproductive rate, net reproductive rate, generation time and the rate of population increase). Age-specific survivorship, life expectancy and fecundity curves were calculated. The following formulae were used for deriving the life history variables (Krebs, 1985):

l_x = Proportion of survivorship per day

m_x = Proportion of offspring produced per female per day

Life expectancy: $$e_x = \frac{T_x}{n_x}$$

where, T_x = number of individuals per day

n_x = number of living individuals at the initiation and the age x (days)

$$\text{Gross reproductive rate} = \sum_{0}^{\infty} m_x$$

$$\text{Net reproductive rate } R_o = \sum_{0}^{\infty} l_x \cdot m_x$$

$$\text{Generation time: } T = \frac{\sum l_x \cdot m_x \cdot x}{R_o}$$

Rate of population increase, Euler-Lotka equation (solved iteratively)

$$\sum_{x=w}^{n} e^{-rx} \cdot l_x \cdot m_x = 1$$

where, r = rate of population increase per day, w = age at maturity (days)

The differences in the life history variables of *M. macleayi* cultured on either *Chlorella* or *Scenedesmus* under different food levels were statistically evaluated using ANOVA and *post hoc* tests as mentioned earlier.

Results

Population Growth Study

The population growth curves of *M. macleayi* showed in general increased abundances with increasing availability of algal food in the medium. When *Chlorella vulgaris* was offered as diet, especially at higher food densities, there was a lag phase of 4-6 days. Under similar conditions, when *Scenedesmus acutus* was used as diet, there was no distinct lag phase. In addition, compared to *Chlorella*, *M. macleayi* grew to higher abundances on *Scenedesmus* (Figure 7.1). The rate of population increase (*r*) derived from the growth data varied from 0.13 to 0.30 per day, depending on the food

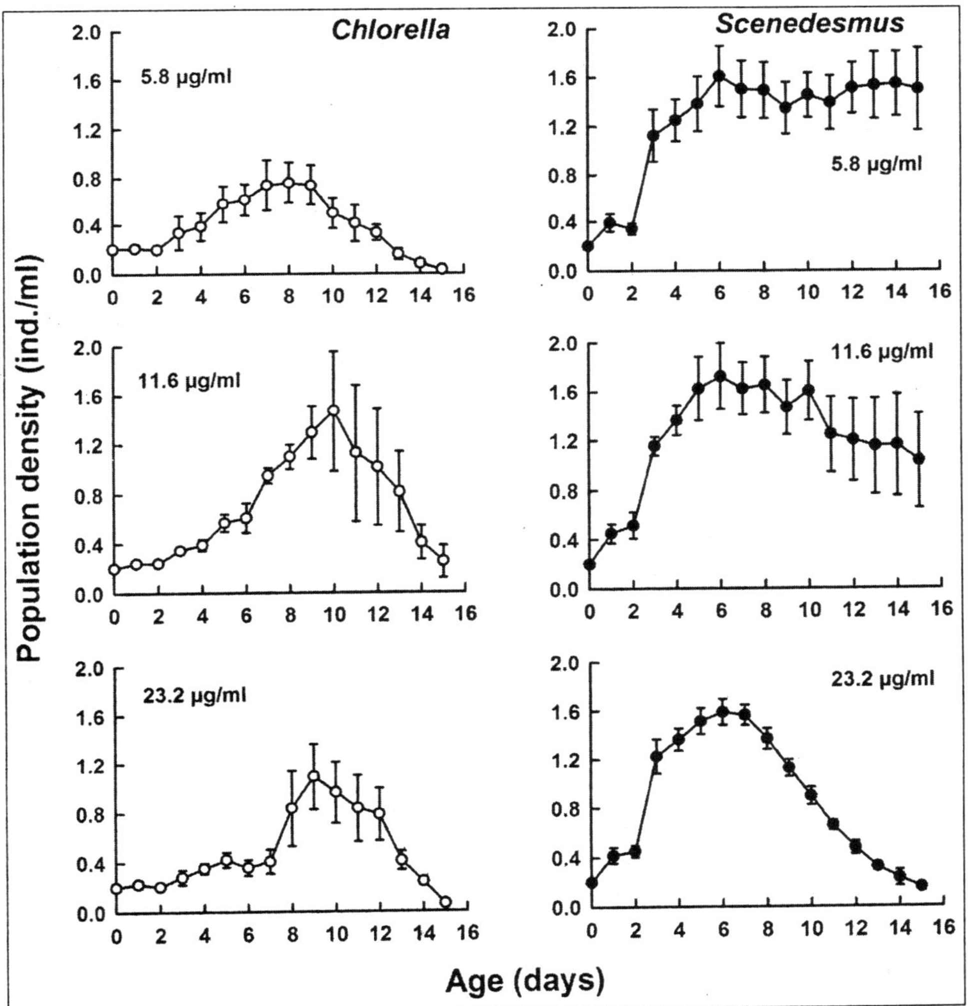

Figure 7.1: Population Growth Curves of *Moinodaphina macleayi* Fed *Chlorella vulgaris* (Open circles) or *Scenedesmus acutus* (Closed circles) Under Different Concentrations (Dry weight, µg/ml). Shown are the mean ± standard error based on four replicates.

type and density but in general, the *r* of *M. macleayi* was higher when fed *Scenedesmus* (Figure 7.2). Statistically, the *r* was significantly influenced by the algal type; and within a given algal type, food level was significant only for *Chlorella* ($p<0.05$; F-test, ANOVA, Table 7.1).

Tabla 7.1: Results of One-way Analysis of Variance (ANOVA) Performed for the Selected Population Level Variables of *Moinodaphina macleayi* Fed *Chlorella vulgaris* or *Scenedesmus acutus* Under Different Concentrations. DF = degrees of freedom, MS = mean square; F = F-ratio. *** = p < 0.001, ** = p < 0.01, * = p < 0.05; ns = non-significant (= p > 0.05).

Variable	*DF Effect*	*MS Effect*	*DF Error*	*MS Error*	*F-raito*
Population growth study					
Rate of population increase					
Food: *Chlorella*	2	0.010	9	0.002	4.27*
Food: *Scenedesmus*	2	0.001	9	0.003	0.64ns
Chlorella vs *Scenedesmus*	1	0.075	22	0.003	22.84***
Life table demography study					
Average lifespan					
Food: *Chlorella*	2	6.28	9	3.42	1.83ns
Food: *Scenedesmus*	2	3.56	9	0.74	4.76*
Chlorella vs *Scenedesmus*	1	3.92	22	2.60	1.50ns
Life expectancy at birth					
Food: *Chlorella*	2	6.28	9	3.41	1.83ns
Food: *Scenedesmus*	2	3.56	9	0.75	4.76*
Chlorella vs *Scenedesmus*	1	3.92	22	2.60	1.50ns
Gross reproductive rate					
Food: *Chlorella*	2	54.17	9	5.57	9.72**
Food: *Scenedesmus*	2	96.81	9	13.65	7.09*
Chlorella vs *Scenedesmus*	1	22.95	22	21.58	1.06ns
Net reproductive rate					
Food: *Chlorella*	2	8.22	9	2.80	2.93ns
Food: *Scenedesmus*	2	42.97	9	2.13	20.16***
Chlorella vs *Scenedesmus*	1	41.61	22	6.67	6.23*
Generation time					
Food: *Chlorella*	2	4.13	9	0.98	4.21ns
Food: *Scenedesmus*	2	2.68	9	0.36	7.56*
Chlorella vs *Scenedesmus*	1	0.13	22	1.16	0.11ns
Rate of population increase					
Food: *Chlorella*	2	0.01	9	0.03	0.32ns
Food: *Scenedesmus*	2	0.01	9	0.002	4.97*
Chlorella vs *Scenedesmus*	1	0.01	22	0.01	1.00ns

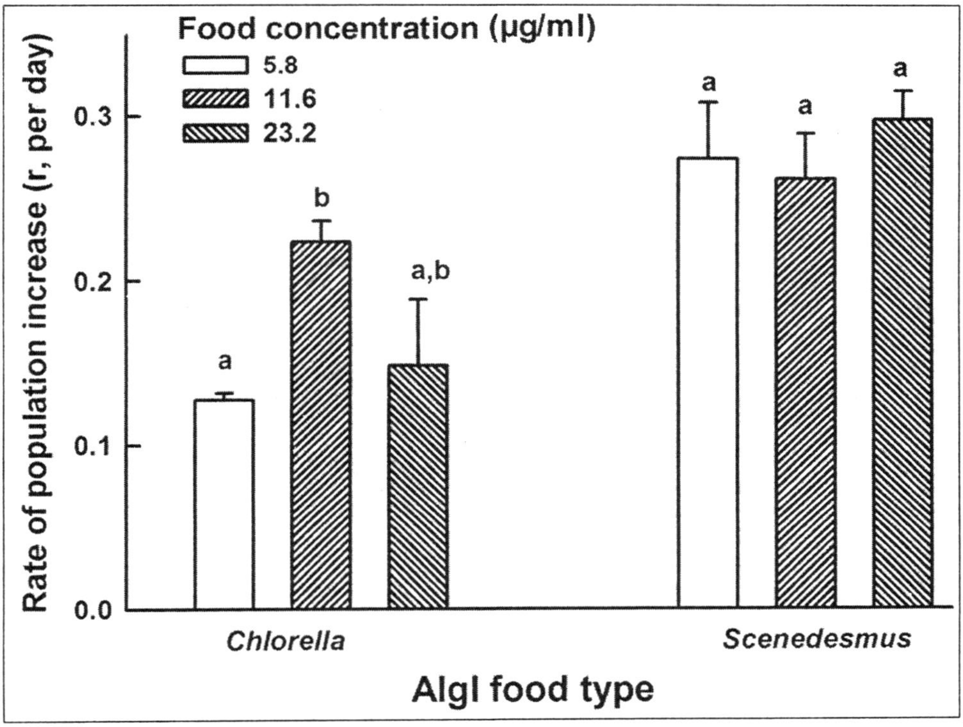

Figure 7.2: Rate of Population Increase (*r* per day) of *M. macleayi* Fed *Chlorella vulgaris* or *Scenedesmus acutus* Under Different Concentrations (Dry weight, µg/ml). Shown are the mean ± standard error based on four replicates. For a given food type, data bars carrying similar alphabets are not significant (>0.5, Tukey tests).

Life Table Demography Study

Age-specific survivorship curves of *M. macleayi* fed *Chlorella* or *Scenedesmus* showed better survival of younger age groups on *Scenedesmus*, although the maximal duration of lifespan was higher in *Chlorella*, especially at low food density (Figure 7.3). Age-specific life expectancy curves showed decreased lifespan with increasing age of the cohort. However this pattern was more pronounced when fed *Scenedesmus* (Figure 7.4). Age specific fecundity patterns were similar on a given algal diet, except that the magnitude varied depending on the concentration. Thus when fed *Chlorella*, the offspring production was lower but spread over a longer period of time. On the other hand, when fed *Scenedesmus*, the offspring production was higher and restricted to two peaks. In both the algal types, increased food availability decreased the offspring production (Figure 7.5).

Data on the selected life history variables (average lifespan, life expectancy at birth, gross and net reproductive rates, generation time and the rate of population increase) of *M. macleayi* fed the two algal types under different concentrations are presented in Figure 7.6. When *Chlorella* was used as diet, except the gross reproductive

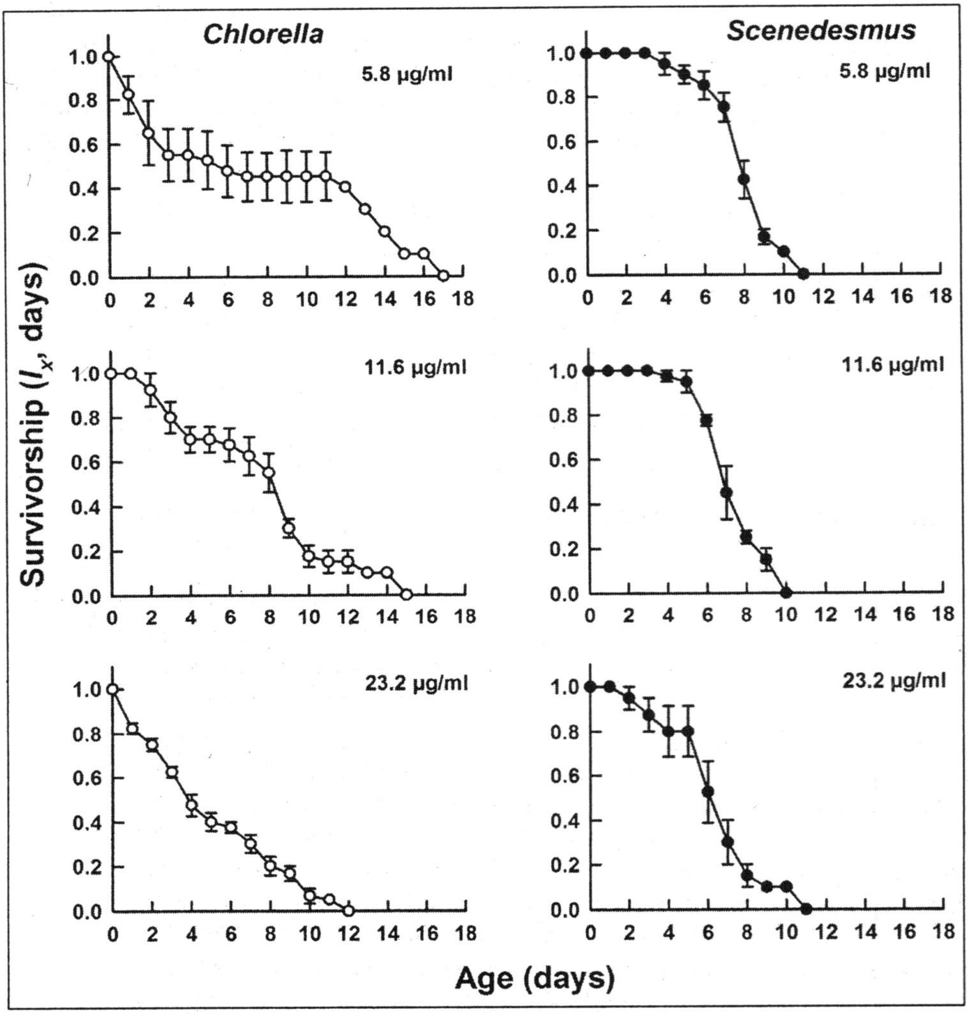

Figure 7.3: Age-Specific Survivorship (l_x) Curves of *Moinodaphina macleayi* Fed *Chlorella vulgaris* (Open circles) or *Scenedesmus acutus* (Closed circles) Under Different Concentrations (Dry weight, µg/ml). Shown are the mean ± standard error based on four replicates (cohorts).

rates, rest of the variables did not show significant ($p>0.05$) differences with food concentration. On the other hand, all the variables were significantly affected by the concentration of *Scenedesmus*. Independent of food concentration, net reproductive rate showed significant differences when fed *Chlorella* or *Scenedesmus* ($p<0.05$) and the rest of the variables were not significantly affected by the algal type ($p>0.05$, F-test, Table 7.1).

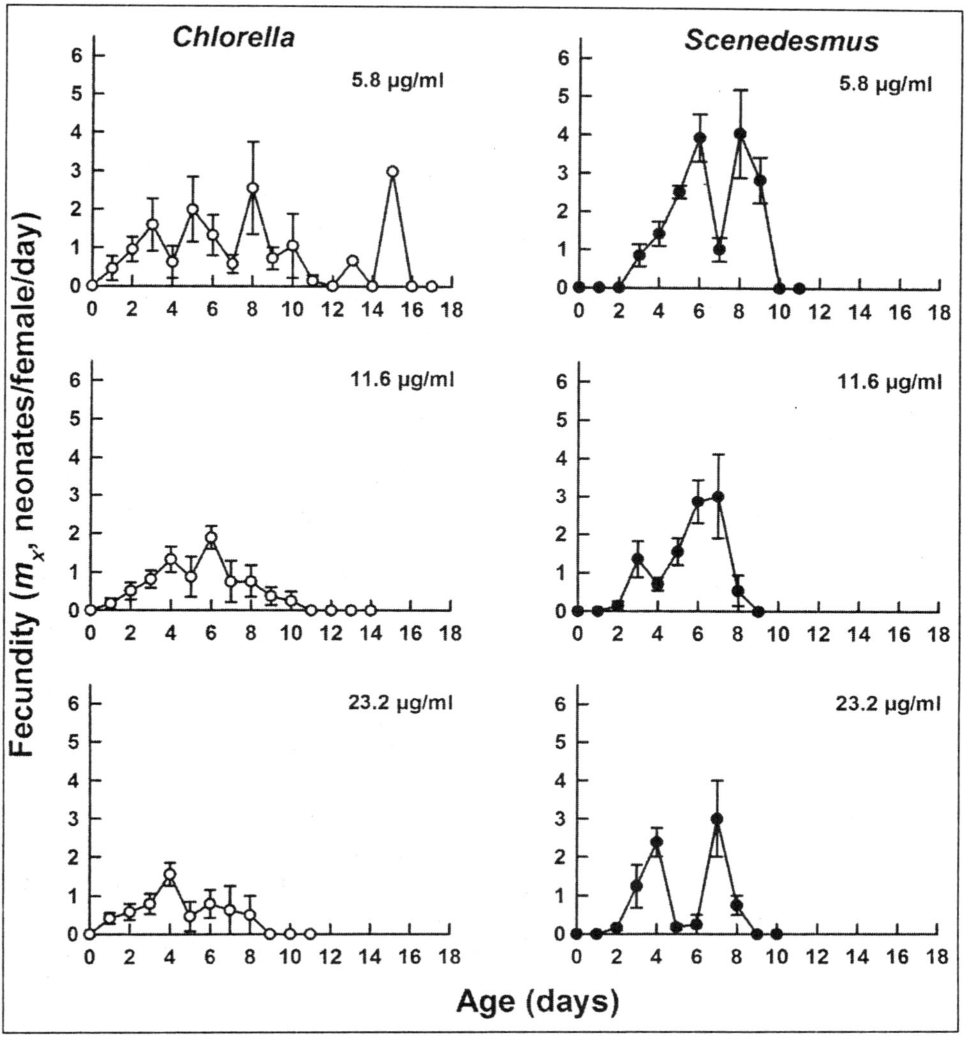

Figure 7.4: Age-Specific Fecundity (m_x) Curves of *M. macleayi* Fed *Chlorella vulgaris* (Open circles) or *Scenedesmus acutus* (Closed circles) Under Different Concentrations (Dry weight, µg/ml). Shown are the mean ± standard error based on four replicates (cohorts).

Discussion

Moinodaphnia macleayi is generally found in eutrophic waters and therefore it may have adapted to live under conditions of high particulate food (López- López and Serna-Hernández, 1999). Therefore the concentrations used represent the natural levels to which the test species is exposed in ponds and lakes. In addition, the natural diet of most cladocerans including *M. macleayi* is alga (Hyne *et al.*, 1993). The two

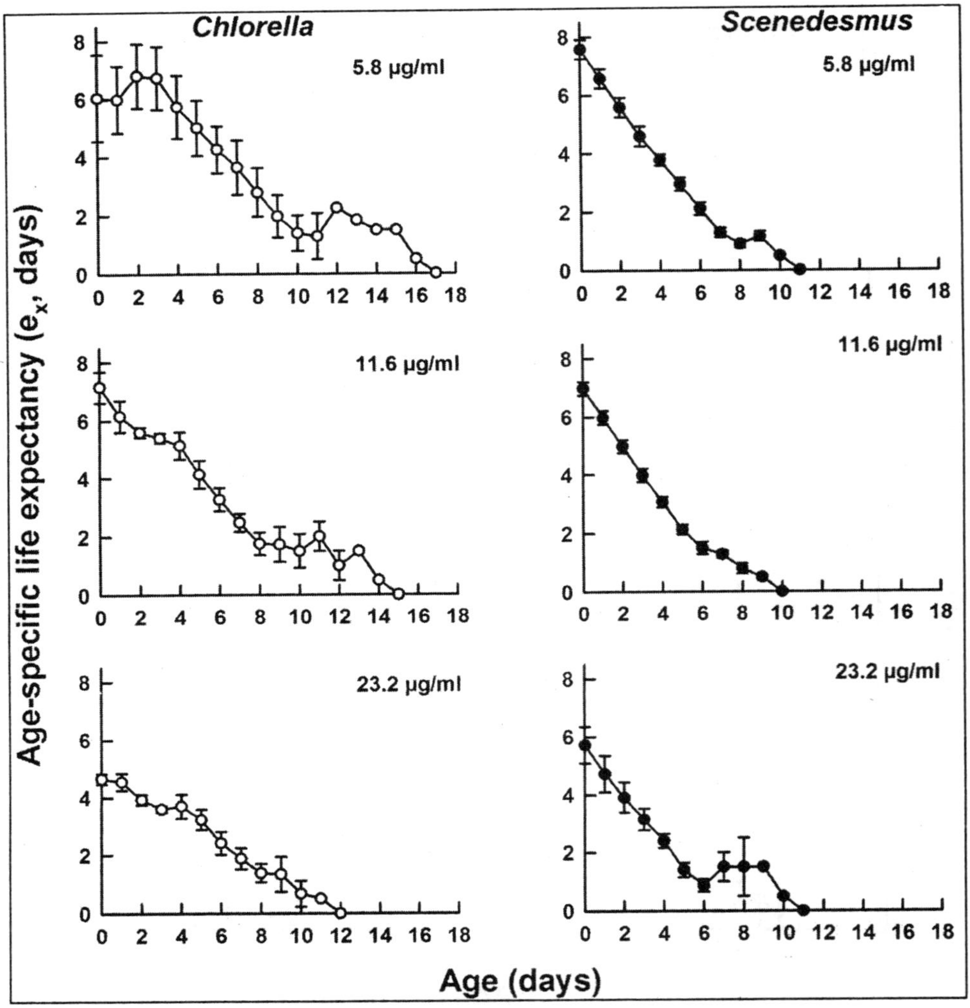

Figure 7.5: Age-Specific Life Expectancy (E_x) Curves of *M. macleayi* Fed *Chlorella vulgaris* (Open circles) or *Scenedesmus acutus* (Closed circles) Under Different Concentrations (Dry weight, µg/ml). Shown are the mean ± standard error based on four replicates (cohorts).

algal species used in this are also common in freshwater bodies in which *M. macleayi* is encountered.

In this study we evaluated effect of two algal diets at different concentrations on the population growth and life table characteristics of *M. macleayi*. Therefore comparisons with other cladoceran genera are possible. Thus, by convention, cladocerans are more preferentially fed *Scenedesmus* than on other algal types such as *Chlorella* (Ovie and Egborge, 2002). There is some indication that *Chlorella* is more suitable to rotifers while for cladocerans *Scenedesmus* is appropriate (Flores-

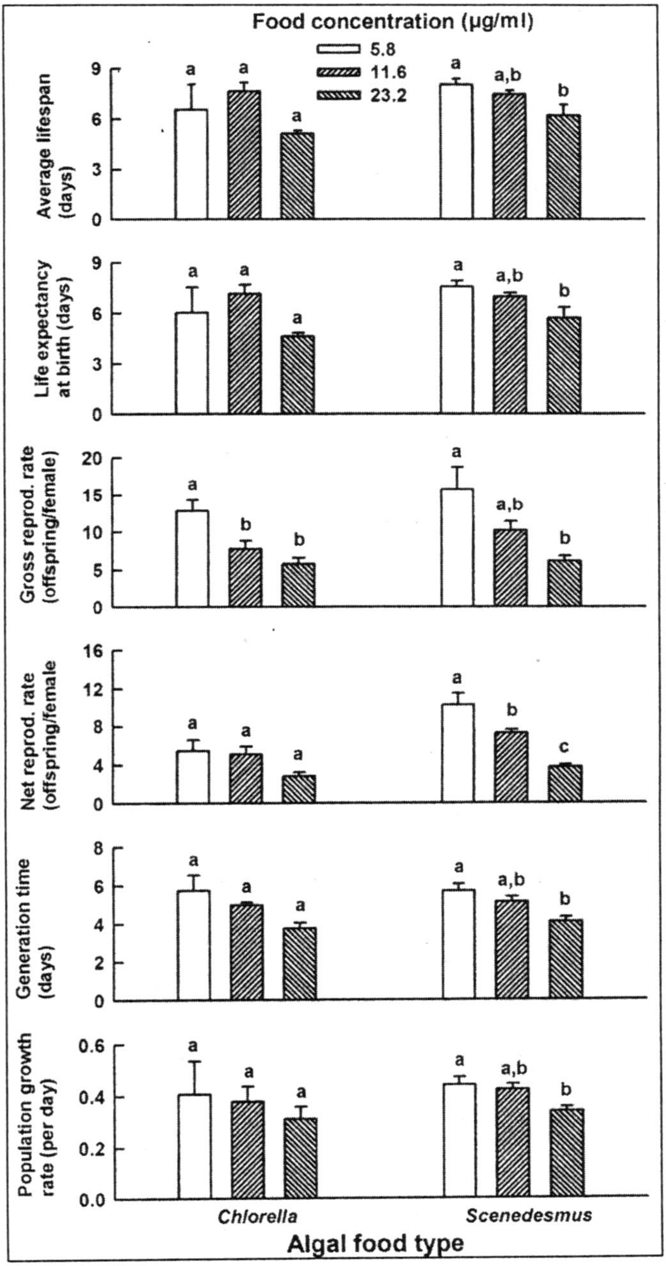

Figure 7.6: Selected Life History Variables (Average lifespan (days), Life Expectancy at Birth (Days), Gross Reproductive Rate (Offspring/female), Gross Reproductive Rate (Survival-weighted offspring/female), Generation Time (Days) and Rate of Population Increase (r, day) of *M. macleayi* Fed *Chlorella vulgaris* or *Scenedesmus acutus* Under Different Concentrations (Dry weight, µg/ml). Shown are the mean ± standard error based on four replicates (cohorts).

Burgos *et al.*, 2003). Our results from the population growth and life table studies showed that for majority of the variables derived here, there was no significant difference between *Scenedesmus* and *Chlorella* within the range of concentrations selected. However, there were significant differences in the life history variables of *M. macleayi*, with reference to food concentrations, especially for *Scenedesmus*. Thus it appears the differences in the life history variables of *M. macleayi* would become reduced with the use of wide food concentrations of these two algal food types.

Regardless of the food type, the data obtained from the population growth and life table of *M. macleayi* are comparable to those reported in literature. For example, Nandini and Sarma (2000) cultured different genera of cladocerans on a wide range of *Chlorella* levels the growth rates reported there (0.01 to 0.28 per day) agree with those obtained in this work. Sarma *et al.* (2005) compared the life history variables of tropical and temperate genera of cladocerans. *Moinodaphnia* was also included there because it is predominantly found in warm waters. Since our experiments were conducted at higher temperature (25°C), it is possible to compare the life history variables of *M. macleayi* with other tropical taxa. Most tropical cladocerans have the lifespan ranging from 2 to 3 weeks, which agrees with the data obtained in this work. In terms of reproduction-related variables, both the gross and net reproductive variables, the reported range from literature for most species of cladocerans is 10-40 and 5-20 offspring per female, respectively. Earlier studies on *M. macleayi* have shown that the net reproductive rates varied from 9 to 28 (offspring per female) (Ferrão-Filho *et al.*, 2000). The present data on the net reproductive rates (3-10 offspring per female), though are on the lower side, agreeable with the published data. The high growth rates on low food levels also explain the high abundances of this cladocerean in the littoral of shallow ponds, under conditions of probable food limitation. The demographic characteristics of this species are also more similar to members of the genus *Moina* than to those of *Daphnia* (Sarma *et al.*, 2005).

Most of the life history variables derived here either did not show significant differences with increasing algal density or showed reduced values with the highest food concentration offered. This suggests that at food concentrations higher than 5.8 µg/ml do not necessarily lead to enhanced survivorship or reproductive output for *M. macleayi*. There is also considerable information that higher food levels actually inhibit reproductive output in cladocerans (Ovie and Egborge, 2002). In this study we did not evaluate the feeding and filtration rates of *M. macleayi* on either *Chlorella* or *Scenedesmus* and therefore, it was not possible to attribute our observations on the basis of assimilation efficiencies. However, there is sufficient information in literature that suggests the decreased assimilation efficiency with increasing food availability in the medium, especially at food concentrations higher than the incipient levels (Downing and Rigler, 1984). This may be responsible for the decreased gross and net reproductive rates and the rate of population increase of *M. macleayi* when fed *Scenedesmus*. When used as *Chlorella* too, there was a similar trend, but statistically was not significant due to strong differences in the replicate values. Our study did not consider lower algal densities because the species we collected was from a eutrophic pond and therefore was expected for having adapted to higher food densities. This was evident too here. Thus, *M. macleayi* in our study continued to survive and reproduce at a food concentration as high as 23.2 µg/ml.

Conclusions

Our study showed that *M. macleayi* was able reproduce on *Chlorella vulgaris* as well as on on *Scenedesmus acutus* within the tested food concentrations of 5.8 to 23. 2 µg/ml. For most the demographic parameters, there were no significant differences when fed *Chlorella* or *Scenedesmus*. Also within the offered food levels of *Chlorella*, except the gross reproductive rates, rest of the demographic variables did not vary significantly. However, all the derived life history variables of *M. macleayi* were significantly affected by the *Scenedesmus* concentration in the medium. In this study though differences in the life history variables of *M. macleayi* fed *Chlorella* or *Scenedesmus* were not significant, it remains unknown if this stands valid when much lower food concentrations (at or near the incipient levels) are used. Thus, the future works need to focus on these aspects for drawing the generalizations on the life history strategies of *M. macleayi*.

Acknowledgements

We thank Prof. Bansi Lal Kaul for critically reviewing our contribution. This was supported by the research projects PAPIIT-IN203107 and PAPIIT-IN201907 (UNAM, Mexico).

References

Alva-Martínez, A.F., Sarma, S.S.S. and Nandini, S., 2004. Population growth of *Daphnia pulex* (Cladocera) on a mixed diet (*Microcystis aeruginosa* with *Chlorella* or *Scenedesmus*). *Crustaceana*, 77: 973–988.

Barata, C., Baird, D.J. and Soares, A.M.V.M., 2002. Demographic responses of a tropical cladoceran to cadmium: effects of food supply and density. *Ecological Applications*, 12: 552–564.

Borowitzka, M.A. and Borowitzka, L.J., 1988. *Micro-algal Biotechnology*. Cambridge University Press, United Kingdom, 477 pp.

DeMott, W.R., 1989. The role of competition in zooplankton succession. In: *Plankton Ecology: Succession in Plankton Communities*, (Ed.) U. Sommer. Springer, New York, p. 195–252.

Dodson, S.I. and Frey, D.G., 2001. Cladocera and other branchiopoda. In: *Ecology and Classification of North American Freshwater Invertebrates*, (Eds.) J.H. Thorp and A.P. Covich. Academic Press, London, p. 850–914.

Downing, J.A. and Rigler, F.H. (Eds.), 1984. *A Manual for the Methods of Assessment of Secondary Productivity in Freshwaters*, 2nd Edn. IBP Handbook 17, Blackwell Scientific Publications, London.

Dutta, S.P.S., Kour, S. and Khajuria, M., 2007. Plankton ecology of a paddy field at Maralia Morh, Miran Sahib, Jammu. In: *Advances in Fish and Wildlife Ecology and Biology*, (Ed.) B.L. Kaul. Daya Publishing House, New Delhi, 4(12): 106–120.

Elmoor-Loureiro, L.M.A., 2007. Phytophilous cladocerans (Crustacea, Anomopoda and Ctenopoda) from Paranã River Valley, Goiás, Brazil. *Revista Brasileira de Zoologia*, 24: 344–352.

Ferrão-Filho, A.S., Azevedo, S.M.F.O. and DeMott, W.R., 2000. Effects of toxic and non-toxic cyanobacteria on the life history of tropical and temperate cladocerans *Freshwater Biology*, 45: 1–19.

Flores-Burgos, J., Sarma, S.S.S. and Nandini, S., 2003. Population growth of zooplankton (rotifers and cladocerans) fed *Chlorella vulgaris* and *Scenedesmus acutus* in different proportions. *Acta Hydrochim. Hydrobiol.*, 31: 240–248.

Hyne, R.V., Padovan, A., Parry, D.L. and Renaud, S.M., 1993. Increased fecundity of the cladoceran *Moinodaphnia macleayi* on a diet supplemented with a green alga, and its use in uranium toxicity tests. *Australian Journal of Marine and Freshwater Research*, 44: 389–399.

Krebs, C.J., 1985. *Ecology: The Experimental Analysis of Distribution and Abundance*, 3rd Edn. Harper and Row, New York, 592 pp.

López-López, E. and Serna-Hernández, J.S., 1999. Variación estacional del zooplancton del embalse Ignacio Allende, Guanajuato, México y su relación con el fitoplancton y factores ambientales. *Rev. Biol. Trop.*, 47: 643–657.

Nandini, S. and Sarma, S.S.S., 2000. Lifetable demography of four cladoceran species in relation to algal food (*Chlorella vulgaris*) density. *Hydrobiologia*, 435: 117–126.

Nandini, S., Sarma, S.S.S. and Ramírez-García, P., 2000. Life table demography and population growth of *Daphnia laevis* (Cladocera) under different densities of *Chlorella vulgaris* and *Microcystis aeruginosa*. *Crustaceana*, 73: 1273–1286.

Ovie, S.I. and Egborge, A.B.M., 2002. The effect of different algal densities of *Scenedesmus acuminatus* on the population growth of *Moina micrura* Kurz (Crustacea: Anomopoda, Moinidae). *Hydrobiologia*, 477: 41–45.

Porter, K.G., Gerritsen, J. and Orcutt, J.D., Jr., 1982. The effect of food concentration on swimming patterns, feeding behaviour, ingestion, assimilation and respiration by Daphnia. *Limnol. Oceanogr.*, 27: 935–949.

Sarma, S.S.S., Nandini, S. and Gulati, R.D., 2005. Life history strategies of cladocerans: comparisons of tropical and temperate taxa. *Hydrobiologia*, 542: 315–333.

Sokal, R.R. and Rohlf, F.J., 2000. *Biometry*. W.H. Freeman and Company, San Francisco: 887 pp.

Vijverberg, J., 1980. Effect of temperature in laboratory studies on development and growth of Cladocera and Copepoda from Tjeukemeer, The Netherlands *Freshwater Biology*, 10: 317–340.

Weber, C.I., 1993. *Methods for Measuring the Acute Toxicity of Effluents and Receiving Waters to Freshwater and Marine Organisms*, 4th Edn. United States Environmental Protection Agency, Cincinnati, Ohio, EPA/600/4–90/027F, xv + 293 pp.

Chapter 8

Bloom Events of Toxin Producing Cyanobacterial Species Associated with Fish Kills in a Tropical Reservoir of Brazil

☆ *N.T. Chellappa and Sathyabama Chellappa*

Abstract

Eutrophication and consequent cyanobacterial bloom formation causes economical damage to fisheries and harmful effects to public health. Mass fish mortality due to toxin producing cyanobacterial blooms was registered during December 2006 in Marechal Dutra Reservoir, Acari, Northeast Brazil. Phytoplankton and fish samplings were carried out on alternate days during the fish kill events and monthly during January to June 2007. The cyanobacterial toxin was identified and quantified from seston samples and liver of dead fish using the standard HPLC method. The results indicate that the toxic blooms of *Cylindrospermopsis raciborskii* and *Microcystis aeruginosa* were persistent for two weeks and represented 90 per cent of the phytoplankton species assemblages. The general nature of the toxic blooms indicates that the liberation of the toxin microcystin accelerates the mortality of a wide range of tropical reservoir fish. The presence and relative abundance of the other algal species during the bloom persistence period was selective and limited compared to the overwhelming dominance of cyanobacterial species. The lethally affected fishes were *Oreochromis niloticus, Plagioscion squamosissimus, Cichla monoculus, Prochilodus brevis, Hoplias malabaricus* and *Leporinus friderici*. The microcystin levels varied from 0.07 to 8.73 µg L^{-1} in seston samples and from 0.01 to 2.59 µg g^{-1} in the liver samples of fish during the bloom period.

***Keywords**: Harmful algal blooms, Anthropogenic activity, Fish mortality, Reservoir, Public health.*

Introduction

The semi-arid freshwater ecosystems of northeast Brazil frequently encounter water level fluctuations because of recurrence of extended drought, irregular rainfalls and high temperatures, with evaporation rates outstripping the yearly precipitation. The seasonality of the freshwater ecosystems of this region is defined by short spells of intense rainfall (March to July) coupled with an extended dry period throughout the rest of the year (Bouvy *et al.*, 2000; Chellappa and Chellappa, 2004).

Eutrophication in tropical freshwater ecosystems is a process that triggers the sporadic occurrence of toxin producing phytoplankton blooms, which results in economical damage to fisheries and harmful effects to public health (Falconer, 2001; Best *et al.*, 2002). The occurrence of cyanobacterial blooms in freshwater is usually accompanied by both toxin and non-toxin producing species (Huisman *et al.*, 2005). Cyanotoxins are responsible for intoxication in wild and domestic animals, contamination of water for human consumption, fish mortality and elimination of other aquatic biota (Carmichael 2001). The commonly encountered toxin producing species of cyanobacteria include *Anabaena circinalis, Aphanizomenon flos-aquae, Microcystis aeruginosa*, and *Cylindrospermopsis raciborskii* (Azevedo *et al.*, 1994; Duy *et al.*, 2000).

In Brazil, the toxic blooms of cyanobacteria have been reported (Azevedo, 1994; Porfirio *et al.*, 1999; Magalhães *et al.*, 2001; Matthiensen *et al.*, 1999), and in northeast reservoirs (Bouvy *et al.*, 1999; Chellappa *et al.*, 2000). It has been reported that the presence of microcystin toxin from the blooms of *Microcystics aeroginosa* in the water storage tank of a haemodialysis clinic resulted 53 deaths in Caruaru city, in the state of Pernambuco (Jochimson *et al.*, 1998). This episode reinforced the importance of toxic blooms on public health and was the starting point of intense monitoring study on water quality of reservoirs and detection of toxic cyanobacterial bloom in order to provide early warning signals to reservoir management programmes. Previous studies conducted in the Marechal Dutra reservoir reported cyanobacterial bloom formation and fish kill episodes in 2000, when *Microcystis aeruginosa* grew exponentially and caused mass mortality of fishes (Chellappa *et al.*, 2000). Ever since the reservoir water has shown persistent symptoms of eutrophication, with the dominance of cyanobacteria species and no further incidence of fish mortality (Chellappa and Costa, 2003).

The purpose of this study was to report an episode of mass fish mortality associated with toxin producing cyanobacteria and its impact on reservoir fish culture.

Study Area

The Marechal Dutra Reservoir is located in Acari city and situated at a distance of 223 Km of Natal, capital of the state of Rio Grande do Norte, Brazil. This reservoir is within the coordinates of 6°26′11″ S and 36°36′17″ W and was built in 1958 by damming the River Acauã, pertaining to the Piranhas–Assu hydrographical basin of the semi-arid northeast zone of Rio Grande do Norte. It has a storage capacity of 40.000.000 m^3, with a drainage area of 2400 Km^2. It is a shallow reservoir with an average depth of 5.2 m and an estimated maximum depth of 21m. The time of water residence was 4.2 years during the study period, with oscillations according to the flux and drainage of water. The reservoir water is used mainly for human consumption, irrigation, fisheries and cage fish culture in the shallow regions of this reservoir by local fishermen.

Methods and Materials

Sampling and Analysis

Sampling was carried out from December 2006 to June 2007, which included both dry (December 2006 to March 2007) and wet seasons (April to June 2007). The study period coincided with the intensification of cage culture practices in the reservoir. Throughout the intense period of bloom formation and fish kill events, which were during the first 15 days of December 2006, triplicate collections were made on alternate days totaling twenty-four samples. The regular phytoplankton samples that were carried out from January to June 2007 were conducted on a monthly basis. Collections were carried out at 4 stations in the reservoir near to the cage cultures from 9:00h to 11:00h. The samples were obtained at surface level (0m) and at depths of 4m, 8m and 11m. The physical-chemical parameters such as, temperature, pH, dissolved oxygen and electrical conductivity of the water, were measured using a Multi-parameter Kit and inorganic nutrients such as, nitrate, phosphate and ammonia were analyzed (Goltermann *et al.*, 1978).

Phytoplankton analysis was carried out using samples of 1L of water fixed in lugol-iodine solution and quantitative enumerations were using inverted microscopy (Lund *et al.*, 1958). Plankton samples were collected from net tows in the water column using a 20μm mesh net and the materials preserved in acetic acid-lugol to determine the species composition. The identification was carried according to Weah and Sheath (2003). Seston samples were collected and filtered through Whatman GF/C acid-washed glass fiber filters and frozen at –20°C until subsequent analysis. Cyanotoxins were analyzed by High Performance Liquid Chromatography technique (Krishnamurthy *et al.*, 1986).

Microcystin extraction from the seston samples was carried out using the combination of solvent methanol: butanol: water (20:5:75 v/v) solution. Afterwards, the extract was centrifuged and the supernatant was evaporated to 30 per cent of its initial volume, with the resultant extract passed in a C-18 cartridge (Bond Elut C-18 variant). The sample was eluted from the cartridge with 20ml of de-ionized water, methanol 20 per cent and methanol 100 per cent, respectively. The 100 per cent methanol fraction was evaporated to dryness, resolubilized in 1ml methanol 50 per cent and filtered in nylon filter (0.45μm).

The microcystin identification and quantification was carried out through HPLC, with a U.V./Vis detector, SOD-10 A, LC-10aS bombs and CR6A integrator, comparing the absorption spectrum with the Microcystin-LR standard. The analysis were held in isocratic conditions and reverse level, using a semi-preparative column (Supercosil LVC-18, 5, 25cm × 10mm), with a mobile phase of acetonitril and ammonium acetate 20mM, pH 5.0 (28:72 v/v), with 1 ml Min^{-1} flow, loop of 100μL and UV detector set at

238nm (Krishnamurthy *et al.*, 1986). Chlorophyll-*a* concentrations were analyzed using acetone extract and corrected to phaeophytin (Goltermann *et al.*, 1978).

Fish sampling in the study area was carried out using dragnets of different mesh sizes, varying from 24 to 100mm, and other types of fishnets, such as, cast nets and hand nets. Fish were identified, numbered, labeled and the biometric data were registered in the Ichthyology Laboratory of the Federal University of Rio Grande do Norte, Brazil.

Results

Table 8.1 presents the summary of physical and chemical characteristics of a vertical profile of the reservoir waters during dry and wet periods. The pH, the bicarbonate alkalinity and dissolved oxygen concentrations gradually declined according to the depth, being high at surface and low at 11m. The temperature did not exhibit drastic variation in both dry and wet seasons, as well as in the vertical profile.

Table 8.1: Mean Values of Physical and Chemical Variables in Relation to Depth Profile in Marechal Dutra Reservoir, Brazil during 2006–2007

Parameters	*Dry 0m*	*Dry 4m*	*Dry 8m*	*Dry 11m*	*Wet 0m*	*Wet 4m*	*Wet 8m*	*Wet 11m*
pH	9.2	8.7	8.6	8.4	8.5	8.2	7.6	7.5
Bicarbonate (mgL^{-1})	131	135	137	138	125	101	114	127
Carbonate (mgL^{-1})	79	39	35	33	7	13	0	0
Conductivity (μscm^{-1})	860	931	855	659	225	221	22	220
Temperature (°C)	27	27	28	26	30	29	29	29
Per cent O_2	85.1	65.1	26.0	16.8	89.8	66.5	34.1	18.2
Dissolved oxygen (mgL^{-1})	6.4	5.1	2.4	1.3	6.4	5.0	2.3	1.2
Total Phosphorus (mgL^{-1})	0.66	0.64	0.70	0.65	0.74	0.79	0.85	0.76
Ammonia (mgL^{-1})	0.17	0.21	0.34	0.48	0.14	0.15	0.18	0.64
Total nitrogen (mgL^{-1})	0.61	0.71	0.66	0.91	1.09	0.85	0.97	0.89
pH	9.2	8.7	8.6	8.4	8.5	8.2	7.6	7.5
Bicarbonate (mgL^{-1})	131	135	137	138	125	101	114	127
Carbonate (mgL^{-1})	79	39	35	33	7	13	0	0

High oxygen values were encountered at the surface with high oxygen saturation and reduced hypolimnetic oxygen values as low as 16 per cent. Conductivity values showed a slight variation along the vertical, with an increase in the surface water during the dry season. The inorganic nutrients were high indicating the degree of eutrophication of the reservoir during the study period.

Presence of toxin producing cyanobacterial species, such as, *Microcystis aeruginosa* and *Cylindrospermopsis raciborskii* were registered during the fish kills. The lethally affected fishes were tilapia, *Oreochromis niloticus, Plagioscion squamosissimus, Cichla monoculus, Prochilodus brevis, Hoplias malabaricus* and *Leporinus friderici,* besides the freshwater shrimp, *Macrobrachium amazonicum.*

Table 8.2 shows the phytoplankton composition during the study period and the reduced species diversity. A total of 90 per cent phytoplankton composition was composed of cyanobacterial species during the bloom period. The toxin producing

Table 8.2: Phytoplankton Species Composition in Marechal Dutra Reservoir, Brazil during 2006–2007

Phytoplankton Species	*Vertical Profile*			
	0m	*4m*	*8m*	*11m*
Dry Season				
Cyanophyceae				
Cylindrospermopsis raciborskii (Woloszinska) Sennayya and Subba Raju	x	x	x	x
Gloeocapsa alpicula (Lyng.) Bornet and Flahaulf	x	x	x	
Lyngbya major Menegh	x			x
Microcystis aeruginosa Küetzing	x	x	x	x
Oscillatoria articulata Gardner		x		x
Oscillatoria lacustris Lemm.	x	x		
Oscillatoria limmetica (Klèb.) Geitler	x	x	x	x
Pseudanabaena catenata Laut.	x		x	
Bacillariophyceae				
Aulacoseira granulata (Ehrenberg) Simonsen	x	x	x	x
Chlorophyceae				
Coelastrum cambricum Archer	x	x	x	x
Nephrocytium lunatum West.				x
Wet season				
Cyanophyceae				
Chroococcus dispersus (V. Keiss) Lemm.	x			
Chroococcus decorticans (A. Br.)	x	x	x	x
Chroococcus giganteus West.	x	x	x	x
Chroococcus varius A. Braun	x	x		x
Gloeocapsa alpicula (Lyng.) Bornet and Flahaulf	x	x	x	
Microcystis aeruginosa Küetzing	x	x	x	x
Bacillariophyceae				
Aulacoseira granulata (Ehrenberg) Simonsen	x	x	x	x
Chlorophyceae				
Coelastrum cambricum Archer	x	x	x	x
Pandorina morum Bory	x	x	x	
Scendesmus quadricauda (Turp) Brébisson	x	x	x	

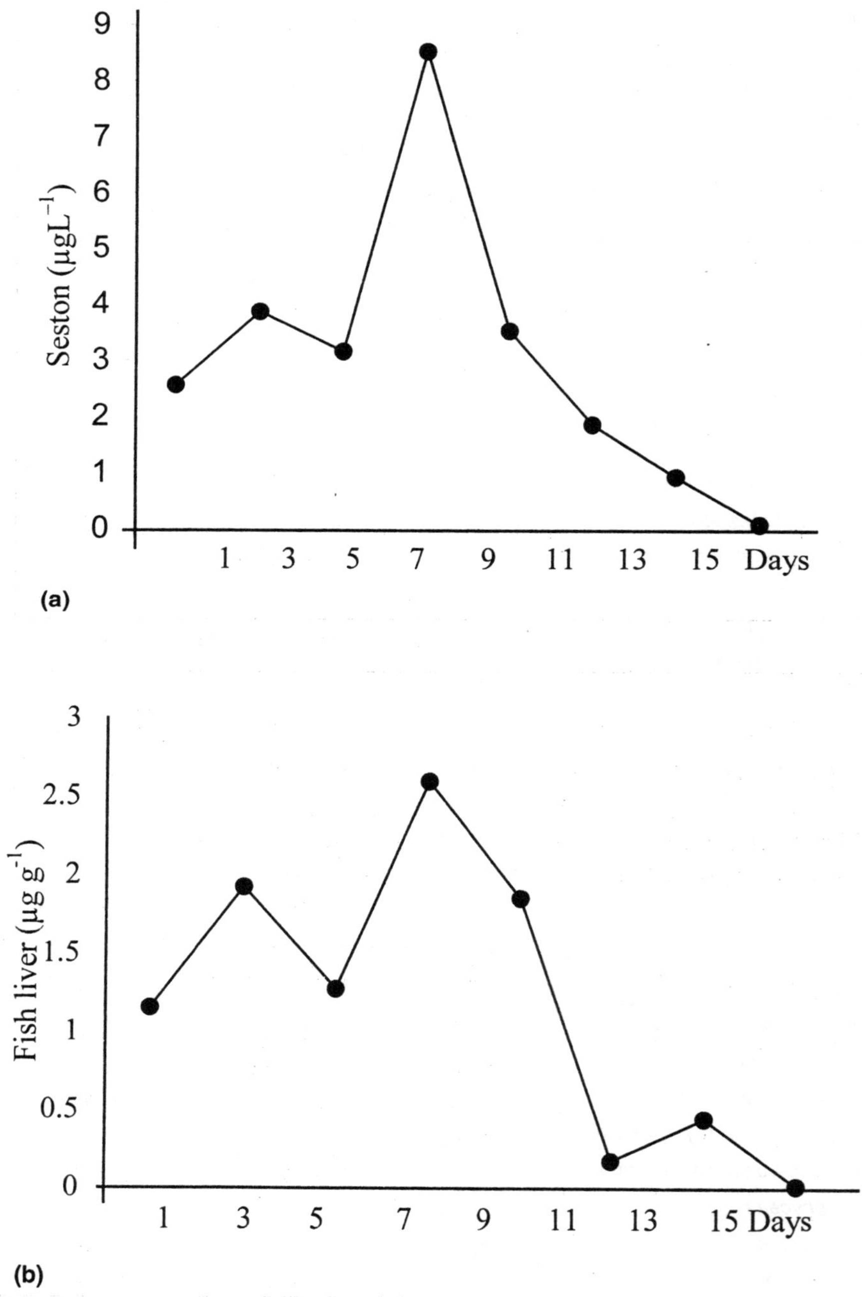

Figure 8.1: Microcystin Levels (µg g^{-1}) in Seston (a) and in Fish Liver (b) during the Bloom Period

species, *Microcystis aeruginosa* and *Cylindrospermopsis reciborskii* alternated dominance during the bloom formation. The first four days of the intense bloom period had a higher number of filaments of *C. raciborskii,* which was gradually substituted by *M. aeruginosa,* and this alternate domination became a characteristic feature. During December 2006, the combined blooms of *Chylindrospermopsis raciborskii* and *Microcystis aeruginosa* produced hepatotoxins. Results of seston and fish liver samples indicated presence of microcystin and the highest concentration was observed on the seventh day of the samples. The ichthyotoxin showed increase concentration until it reached the maximum in liver in the first week and decreased gradually during the second week of bloom formation (Figure 8.1).

Results of chlorophyll biomass indicated that it doubled from the first to the seventh day (exponential phase), and collapsed after the seventh day of the bloom formation (collapse phase). The values of chlorophyll *a* and its degraded component, phaeophytin, were inversely related (Figure 8.2).

The ichthyofauna of the reservoir is composed of native fish species as *Prochilodus brevis* (6 per cent), *Hypostomus pusarum* (4 per cent), *Synbranchus marmoratus* (3 per cent), *Hoplias malabaricus* (2 per cent), and introduced fish species, such as, *Oreochromis niloticus* (44 per cent), *Plagioscion squamosissimus* (22 per cent), *Cichla monoculus* (9 per cent), *Leporinus friderici* (6 per cent) and *Astronotus ocellatus* (4 per cent).

The composition of fish fauna of Marechal Dutra reservoir revealed *Oreochromis niloticus* as the dominant component. Mass killing of diverse species of the fish community in December 2006 (Figure 8.3) coincided with the heavy toxin producing bloom of *Microcystis aeruginosa* and *Cylindropsemopsis raciborskii* in the reservoir. In the current study, liver samples from *Oreochromis niloticus* and *Plagioscion*

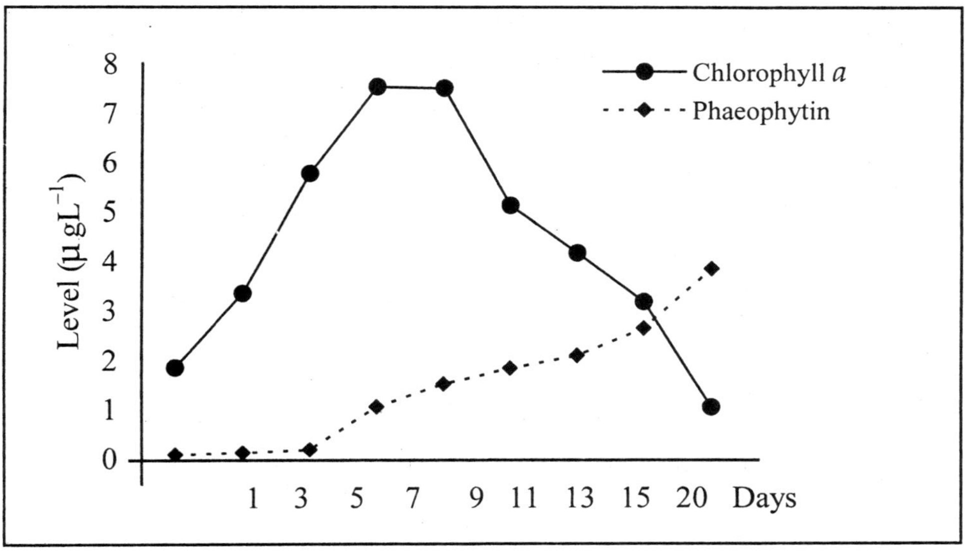

Figure 8.2: Levels of Chlorophyll *a* and Phaeophytin (mg L^{-1}) during December 2006

Figure 8.3: Fish Mortality During the Bloom Period in Marechal Dutra Reservoir During 2006

squamosissimus were contaminated with hepatotoxins, which caused the mass mortality of the fishes.

Discussion

The results from the phytoplankton species composition of eutrophicated impounded waters of the Marechal Dutra Reservoir, Acari, Rio Grande do Norte, recorded in this study demonstrated an overwhelming dominance of cyanobacteria during both dry and wet seasons. Water quality depends on a variety of biotic and abiotic factors, seasonality, as well as anthropogenic activities, such as, dam construction and increased soil erosion. As a result, eutrophication in water bodies frequently occurs. The development of a bloom is based on the assumption that a species or a species assembling becomes dominant in density by possessing mechanisms that enable a competitive advantage in comparison to other species present in the water body (Codd, 2000). Cyanobacteria have some characteristics that possibly explain its success under certain environmental factors, such as, macronutrients like nitrogen and phosphorus (Reynolds *et al.*, 2000), temperature, pH, and light (Lee and Rhee, 1999).

In the current study, total nitrogen and total phosphorus increased considerably during the wet season, probably due to increased turbidity and to the artificial feeds added to the maintenance of cage culture of fish suspended in the reservoir. Therefore, anthropogenic stresses may result in modifications of the physical and chemical water composition and phytoplankton composition, particularly in favor of the cyanobacterial population in dry season and chlorophycean population in wet season.

In relation to the chlorophyll/phaeophytin concentrations, high chlorophyll *a* and low phaeophytin degraded pigments concentrations were observed during the first two weeks of December 2006, which should be addressed to the intense bloom formation. During January to June 2007, the concentration of these substances could be attributed to the turbid water condition, which was likely due to the presence of large suspended organic matter, particularly during the wet season.

Although eutrophication has been recognized as worldwide growing concern, only recently cyanobacterial toxins have been studied in detail and widely recognized as a public health issue (Jochimson *et al.*, 1998; Chorus and Bartram, 1999). The cyanobacterial dominance in eutrophicated freshwaters has been studied since 1990 in the state of Rio Grande do Norte (Chellappa and Chellappa, 2004). The need for toxin analysis originated from the association of fish mortality with the toxic blooms of *Microcystis aeruginosa*, which occurred in Marechal Dutra reservoir of Rio Grande do Norte in January 2000 (Chellappa *et al.*, 2000). This study registered the presence of microcystin and its coincidence with the bloom of *Cylindrospermopsis raciborskii* and *Microcystis aeruginosa*, thus suggesting the dominance of potential toxin producing species in this reservoir. The production of saxitoxin, microcystin, nodularin and cylindropsermopsin has been observed both in natural populations and in unialgal cultures of toxic species (Falconer, 2001; Jochimsen *et al.*, 1998; Huisman *et al.*, 2005). A previous study has observed the production of saxitoxin by *Aphanizomenon, Anabaena, Lyngbya* and *Cylindrospermopsis* from unialgal cultures, with dominance of

C-toxins, neoSTX and GTX, respectively (Chorus and Bartram, 1999). A similar such trend in relation to the production of hepatotoxic microcystin was encountered in this study.

The maximum level for microcystin accepted in potable water is 1μg L^{-1} in Brazil (Azevedo, 1998). However, the results revealed microcystin levels as high as 8.73μg L^{-1} in seston materials of reservoir, thus endangering consumer's health. Reservoirs with planktivorous fishes that are not controlled by piscivorous fish communities have higher macrozooplankton biomass. Fish assemblage composition effects may cascade down the food web in reservoirs, although bottom-up and top-down effects may be unsymmetrical in character. For reservoirs with balanced fish assemblages, it is expected that efficiency of nutrient transference from the bottom up in the trophic web should diminish from algae to piscivorous fishes (Quirós and Boveri, 1999).

In the study reservoir, fish community composition is usually dependent on human actions, such as, fish introductions and stocking densities by local fisherman. Presence of facultative planktivorous fish that depend on benthic resources should be considered. Changes in fish assemblage composition may affect trophic relationships in the reservoir. Careful measures of fish introduction and fish stocking densities are required. The overstocking of planktivorous fishes could result in low water quality through increased microalgal blooms and diminished water transparency. The concurrent stocking of piscivorous fishes may produce more balanced trophic interactions at the bottom of the food webs of the reservoir. Cage culture practice was introduced into Marechal Dutra in 1999 to provide a low cost alternative for increasing fish food to the local population. One of the unpleasant consequences of the eutrophication trend and the formation of cyanobacterial bloom, though seldom, is its association with fish kills (Beveridge, 1987). The first formation of *Microcystis* bloom and fish kill incidence in Marechal Dutra was registered in January 2000 (Chellappa *et al.* 2000). Since then cyanobacterial species dominance are common but the toxin production was occasional.

Soares *et al.* (2004) demonstrated the presence of 2.8μg g^{-1} body weight of microcystins in the liver cells of *Tilapia rendalli* from laboratory experimental studies with different feeds including toxic cyanobacteria and indicated the calculated risks involved in aquaculture practices in ponds. The occurrence of 2.59μg g^{-1} body weight of microcystin in liver samples shown in the present study exposes how the local population and reservoir fisheries are vulnerable to ecotoxicological nature of freshwater ecosystems of water scarce semiarid North-eastern Brazil. The expanding aquaculture practices in Marechal Dutra, inappropriate use of fish feed, installation of many fish cages in the shallow regions of the reservoir and improper stocking densities of fish in the cages lead to eutrophication of the waters and harmful phytoplankton bloom. As a consequence, toxic blooms cause fish kills, contamination of potable water and harmful effects to public health.

Acknowledgments

The authors wish to thank the National Council for Scientific and Technological Development of Brazil (CNPq) for the Research grants awarded (N. T. Chellappa, Grant No. 473205/2007-6 and S. Chellappa, Grant n°. 307497/2006-2).

References

Azevedo, S.M.F.O., Evans, W.R., Carmichael, W.W. and Namikoshi, M., 1994. First report of microcystis from a Brazilian isolate of the cyanobacterium, *Microcystis aeruginosa*. *J. Appl. Phycol.*, 6: 261–265.

Best, J.H., Pflugmacher, S., Wiegand, C., Eddy, F.B., Metclf, J.S. and Codd, G.A., 2002. Effects of enteric bacterial and cyanobacterial lipopolysaccharides, and of microcystin–LR, on glutanione S-transferase activites in zebra fish (*Danio rerio*). *Aquatic Toxicol.*, 60: 223–231.

Beveridge, M.C.M., 1987. *Cage Culture*. Fishing News Book, Blackwell Science Publication, Oxford.

Bouvy, M., Falcão, D., Marinho, M., Pagano, M. and Moura, A., 2000. Occurrence of *Cylindrospermopsis* (Cyanobacteria) in 39 Brazilian tropical reservoirs during the 1998 drought. *Aquat. Microbiol. Ecol.*, 23: 13–27.

Carmichael, W.W., 2001. Health effects of toxin-producing cyanobacteria, "The CyanoHABs". *Human and Ecological Risk Manag.*, 7(5): 1359–1401.

Chellappa, N.T., Costa, M.A.M. and Marinho, I.R., 2000. Harmful cyanobacterial blooms from semiarid freshwater ecosystems of Northeast Brazil. Australia. *Aust. Soc. Limnol.*, 38(2): 45–49.

Chellappa, N.T. and Costa, M.A.M., 2003. Dominant and co-existing species of Cyanobacteria from a semi-arid reservoir of Northeast Brazil. *(Acta Oecologica) International Journal of Ecology*, 24: 3–10.

Chellappa, S. and Chellappa, N.T., 2004. Ecology and reproductive plasticity of the Amazonian cichlid fishes introduced to the freshwater ecosystems of the semi-arid Northeastern Brazil. In: *Advances in Fish and Wildlife Ecology and Biology*. (Ed.) B.L. Kaul. Daya Publishing House, New Delhi, pp. 49–57.

Chorus, I. and Bartram, J., 1999. *Toxic Cyanobacteria in Water: A Guide to their Public Health Consequences, Monitoring and Management*. Spon, London, UK.

Codd, G.A., 2000. Cyanobacterial toxins, the perception of water quality, and the priorisation of eutrophication control. *Ecol. Engineering*, 16: 51–60.

Duy, T.N., Lam, P.K.S., Shaw, G.R. and Connell, D.W., 2000. Toxicology and risk assessment of freshwater of freshwater cyanobacterial (blue-green algal) toxins in water. *Reviews Environmental Contamination and Toxicology*, 163: 113–186.

Falconer, I.R., 2001. Toxic cyanobacterial bloom problems in Australian waters: Risks and impacts on human health. *Phycol.*, 40: 228–233.

Goltermann, H.L., Clymo, R.S. and Ohnstad, M.M.M., 1978. *Methods of Physical and Chemical Analysis of Freshwaters*. Blackwell Sci. Publ., Oxford.

Huisman, J., Matthijs, H.C.P. and Visser, P.M., 2005. *Harmful Cyanobacteria*. Aquatic Ecology Series, Springer, 241 p.

Jochimson, E.M., Carmichael, W.W., An, J., Cardo, D.M., Cookson, S.T., Holmes, C.M.D., Antunes, M.B.D., De Melo, D.A., Lyra, T.M., Barreto, V.S.T., Azevedo, S.M.F.O.

and Jarvis, W.R., 1998. Liver failure and death after exposure to microcystins at a hemodialyses center in Brazil. *New England J. Med.*, 338: 873–878.

Krishnamurthy, T., Carmichael, W.W. and Sarver, E.W., 1986. Toxic peptides from freshwater cyanobacteria (blue–green algae). I. Isolation, purification and characterization of peptides from *Microcystis aeruginosa* and *Anabaena flosaquae. Toxicon.*, 24: 865–873.

Lee, T.J. and Rhee, G.Y., 1999. Kinetics of growth and death of *Anabena flosacquae* (Cyanobacteria) under light limitation and supersaturation. *J. Phycol.*, 35: 700–709.

Lund, J.W.G., Kipling, C. and Le Kren, E.D., 1958. The inverted microscope method of estimating algae number and the statiscal basis of estimating by counting. *Hydrobiol.*, 11: 143–170.

Magalhães, V.F., Soares, R.M. and Azevedo, S.M.F.O., 2001. Microcystin contamination in fish from the Jacarepaguá Lagon (Rio de Janeiro, Brazil): Ecological implication and human health risk. *Toxicon.*, 39: 1077–1085.

Matthieson, A., Yunes, J.S. and Codd, G.A., 1999. The occurrence, distribution and toxicity of cyanobacteria in the estuary of Lagoa dos Patos, Rio Grande do Sul. *Rev. Bras. Biol.*, 59(3): 361–376.

Porfirio, Z., Ribeiro, P.M., Estevan, C.S., Ricardo, L.S. and Santana, A.E.G., 1999. Hepatosplenomegaly caused by an extract of Cyanobacterium *Microcystis aeruginosa* bloom collected in the Manguaba Lagoon, Alagoas-Brazil. *Rev. Microbiol.*, 30: 278–285.

Quirós, R. and Boveri, M.B., 1999. The relation between fish yield and stocking density in reservoirs from tropical and temperate regions. In: *Theoretical Reservoir Ecology and its Applications*, (Eds.) J.G. Tundisi and M. Straskraba. Backhuys Publishers, The Netherlands, pp. 67–83.

Reynolds, C.S., Reynolds, S.N., Munawar, I.F. and Munawar, M., 2000. The regulation of phytoplankton population dynamics in the worlds largest lakes. *Aquat. Ecosystems Health Manag.*, 3: 1–21.

Soares, R.M., Magalhães, V.F. and Azevedo, S.M.F.O., 2004. Accumulation and depuration of microcystins in *Tilapia rendalli* (Cichlidae) under laboratory conditions. *Aquatic Toxicol.*, 70: 1–10.

Weah, J.D. and Sheath, R.J., 2003. *Freshwater Algae of NorthAmerica: Ecology and Classification*. Academic Press, London.

Chapter 9
Bhoj Wetland: A Review

Ashwani Wanganeo and Rajni Wanganeo

Abstract

The present chapter is a review of the Bhoj Wetland which is one the twenty one wetlands of national importance. Bhoj wetland is located at Bhopal, capital of the Indian state of Madhya Pradesh. Among the dozen or so water bodies in the city the most important are two lakes *i.e.* the Upper Lake (Bera Talao) and the Lower Lake (Chota Talao). Unfortunately, however, the quality of water of these twin lakes has deteriorated due to increase in population, human interference, agricultural activities, inflow of wastes and sewage from human settlements and siltation. Consequently an ecological damage has resulted and there is an urgent need to reverse the process of deterioration of this important wetland of national importance.

Keywords: *Catchment, Siltation, Weed infestation, Aquatic fauna, Phytoplankton, Zooplankton.*

Introduction

Bhopal shot into prominence the world over on the early noon of December 3, 1984 on account of the Bhopal gas tragedy that claimed several hundred lives and many times more in injuries caused by the inhalation of the deadly MIC gas that leaked from the Union Carbide Factory on that fateful day. Bhopal was chosen as the capital of new State of Madhya Pradesh in 1956. Situated on the eastern side of the Malwa plateau amidst rolling downs of yellow grass, interspersed with rich fields, the original city was situated on a sandstone ridge (1652 ft) at Lat. 23°16′ North, Long. 77°25′ East. Though having nearly a dozen lakes, the best known of these are two lakes, the larger known of Bara Talao (upper lake) and the smaller one as the Chota Talao (lower lake).

These two lakes of Bhopal have been recognized by the Ministry of Environment and Forests of Government of India as wetlands of National importance and as Bhoj Wetland in 1988. The Ministry of Environment and Forests have recognized twenty-one wetlands of national importance in the country, of which six have been elevated to wetland of international importance under the Ramsar convention. The National committee on Wetlands, Mangroves and Coral Reefs of the Ministry of Environment and Forests of the Government of India has been constituted to suggest specific sites for conservation, action and to identify research and training priorities. Of these twenty-one wetlands of national importance, management action plans for ten have been prepared. The Bhoj wetland, located at Bhopal is one among those.

The Bhoj Wetland

Bhopal, where the Bhoj wetland is located is the capital of Madhya Pradesh sprawled across the Vindhya and Singarcholi mountains. It is a fast developing city with well planned wide roads, luxuriant verdure and modern dwellings nestling in an undulating terrain, in company with unplanned and unhygienic congested slums, which have sprung up in most of the vacant riparian zone. In fact this capital of Madhya Pradesh is the best example of man-made squalor thriving in most beautiful surroundings provided by low hills and expansive water bodies.

Among the dozen or so water bodies in the city the most famous are two lakes the Upper Lake (Bara Talao) and the Lower Lake (Chota Talao), the two together have been designated as the Bhoj wetland, one of the twenty one wetlands of National importance, selected by National committee on wetlands, Mangroves and Coral Reefs of the Ministry of Environment and Forests of the Government of India.

The two water bodies constituting the Bhoj wetland the Upper and Lower lakes are separated by an ancient massive earthen bund. Though adjacent to each other, these two water bodies of Bhoj Wetland are separate entities that came into existence at different times, and apparently for different purposes. The larger of the two, *i.e.*, the Upper Lake was a satellite of a much larger (almost like a sea) lake that was created by the Parmar King Raja Bhoj between 1010 A.D. and 1053 A.D. near Bhojpur in what is now known as the Tal tehsil of Bhopal district. About 250 sq. miles in area, this massive water body did not hold sufficient water, hence Raja Bhoj directed his minister to divert some more water into the lake from surrounding localities. The site chosen for the purpose was near the present Bhopal city (which came into existence several centuries later) where the Kolans river flowed through a narrow rift valley surrounded by a natural wall of hills. A massive earthen bundh was built across this valley on the site currently known as Kamla Park creating a large impoundment (the Bara Talao or Upper lake) which on its south side flows into the Kaliasot river (after Bhadbhada) which turned from its natural course to feed the huge reservoir near Bhojpur. The Upper lake of Bhopal, now constituent of the Bhoj wetland, is as old as the Bhojpur lake (called Hauze-e-Bhim or Bhim Kund by the later muslim rulers) and was created to supply more water to fill the latter. The king Husan (Shah of Mandu 1405-1434 A.D) destroyed the huge Bhojpur lake by cutting through the smaller dam over the Betwa river either intentionally or in a fit of destructive rage. The climate of Malwa is said to have been materially altered by removal of this vast sheet of water as a result of the thoughtless and stupid action of Hoshang Shah. The upper lake (Bara Talao) of Bhopal is the sole survivor of an era of a prosperous Malwa and a historical relict of an ancient heritage. It is our firm conviction that the Bara Talao deserves to be declared as a heritage site.

The other constituent of Bhoj wetland the Lower lake (or Chota Talao), is a satellite of the Upper Lake. It was built in 1794 by Chhote Khan, then Minister of Nawab Hyatt Muhammad, by building a dam called pul-pukhta (bridge of stones) across the Banganga river holding waters of the lower lake. The purpose of constructing the Chota Talao was to beautify the city of Bhopal which has grown on the eastern side of the upper lake, and also to meet water requirements of the Afghan army stationed there. Besides, impounding waters of Banganga river, the lower lake also held a seepage of water from the upper lake through the massive bundh at Kamla Park.

Geology

The principal rock formations of Bhopal district are Deccan traps basalt and Vindhyan sand stones. The Deccan traps formed out from cretaceous to Eocene period, weathering of these volcanic rocks gave rise to heavy black soils, a wide spread rock type, known as black cotton soils, suitable for cotton cultivation. The trap rocks have low porosity and permeability, not ideal for ground water storage. The Vindhyan sand stones of this area belong to Bhander series (Upper Precambrian).

Bhopal area is covered by various lava flows of Daccan traps. Lava flows are generally confined to the valleys of small hills and the surrounding low-lying area is composed of black cotton soil. Rainfall and climatic conditions have played an important role in its formation. The Bhoj wetland and its catchments area depicts sand stone formation towards the eastern flanks while, rest of the catchments area extending in east-west direction is mainly basaltic. The reservoir rests on sand stones.

Climate

Though located in the tropical belt the presence of large water bodies, besides black cotton soil (having high retention capacity for moisture) have together exercised moderating influence on the climate of the region and as such, the climate of Bhopal is of sub-tropical type.

Earlier *i.e.*, up to 15th century the climate of Bhopal region might have been similar to that of a moderate zone, on account of the locality of the city on the banks of a vast artificial lake. The destruction of the vast Bhojpur lake in the fifteenth century affected the climate materially (The winds from the south blowing over its area of 250 sq. miles must have appreciably affected conditions in the plateau to its North and assisted to maintain the forests in flourishing state).

The summer season begins from mid February and continues up to June, the maximum temperature reaching above 45.1°C during May and June. Due to rocks around the city, the mean temperature during the day remains very high. In winter season, the months of December and January are the coldest. Sometimes the minimum temperature goes down below 4°C. Thus an extreme variation in temperature occurs through summer and winter season. Four seasons of varying duration have been encountered in this region of Madhya Pradesh. The duration of different season is given below:

1. Summer: Mid February-Mid June
2. Monsoon : Mid June-August

3. Post monsoon: September-November
4. Winter: December-February.

Relative humidity varies from 15 per cent to 86 per cent. During the south-west monsoon season the humidity is usually above 70 per cent. In the summer months the relative humidity is less than 20 per cent.

Average rainfall at Bhopal is about 919 mm and about 92 per cent of the precipitation is falls during June to September.

Catchment

About 361 km^2, Catchment area of Bhoj wetland spreading southeastwards up to Sehore district, drains into Kolans River while, the catchment area of lower lake (9.6 km^2) is limited within the city area.

Wanganeo (1996-1998) divided the whole catchment area of the upper lake into eleven sub watersheds (based on their drainage pattern) for the micro level analysis (Figure 9.1). Watershed no's I, II, III and XI are of the urban nature. Watershed no's IV, V and X are representing country side, while watershed no's VI, VII, VIII and IX come

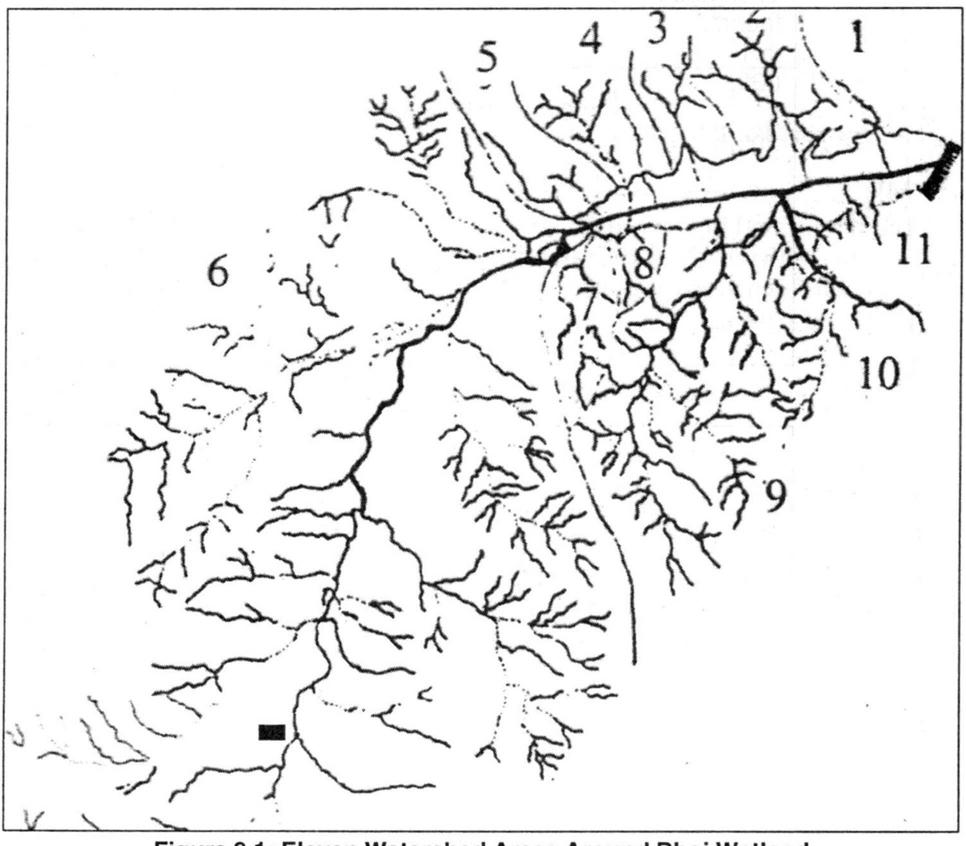

Figure 9.1: Eleven Watershed Areas Around Bhoj Wetland

under rural areas. The watershed number VI is the largest in area (236.10 km^2) and basically it is agricultural in nature. There are twenty two defined nallas in the eleven watersheds bringing in sizeable amount of sludge, silt, agricultural residue etc. The landscape of each watershed is of different nature. The land use/land cover changes in these watersheds from the past ten years (1984-1994) have been recorded in Table 9.1. The study reveals that built up land is increasing thereby exposing the system to anthropogenic pressure. Besides residential buildings, tourist infrastructures have also experienced mushroom growth along the fringe areas of the wetland. Barren rocky area is considerably decreasing on account of quarrying activity in watershed no's II and IX and in human settlements. Decrease in cropland and simultaneously increase in fallow land has also been recorded. Forest cover in watershed no's II and IX has also started declining. Thus the change in land use land pattern of the wetland catchments area indicates further degradation which will decrease life span of Bhoj wetland. The catchment area activities of upper lake may be divided in three sectors namely the urban, semi-urban and rural. The north side of the lake is having major thrust of urban and semi-urban activities which contribute the solid and liquid waste. Towards the southern side of the lake a national park and natural museum of man are located. Sewage inflow from this are is insignificant contrary to the eroded material. The north west and south west portion of the catchment area is basically of rural in nature and is under agricultural practices. Towards the north east side a link road from retghat to lalghati has been constructed covering a distance of 4900 mts.

Table 9.1: Change in Landuse/Landcover in Different Watershed of Bhoj Wetland during Ten Years

Water-shed No.	*Area Km² 1974*	*Classification*	*Percent Change*			*Phase I*	*Phase II*	*Total % Change*
			84 (A)	*88 (B)*	*94 (C)*	*B–A*	*C–B*	*C–A*
I	1.0428	Built up land	58	59	90	1	31	32
		Crop land	4	22	10	18	–12	6
		Barren Rocky	38	19	–	–19	–	–
II	6.4864	Built up land	18	21	31	3	10	13
		Crop land	43	65	59	22	–6	16
		Decidous (open)	3	–	2	–	2	–1
		Fallow land	23	7	–	–16	–	–
		Barren Rocky	10	7	8	–3	1	–2
		Water logged	3	–	–	–	–	–
III	5.523	Built up land	56	66	8	–2	6	18
		Crop land	24	24	36	0	12	12
		Fallow Land	13	9	–	–4	–	–
		Barren Rocky	5	1	–	–4	–	–
IV	8.94	Built up land	23	16	26	–7	10	3
		Crop land	70	65	63	–5	–2	–7
		Fallow Land	7	12	11	5	–1	4
		Water Logged	–	7	–	7	–	–

Contd...

Table 9.1–Contd...

Water-shed No.	*Area Km² 1974*	*Classification*	*Percent Change*			*Phase I*	*Phase II*	*Total % Change*
			84 (A)	*88 (B)*	*94 (C)*	*B–A*	*C–B*	*C–A*
V	5.265	Built up land	3	7	8	4	1	5
		Crop land	66	75	75	9	–	9
		Fallow Land	23	11	14	–12	3	–9
		Barren Rocky	6	7	3	1	–4	–3
		Water Logged	2	–	–	–	–	–
VI	255.912	Built up land	0.01	0.28	0.23	0.26	–0.04	0.218
		Crop land	83	77.9	67.5	–5	–10.39	–15.45
		Fallow Land	14	19.5	30.8	5.5	11.27	16.79
		Barren Rocky	2.41	2.11	1.21	–0.3	–0.84	–1.154
		Water Logged	0.57	0.04	–	–0.53	–	–
		Plantation	–	0.13	0.17	–	–	–
VII	10.41	Built up land	0.82	2.05	0.97	1.23	–1.08	0.15
		Crop land	86	92.2	81.1	5.15	–11.09	–5.94
		Fallow Land	10	2.71	16.6	–7.29	13.93	6.64
		Barren Rocky	2	3.07	1.78	1.07	–1.29	–0.22
		Water Logged	0.18	–	–	–	–	–
VIII	2.99	Crop land	97	79	84	–18	5	–13
		Fallow Land	3	17	13	14	–4	10
		Barren Rocky	–	4	3	4	–1	3
IX	39.91	Built up land	0.17	0.2	0.26	0.034	0.06	0.09
		Crop land	84	72	55	–12	–17	–29
		Fallow Land	10	22	40	12	18	30
		Barren Rocky	4	5.8	3.8	1.8	–2	–0.2
		Water Logged	1.83	–	–	–	–	–
X	9.075	Crop land	91	66	60	–25	–6	–31
		Fallow Land	0	25	30	25	5	30
		Barren Rocky	5	9	10	4	1	5
		Water Logged	4	–	–	–	–	–
XI	15.394	Built up land	18	23	35	5	12	17
		Crop land	6	–	–	–	–	–
		Decidous (open)	68	77	65	9	–12	–3
		Fallow Land	8	–	–	–	–	–

Inflow Points

There are mainly 22 inflow nallas which join the upper lake at various points. These nallas are responsible for the inflow of silt and wastewater from 47 sub basins. From these 47 sub basins untreated human waste, large volume of silt and humus material, agricultural residues flow into the lake. Further the drainage pattern reveals that most of the streams are of 1st and 2nd order. In addition to this drainage density and infiltration number clearly support the fact that Bhoj watershed areas indicate moderate nature of flash floods.

Inflow of Silt from the Catchment Area

Vast catchment area adds to the silting of lakes. Since most of the catchment area is used for agriculture the top soil is in loose form due to repeated ploughing (towards Bhadbhada and Bairagarh ends). This loose top soil is easily eroded by surface runoff during monsoon months. When silt-laden rivulets enter the lake the suspended matter settles in the bed due to sudden decrease in velocity.

Watershed no. VI has been found to contribute the maximum amount of silt, which has resulted in considerably shallowing of the water body near its entry point. Besides watershed no's 7,8,9 have also been found to contribute some amount of silt but not to the extent watershed no. 6 has contributed, owing to the flow patterns and the presence of Bhadbhada spillway maximum silt deposit has been found towards south west direction. Rest of the watersheds *viz.* II, III, IV and V and X has also been found to contribute to this problem but not to the extent watershed no VI has been found. The reason for this magnitude of silt input from these watersheds is that the entry points of silt have been found to culminate into the armpits of Bhoj Wetland.

Besides eroded soil, residues from agricultural lands, surrounding households and other human activities have supplemented to this phenomenon of siltation.

Based on the lake bed survey (SAPROF 1994) about 6.6 km^2 area in the deeper part of the lake near Kamla Park is deposited with silt and sediments to an average thickness 0.75 m (maximum depth 1.5 m), whereas in the shoreline towards the west end of the lake an area of about 10 km^2 is silted to an average thickness of 0.5 meters.

Hydrological Inputs from Eleven Watersheds into Upper Lake

Maximum hydraulic input is from watershed no. VI (64.244 m^3 × 10^6) followed by watershed no IX (11.74 m^3 × 10^6) and no. II (5.9 m^3 × 10^6). Area wise also watershed no's VI and IX are big and their hydraulic contributions is maximum, while watershed no. II which is small in area, yet contributes good amount hydraulic input. This may be on account of increased degree of human activity. Watershed nos. I, II and IX record a continuous input in almost all seasons into the wetland. This may be on account of the urban nature of these watersheds which continuously discharge human generated waste into the system.

Human Population in the Catchment Area

A total no. of 1,62, 269 individuals (both male and female) are residing in the whole catchement are of Bhoj wetland. With respect to the distribution of human population, the watershed no. I show a high population density of 12,899 persons/sq.km., though it is the smallest watershed among others. Density wise the watershed no's II and III stand in the same position *i.e.*, 2nd and 3rd. They also recorded higher densities (5033 and 3345 persons/sq.km.). After watershed no. 1 all these three watersheds cover the maximum urban area of the Bhoj catchment, from where as perennial source of input channel enters into the water body bringing direct human generated waste from the settlements.

Animal Population

Based on distribution of animals /sq.km. area, watershed no's III, IV and IX occupy the 1st, 2nd and 3rd position while the biggest watershed area (VI) occupies seventh position. As such, the maximum input of animal generated waste coming from watershed no's III, IV and IX drains into Bhoj wetland.

Flow of Untreated Sewage from Human Settlements from the Catchment

4.55 lakh population of 29 wards of city falling under the catchment area of the Bhoj wetland exerts direct and indirect pressure on the system from northern side while, pollution load from 2 wards with the population of 28262 is being received from southern side.

Similarly 55 villages falling on the southwest and northwest part of the lake with a human population of 55,000 and cattle population of about 15,000 contribute pollution load into the lake. In general more than sixty percent of the pollution in the catchment area is using open land for their lavatory purpose and this untreated human waste finds its way into the lake very conveniently.

Agriculture Waste Input due to Agricultural Practices

Agricultural wastes mainly composed of organic matter enter the lake during monsoon month.

Pesticides, Insecticides and Fertilizers

Increased use of pesticides, insecticides and fertilizers in the catchment area is also a threat to the lake.

Problems of Environmental Deterioration of Bhoj Wetland

Bhoj wetland faces number of problems on account of opening of its catchment area due to large scale of human activities. The major threats to the lake are briefly mentioned in subsequent sections.

Siltation

As there is hardly any forest or vegetation other than agricultural crops which are mainly grain types, the catchment area having mainly black cotton soils, is subject to severe erosion. As a result, a large volume of silt and humus material has been carried into the lake over the years by Kolans river and other rivulets entering the lake. Besides this the agricultural residues from the village areas and solid wastes including construction debris etc. from the residential and commercial areas also find their way into the lake through the drains and streams particularly during the rainy season. The inorganic residues after decomposition adds to the sediment load in the lake. Exposure of catchment devoid of any vegetation due to housing and excessive construction activities on the higher slopes, particularly in the urban areas such as Kohe-fiza and Khanoogaon area on the northern bank have advanced the rate of soil erosion from these areas. As a result a large amount of silt deposits are formed within the lake thus reducing the storage capacity and water spread area of the lake. The silting rate of the lake is estimated to be 1 cm to 2.58 cm per year on an average. The estimated sedimentation rate from the catchment area is to the tune of 3.67 ha.m/100km^2/year. The spillway channel position of the lake upto Bhadhbhada weir covering further distance of 4.4 km has become shallow and narrow partly as a result of siltation and its discharge capacity has also reduced. The storage capacity of the lake has reduced by about 5 million m^3. Further, it has been estimated that about 6.6 km^2 area in the deeper part of the lake near Kamla Park is deposited with silt and sediments to an average thickness of 0.75 m (maximum depth 1.5m) whereas, in the shoreline towards the west end of the lake an area of about 10 km^2 is silted to an average thickness of 0.5m. Estimated volume of sediments in Upper lake is about 5 millions m^3 considering an area of 6.6 k m^2 deposited to an average thickness of 0.75m.

Inflow of Sewage and Wastewater from the Catchment Area

Untreated human waste besides the waste from the cattle and other domestic animals finds its way into the lake directly or indirectly. Flow of sludge and sewage into the lake is predominant from the area around the Medical College Hostel, Kohe-fiza settlements and Vanvihar area. It is estimated that 7500 m^3 per day of sewage water and 360 m^3 per day of animal liquid discharge into the Upper lake. Further, about 2000 persons are bathing per day in the lake besides washing and cleaning their clothes and vehicles. These activities lead to large inputs of nutrients and pollutants into the system which in turn promotes eutrophication and growth of algal and aquatic plants besides deteriorating the quality of potable water. This problem is further compounded by addition of manure used for the cultivation of singhara above 2 km^2 of the lake area near Bairagarh.

Encroachment

The expansion of urban areas and development of unorganized human settlements in the catchment area has considerably reduced the lake area torwards the northern fringe of the lake.

Weed Infestation

The upper lake is severally infested with macrophytes towards the northwest and south west areas. Among the dominant weeds, *Polygonum glabrum, Ipomea festulosa* and *Eichhornia crassipes* have assumed nuisance proportion. Added to this problem is the release of nutrients on desiccation of plants (which are left open on the bank due to reducing water during summer period). In some part of the lake where maximum growth of the *Polygonum* and other weeds has been detected, the lake is said to have become shallow which has probably resulted in the water flow from the deeper Kamla Park area to the Bhadhbhada area. Besides this, fertilizers and pesticides used for the cultivation of *Trapa* has further compounded this problem by way of not only adding nutrients on but also adding to siltation problem.

Immersion of Tadjias and Idols during Festivals

A number of religious activities take place on the catchment area during festival seasons. During the festival of Ganesh and Durga Puja nearly 2000 idols of different sizes are immersed in the lake near the Kamla Park intake and it is estimated that each year about 70 tones is added this way. During Muharram again several Tadjias are immersed every year near Kamla intake point. As a result a variety of materials *viz.*, clay, hay, clothes, papers, bamboo, wood, adhesive material, water soluble and insoluble paints containing harmful substances are released into the lake water. "Bhujaria" (wheat grown in covered earthen pots) are immersed in sizable quality into this upper lake. Besides, siltation due to clay from the idols, decomposition of biodegradable matter in the idols adds to pollution of the lake. It is reported that cadmium, chromium, zinc etc. have been detected in the lake bed material above environmentally tolerable limits and it is believed that the paints used in the idols and tadjias which are likely to contain heavy metals and toxic elements are the possible cause. Further, during the festival a large volume of flowers are also thrown in to the lake closed to the intake points and there subsequent decay adds to the pollution of lake water.

Reduction of the Storage Capacity of the Lake

Effective capacity of the lake has largely reduced due to siltation and this is estimated to be about 5 million m^3. Reductions in the depth of lake due to deposition of silt and organic matter also have a direct effect on increasing pollution in the lake. If soil erosion, adverse human activities in catchment area and inflow of silts and waste etc. into the lake are left unchecked, siltation of the lake is likely to increase at an alarming rate.

Table 9.2: Human Actions or Direct Causes of Wetland Loss

Sl.No.	Causes	Loss
1.	Roads	4.6 km
2.	Residential and Commercial settings	4.105 km^2
3.	Conversion of Agricultural land into Fallow land	51.82 km^2
4.	Loss of evaporation	0.211 m cu
5.	Water supply	27 MGD
6.	Siltation since two decades	200 m cu
7.	Tourist influx	Increase in number of boats plus water sports.
8.	Removal of macrophytes	Loss of: biodiversity, breeding ground of fishes and oxygen replenishment

Natural Actions

1. Draught
2. Erosion

Threat to Aquatic Life

Some specimens of dead fishes had been spotted at the shore of the lake in Dec. 1993 and upon examination by the concerned authorities, they were found to have ulcer marks over their body which denotes that the fish might have died due to epizootic ulcerative syndrome disease. The fishes affected by disease include *Mastocemblus armatus, Mystus vittatus, Puntus sarana, Puntus ticto, Cirhinus ruba, Chanda ranga, Chanda nama, Channa gachua* and *Lapidocephalichthys guntea*. Some other living fishes namely *Notopterus notopterus, Heteropneustes fossilis* and *Calarius batrachus* had also been examined but without symptoms of the disease. Also the major carps *Catla, Rohu* and *Mrigala* were not found infected with the disease. The cause of epizootic ulcerative syndrome is believed to be due to some virus, bacteria, fungi and protozoan's although further investigations by the fisheries research laboratory are in progress. However, this situation is alarming as it can lead to health problem among people who consume water and fish from the lake.

Health Hazards

Use of catchment area for open defecation and the disposal of solid and liquid wastes directly into the lake lead to serious health hazards. This is further aggravated by the poor sanitary condition particularly of the lower income groups and slum dwellers that occupy the catchment area. Though detailed statistical data are not available, it is reported that 86 per cent of the population using water from the upper lake have undergone abdominal disorders mostly due to tape worms and amoebic infection.

Table 9.3: Incidence of Human Diseases in Bhopal

Sl.No.	*Types of Disease*	*Percentage of Patients Affected*
1.	Pulmonary tuberculosis	0.8
2.	Respiratory infection	0.2
3.	Leprosy	0.08
4.	Water borne diseases	86.0 (include 0.5 per cent deaths)
5.	Hypertension	2.0
6.	Diabetes	1.0
7.	Skin disease and STD	2.0
8.	Gynecological disease	2.9
9.	Malaria and others	5.0

Source: Bhopal Hospital Records.

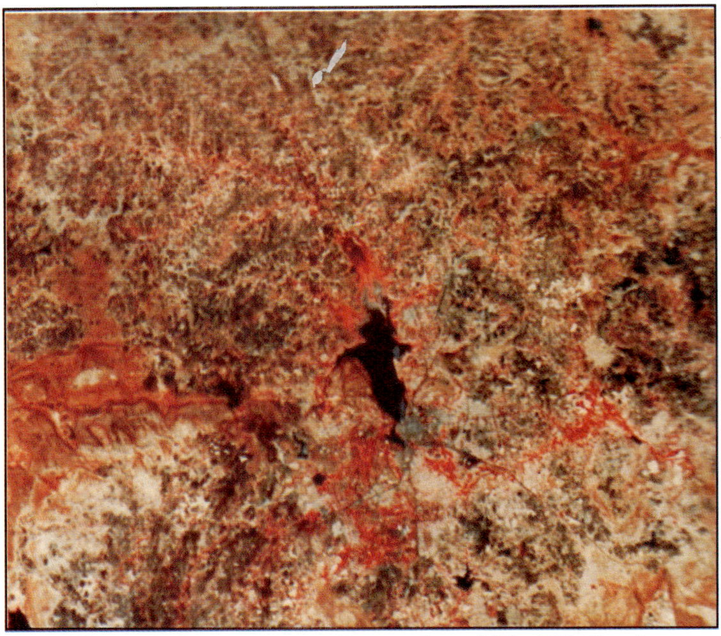

Plate 9.1: Pre-monsoon Expansion of Bhoj Wetland

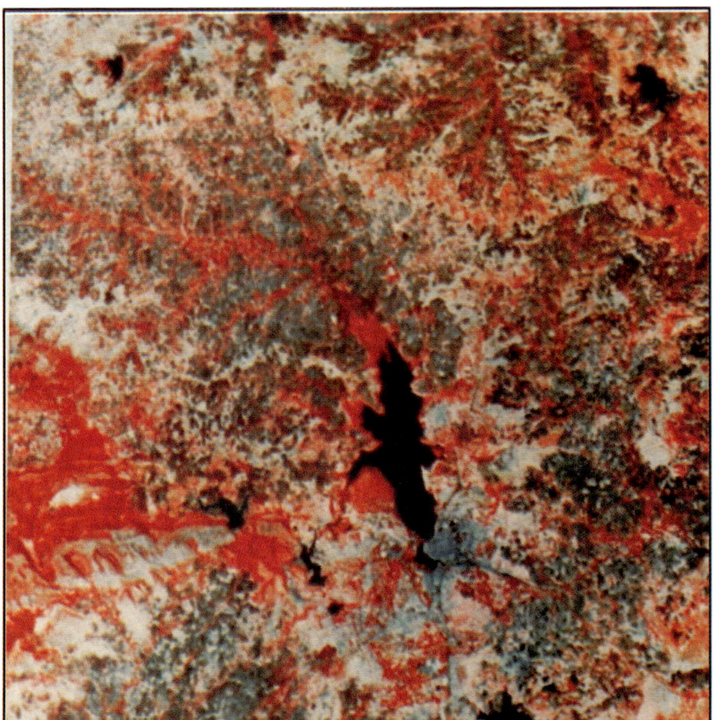

Plate 9.2: Post-monsoon Expansion of Bhoj Wetland

Water Quality

General Status of Upper Lake

Upto the middle part of 20th century, the Upper Lake water has been supplied as potable water without any treatment. However, gradual pollution of the lake has become noticeable, thereafter, as evident not only from the visible changes to the lake environment, but also from the water quality tests conducted on the routine or on adhoc basis. Natural phenomena such as soil erosion and siltation bringing in soil and other material by the inflowing river and steams; increase in pollution in the catchment area and particularly along the periphery of the lake. Cultural activities such as immersion of idols and Tadjias during the festival and inflow of untreated sewage and waste water from the human settlements, etc. have changed the trophic status of the water body considerably. It was during the time of Mr. Cook, the then water works engineer of Bhopal state that the masonry weir was constructed on the sandstone ledge in the Kaliasot nala about 4.8km., below the escape head of the lake in order to increase the capacity of the lake to the extent required at the time. The weir is popularly known as Bhadbhadha weir and the site is a picnic spot during the time the weir is spilling. The earlier weir was 93.2 m long with 6.1m flat crest and the gentle balter of 1 in 10 on downstream side and vertical face on the upstream side. The flat crested weir is slopping from both ends towards the middle the highest level at the ends being R.L. 505.75 m, the lowest being R.L. 504.50 m. The weir was constructed in the stone masonry in lime mortar.

Since then the water body (Upper Lake) is experiencing fluctuation in its total area on account of changing pattern in climatic regimes *viz.*, monsoon evapo-transpiration draining of the water for potable purpose. Beside, irrigation, siltation and agriculture encroachment there is no doubt about the fact that owing to change in climatic regime and other related factors of water body (Upper lake) is greatly subjected to number of stresses which get quickly reflected by the changing morphometric characteristics.

On account of hydraulic inputs from 11 watershed bringing sizable, colloidal and detritus material with it, has been found to affect both water area and lake area as well.

Wanganeo (1996-1998) records the continuous reduction in the water spread area during both the seasons (*i.e.* pre-monsoon and post-monsoon). Similar trend have been recorded for the lake surface areas as well. The fluctuation in water spread area exerts its major impact on the vegetation and its nutrient cycling and hydraulic function.

It has been found that the enhancement in the growth of aquatic vegetation for the last ten years by 25 per cent (Wanganeo 1996-1998) led to the death of this once beautiful lake. The present status of the lake is that it is highly polluted due to eutrophication and a vast area of the former water spread is either replaced with silt landmass or covered with aquatic weeds and this situation is prominent during the drought seasons. This has posed a serious threat to the quality and the effective usable quantity of water from the lake for Bhopal city dwellers. Water supply scheme is already handicapped due to the absence of alternate inexpensive water sources.

The time series changes on the present water body have been worked out by Wanganeo (1996-1998) with the help of Land Sat imageries. It has been reported that Upper Lake is experiencing reduction in its total lake area by 3.175 sq.km., during pre and post-monsoon period for the last ten years respectively.

The significant variation observe in the depth parameter may be on account of the variation in its measurement which arises due to no fixed place of observation beside the periodical fluctuation related to a climatic conditions and the exploitation of Upper Lake water resource for potable purpose. While scanning the literature difference in maximum length, breadth and depth in Upper lake has been recorded which is obvious to a certain extent, as the capacity of the water body has been increased after 1963. For the last one decade the Upper Lake water body has been subjected to the various types of anthropogenic pressures which have been found to be responsible for bringing about change in its morphometric characteristics. General physical features of the water body are given in Table 9.4.

Table 9.4: Physical Characteristics of Two Basins of Bhoj Wetland

Sl.No.	Item	Upper Lake	Lower Lake
1.	Constructed in	11th century	Late 19th century
2.	Type of dam	Earthen Dam	Earthen dam
3.	Longitude	77°18'–77°24'	77°24–77°26'
	Latitude	23°30'–23°16'	23°14'–23°15'
4.	Catchment area sq.km (sq.miles)	361	9.6
5.	Submergence area of FTL km^2	30.72	1.29
6.	Maximum water level Rlm(ft)	508.65	
7.	Dead storage level Rlm (ft)	503.53	
8.	Storage Capacity 10^6 m^3 (Mft^3)	101.6	8.0
9.	Deepest bed level Rlm (ft)	499.26	
10.	Maximum depth m (ft)	11.7	
11.	Mean Depth m (ft)	6.0	
12.	Total length of weir m(ft)	102.1	
13.	Crest level of spillway Rlm(ft)	504.38	
14.	Designed flood discharge m^3/sec (cusecs)	2208	
15.	Moderate flood discharge m^3/sec. (cusecs)	538.02	
16.	Sewage water inflow m^3/day (MGD)	330.80	
17.	Source of water	Rain water	Seepage from upper lake sewage
18.	Main use of water	Potable water	Cloths, Washing, boating

Physico-chemical Characteristics

Water Temperature

The Lake under consideration depicted a steady increase in water temperature from 1975 onwards. The increase in water temperature remained within the limit of

10°C from 1975 to 1996. The variation in rise in temperature was pronounced in between 1990-1995, when the maximum rise in water temperature was recorded. With the increase in atmospheric temperature a corresponding increase in surface water temperature has also been recorded. Seasonal variation in water temperature shows an increase in temperature during summer months almost at all the watersheds. Whereas, a decrease in value was encountered during winter months.

The water body under study has not been found to have any thermal stratification although the lake depicted maximum surface water temperature during summer months. Lack of depth prevents the water body to have any summer stratification.

Among 11 watersheds, the minimum (20.6°C) temperature was recorded at water shed No. I while the maximum value (33.6°C) was encountered at watershed No: XI (Table 9.5).

Table 9.5: Water Temperature (°C) at the Inlet Sites from Different Watersheds

Watershed No.	*Minimum*	*Maximum*	*Mean*	*Sd.*
I	20.6	31.4	27.1	3.57
	XII	VIII		
II	20.9	32.3	26.98	3.71
	XII	VIII		
III	25.3	31.8	28.51	3.76
	X	VIII		
IV	22.1	31	26.96	3.83
	XI	VII		
V	20.2	31.6	27.42	4.35
	XI	VIII		
VI	21.5	31	27.72	3.21
	XI	VI, VII		
VII	22.3	31.2	28.20	3.35
	XI	VI		
VIII	29.8	32.2	30.6	0.81
	IX	VII		
IX	30	32.2	30.9	1.11
	XI	VII		
X	24.3	33.5	28.6	3.35
	II	VII		
XI	21.4	33.6	27.88	3.41
	XII	V		

Turbidity

The present lakes where number of treatment plants are installed for supplying water for domestic purpose therefore invites attention for turbidity study.

The lake water during first decade (*i.e.* from 1960-1970) was more turbid in nature, which could be due to increase in entry of clay, silt, and carbonate particles etc. from its catchment area which had not been protected from any side at the very beginning of this study. The turbidity of water afterwards depicted a considerable decrease from 1975 onwards *i.e.* till 1995.

pH

Hydrogen ion concentration in out let waters varied from 7.2 unit l^{-1} to 8.4 unit l^{-1}. The maximum value was recorded during summer month. (April). In general pH values during different months depicted that the water was slightly alkaline in nature.

The hydrogen ion concentration in upper lake depicted a steady increase from 1960 onwards. The lake water always recorded pH values higher than 7. The higher pH concentration in the lake water signifies that the water body is becoming more productive with the changing time.

Conductivity

Conductivity varied from 206 m mhos to 340 m mhos. The maximum value was recorded in July while the minimum value was encountered in the month of April.

Chloride

Concentration of chloride in natural water lakes play metabolically active role in photolysis of water and photophosphorylation reaction in autotrophs There high concentration are considered to be the indications of pollution which could be either due to organic waste of animal origin or industrial effluents.

The chloride content of the lake under study depicted an increasing trend from 1965 onwards thereby indicating an increase of organic matter in lake water.

Water shed No.II recorded maximum value (45.99 mg l^{-1}) of chloride concentration over others (*i.e.* rest of the water sheds while the minimum value (9.99 mg l^{-1}) was encountered at water shed No. II.

Seasonal variation of chloride concentration depicts higher values of chloride content during March, April, and May at water shed No. I and watershed No: XI. In rest of the watershed *i.e.* III, IV, V, VI, VII VIII, the chloride content of the water didn't depict any significant trend when the flow was regular (Table 9.6).

Chloride content in outgoing waters varied from 20.97 mg l^{-1} (December) to 32.97 mg l^{-1} (July).

Calcium and Magnesium

Table 9.7 and 9.8 show the monthly variation in calcium and magnesium concentration from various watershed in puts in Bhoj wetland. The minimum (24 mg l^{-1}) concentration of calcium was recorded at watershed numbers XI. Maximum value (80.5 mg l^{-1}) of calcium was recorded at both watershed numbers II and VI.

Magnesium concentration of the water at almost all the watersheds also recorded a similar trend as that of calcium content. However higher value of magnesium content was recorded at watershed No. 1 during the month of September.

Table 9.6: Chloride (mg l^{-1}) at the Inlet Sites from Different Watersheds

Watershed No.	*Minimum*	*Maximum*	*Mean*	*Sd.*
I	13.99	38.99	28.41	7.39
	VIII	XI		
II	9.99	45.99	28.99	8.8
	X	VII		
III	14.99	22.99	12.82	6.64
	XI	VI		
IV	10.99	23.99	15.59	4.67
	XI	VIII		
V	12.99	33.99	30.1	4.66
	XI	X		
VI	8.99	30.99	19.15	8.0
	IX	X		
VII	15.99	36.99	24.8	7.17
	VIII	XI		
VIII	10.21	38.59	22.74	9.42
	VIII	VII		
IX	25.99	37.99	33.32	5.18
	VII	VIII		
X	32.99	33.99	32.54	3.79
	VIII	VII		
XI	16.99	30.32	23.38	5.61
	IX	III		

Table 9.7: Calcium (mg l^{-1}) at the Inlet Sites from Different Watersheds

Watershed No.	*Minimum*	*Maximum*	*Mean*	*Sd.*
I	36.3	72	52.26	9.75
	VI	VII		
II	30.70	80.05	43.99	15.25
	XI	VI		
III	27.4	68	41.83	13.67
	XI	VI		
IV	20	41.2	35.02	8.07
	X	VII		
V	41.3	68	47.66	10.59
	X	XI		
VI	31.8	80.05	59.54	21.05
	X	IX		

Contd...

Table 9.7–Contd...

Watershed No.	Minimum	Maximum	Mean	Sd.
VII	33.6	56	42.7	6.86
	X	IX		
VIII	42	68.4	54.16	9.68
	X	IX		
IX	27	36.3	29.36	6.87
	VIII	IX		
X	35.66	57.12	46.99	9.84
	VII	IX		
XI	24	46.20	42.7	12.12
	IX	II		

Table 9.8: Magnesium (mg l^{-1}) at the Inlet Sites from Different Watersheds

Watershed No.	Minimum	Maximum	Mean	Sd.
I	7.50	28	12	6.38
	II	IX		
II	6.99	21	11.1	4.21
	V	VII		
III	7.29	12.34	10.84	1.81
	VI	VII		
IV	6.86	21	10.94	5.15
	X	XI		
V	9.0	18	10.8	6.49
	VIII	IX		
VI	3.79	14.38	11.38	3.5
	VI	VIII		
VII	4.61	9.89	7.57	1.67
	VI	VIII		
VIII	4.8	12.16	9.64	1.43
	VII	IX		
IX	9.0	11.3	9.78	1.1
	IX	VII		
X	6.52	8.2	7.41	0.68
	VII	IX		
XI	5.23	21	9.86	4.11
	V	IX		

Calcium hardness remained within the range 32.89 mg l^{-1} to 47.26 mg l^{-1}. In general higher values of calcium concentration was recorded during monsoon months. Higher values of magnesium content was also recorded during monsoon months.

Dissolved Oxygen

D.O. is of paramount importance for all the living organisms and is considered to be lone factor which to a great extent can reveal the nature of the whole aquatic system at a glance, even when information on other chemical, physical and biological parameter is not available. Dissolved oxygen in lake water remained almost stable over a decade *i.e.* from 1980-1990, thereafter depicted a considerable fall.

Dissolved oxygen values in outgoing waters remained within the range of 6.6 mg l^{-1} to 8.4 mg l^{-1} during the study. The highest value was encountered during summer (May) while the lowest value was recorded during winter (January).

Table 9.9: Orthophosphate (mg l^{-1}) at the Inlet Sites from Different Watersheds

Watershed No.	*Minimum*	*Maximum*	*Mean*	*Sd.*
I	0.017	0.087	0.047	0.017
	IX	VIII		
II	0.021	0.093	0.054	0.021
	XI	VI		
III	0.016	0.116	0.064	0.031
	X	VII		
IV	0.044	0.144	0.75	0.035
	X	VII		
V	0.017	0.093	0.046	0.025
	VII	IX		
VI	0.021	0.116	0.07	0.29
	X	VII		
VII	0.079	0.163	0.09	0.039
	IV	IX		
VIII	0.049	0.089	0.07	0.014
	II	IX		
IX	0.05	0.057	0.08	0.042
	VIII	VII		
X	0.021	0.046	0.036	0.011
	VIII	VII		
XI	0.036	0.318	0.076	0.074
	IV	VII		

Orthophosphate

Orthophosphate at different watershed shows a gradual increase from February onwards reaching the maximum (0.163 mg l^{-1}) during post monsoon months. Maximum value of orthophosphate was encountered at watershed no XI. While the minimum value (0.015 mg l^{-1}) was recorded at watershed No. V. In rest of the watersheds where the flow was irregular, the orthophosphate values never depicted any significant trend throughout the study period (Table 9.9).

Orthophosphorus values ranged from 0.028 mg l^{-1} to 0.052 mg l^{-1}. The highest value was registered during July while the lowest value was encountered during December.

Total Phosphorous

Maximum concentration of total phosphorous values were recorded during post monsoon months at all the watersheds. Both the highest (0.634 mg l^{-1}) as well as the lowest (0.163 mg l^{-1}) values however were encountered at watershed No. VII. Rest at the watersheds *viz.* III, IV, V, VI, VII, VIII, IX, and X didn't show any significant trend during the course when the flow was regular (Table 9.10). Total phosphorus values remained within the range of 0.121mg l^{-1} (November) to 0.265 mg l^{-1} (July).

Table 9.10: Total Phosphorus (mg l^{-1}) at the Inlet Sites from Different Watersheds

Watershed No.	*Minimum*	*Maximum*	*Mean*	*Sd.*
I	0.164	0.434	0.274	0.101
	I	VII		
II	0.21	0.459	0.236	0.096
	I	VII		
III	0.181	0.612	0.327	0.0143
	IX	VII		
IV	0.289	0.616	0.412	0.0144
	I	IX		
V	0.289	0.459	0.381	0.079
	VIII	XI		
VI	0.169	0.502	0.321	0.108
	XI	VII		
VII	0.163	0.634	0.347	0.147
	IX	VII		
VIII	0.214	0.514	0.34	0.106
	IX	VII		
IX	0.163	0.413	0.28	0.103
	VIII	VII		
X	0.263	0.612	0.42	0.144
	VIII	IX		
XI	0.064	0.279	0.181	0.074
	II	IX		

Nitrate

Nitrate is most oxidized form of nitrogen and is an important plant nutrient. Due to its higher mobility as compared to other vital nutrients, its concentration is freshwater apart from autochthonous production is largely regulated by waste water loading, agricultural runoff and ground water inputs. Nitrate in lake water under consideration showed an increasing trend from 1985 onwards.

Monthly variation in nitrate values at all the input points from various watersheds has been depicted in Table 9.11. Maximum value (0.96 mg l^{-1}) of nitrate content was encountered in watershed no XI. The minimum (0.038 mg l^{-1}) value was recorded at watershed on III. In general both at watershed No. I and II the nitrate values depicted an increase during first half of the course of investigation. In rest of the watersheds particularly IV, VI and VIII, no definite trend was noticed. Nitrate nitrogen content in outgoing waters remained within 0.118 mg l^{-1} (March) to 0.386 mg l^{-1} (July).

Table 9.11: Nitrate (mg l^{-1}) at the Inlet Sites from Different Watersheds

Watershed No.	*Minimum*	*Maximum*	*Mean*	*Sd.*
I	0.092	0.9	0.296	0.244
	VI	VIII		
II	0.072	0.83	0.264	0.21
	VII	X		
III	0.038	0.191	0.089	0.048
	XI	VIII		
IV	0.064	0.194	0.10	0.056
	XI	VIII		
V	0.09	0.9	0.339	0.305
	XI	VII		
VI	0.067	0.83	0.247	0.262
	X	VII		
VII	0.088	0.7	0.250	0.231
	VI	VIII		
VIII	0.121	1.6	0.812	0.65
	VII, IX	VIII		
IX	0.064	1.0	0.376	0.44
	IX	VII, VIII		
X	0.128	0.412	0.243	0.122
	VII	VIII		
XI	0.092	0.96	0.447	0.711
	VIII, IX	X		

Higher values of nitrate nitrogen generally encountered during monsoon months. Phosphorous and Nitrogen budgeting has been depicted in Figures 9.2 a, b. An

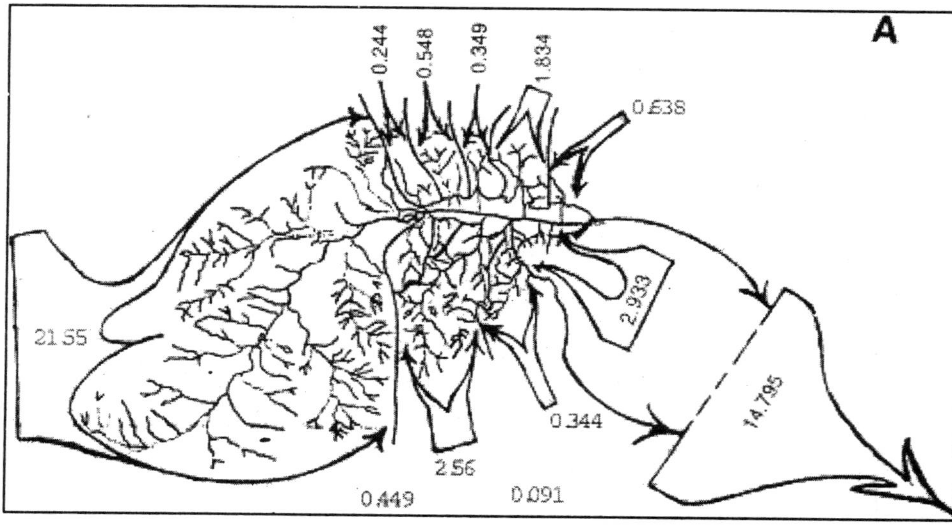

Total Phosphorus (Tonnes per year)

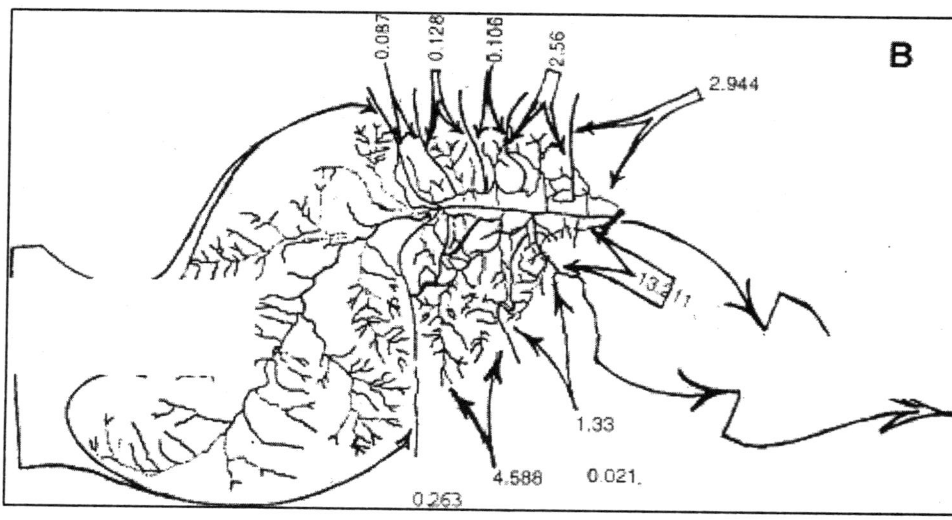

Total Nitrogen (Tonnes per year)

Figure 9.2: Phosphorus and Nitrogen Budgeting in the Bhoj Wetland

amount of 14.2 tones/year of total phosphorous has been found to be present in the system on yearly basis while on subtracting the output from input an amount of 16.735 tones / yr is supposed to be present in the system. The less amount present in the system (14.2 tones / yr) suggest that around 2.535 tones of phosphorous are trapped in different components of the system.

An amount of 19.757 tones / year of Nitrogen are left in the system (while subtracting the output from input values). In actual an amount of 23.30 tones / yr of NO_3-N are present in the system, suggesting that an additional amount of 3.543 tones/yr of NO_3-N is being added to the system from some other source.

Total Iron

Table 9.12 records monthly variation in total iron concentration from 11 watershed inputs of Bhoj wetland. Iron content was maximum (0.79 mg l^{-1}) at watershed No: I while it was minimum (0.045 mg l^{-1}) at watershed No. II.

Table 9.12: Total Iron (mg l^{-1}) at the Inlet Sites from Different Watersheds

Watershed No.	*Minimum*	*Maximum*	*Mean*	*Sd.*
I	0.064	0.79	0.297	0.20
	XI	V		
II	0.045	0.55	0.181	0.12
	VI	IV		
III	0.067	.268	0.147	0.06
	IX	VI		
IV	0.092	0.288	0.213	0.08
	IX	X		
V	0.18	0.48	0.31	0.11
	IX	VII		
VI	0.08	0.560	0.283	0.162
	VII	X		
VII	0.39	0.812	0.504	0.174
	VII	VI		
VIII	0.16	0.81	0.322	0.27
	X, XI	VII		
IX	0.19	0.30	0.253	0.046
	VIII	IX		
X	0.30	0.56	0.393	0.118
	VIII	VII		
XI	0.071	0.39	0.191	0.094
	V	IX		

Major Ionic Concentration at Different Watersheds of Bhoj Wetland

Inlet Waters

The major ionic concentration of inlet waters at different watersheds. Among major anions, concentration of bicarbonate was more at watershed No. 9, followed by watershed No. 10. The concentration of chloride was also recorded high at watershed No. 9. In general the sequence of dominance of various anions is as follows:

Bicarbonate > Chloride > Sulphate > Carbonate.

Among cations, calcium was more abundant at watershed No. 6, followed by watershed No. 8. The presence of rest of the cations was not of that significance.

The sequence of dominance of cations is as follows:

Calcium > Magnesium > Sodium > Potassium

Ambient Water

In ambient water, among major anions, concentration of bicarbonate was slightly more at surface water than at middle and bottom waters, while concentration of chloride was more at bottom waters than at surface and middle waters.

Of cations calcium was slightly high at surface and bottom waters.

The sequence of dominance of various cations is as follows:

Calcium > Magnesium > Sodium > Potassium

Outgoing Waters, Intake Point of Potable Pumping Water and Seepage Water

Among cations higher values of calcium, magnesium, sodium was recorded at outgoing (Bhadbhada site) waters than seepage and pumping water.

Among anions bicarbonate was also recorded to be more in outgoing water than in seepage and pumping water.

Biological Population

Plankton Population

Phytoplankton Population

In all 151 species of phytoplankton have been reported by Agarker *et al.* (1994). Out of these 72 species belonged to Chlorophyceae, 29 species to Cyanophyceae, 23 species each to Euglenophyceae and Bacillariophyceae and 4 species to Dinophyceae. Phytoplankton population attains maximum density during early summer period thus depicting positive role of temperature, on the growth of various algal forms. Second maximum density was attained during post monsoon period after the entry of nutrients from catchment area.

On the other hand Anonymous (1995-1996) recorded 49 species only of phytoplankton out of which 21 belonged to chlorophyceae and 11 belonged to Bacillariophyceae. Cyanophyceae was represented by 13 species whereas, Euglenophyceae and Dinophyceae was represented by two and one species respectively.

Zooplankton Population

A total of 78 zooplankton representing Protozoa (11 species), Rotifera (43 species), Cladocera (14 species), Copepoda (5 species) and Ostracoda (6 species) were identified in Bhoj wetland. The percentage of rotifers among the zooplankton varied from 28.98 per cent to 52.17 per cent. The difference of rotifers community structures in the water body attributed to the intense variability in physio-chemical, geographical conditions, human interference and presence of macrophytes (Agarker *et al.*, 1994). Anonymous (1995-1996) recorded only 27 species of zooplankton out of which 14 belonged to rotifera, 6 belonged to cladocera and 7 belonged to copepoda and 4 belonged to ostracoda.

Aquatic Insects

69 species of aquatic insects were identified from Upper lake. 5 species under Ephemeroptera, 9 species under Odonata, 29 species under Hemiptera, 11 species under Diptera and 15 species under Coleoptera with dominance of Hemipteran insects.

Primary Production

Seasonal variation in primary production has been depited in Table 9.13. In general carbon fixation at the surface water of littoral site always depicted higher values than pelagic site.

Table 9.13: Seasonal Variation in Primary Production Values at the Pelagic and Littoral Site (mg Cm^{-3} day^{-1})

Seasons	*Littoral Site*	*Pelagic Site*
Summer	448.92	137.86
Mansoon	68.53	41.26
Post mansoon	562.91	116.76
Winter	326.41	96.75

At littoral site the production varied from 68.53 mgCm^{-3} day^{-1} (Monsoon) to 562.91 mg Cm^{-3} day^{-1} (Post monsoon) while at pelagic site the production varied from 41.26 Cm^{-3} day^{-1} (Monsoon) to 137.86 mgCm^{-3} day^{-1} (Summer).

Management Strategies Adopted

Under the conservation process number of works has been carried out by the Lake conservation authority of Bhopal which are as:

Construction of Check Dams, Silt Traps, Toe Walls and Cascading for Controlling Silt Entry

The aim of the catchment area treatment of the lakes was to mitigate inflow, of silt, agricultural residues and other wastes etc. into the lake through inlet channels discharging into the lake. This has been achieved by constructing gabion structures, check dams silt traps and toe walls on all the inlets channels leading to the Upper and Lower lake. Under this sub-project at 3 Nallas (Banganga, Hindi Granth Academy and MVM) into the catchment area of lower lake and 12 Nallas (Halallpura, Laukhedi, Bairagarh Kalan, Bhainsakhedi, Intkhedi, Sirsa Nala, Mugaliya Chhap, Gora Nala, Kotra Sultanabad, SAF Nalla, Bhauri and Jamunia Chhir Nala) into the catchment area of Upper lake, a total number of 75 check dams, high level gabion structure, low level gabion structure and 2 silt traps has been executed.

Desilting and Dredging of Lakes

The prime objective of desilting and dredging was to remove volume of silt and sediments that have accumulated in the lakes over a long period of time. At those places where nallahs and rivulets meet the lakes and silt has accumulated generally at the peripheral area causing the lakes to become shallow and also resulted in growth of weeds. To avoid this desilting and dredging work has given importance.

Deepening and Widening of the Bhabhada Spill Channel

The deepening and widening works have been carried out in 2.6 km long spill channel to increase the discharge capacity and about 9.87 lakh cubic meter silt has been removed. The channel has been given a proper shape to attain the required discharge.

Construction of Idols/Tajias Immersion bay at Prempura (Bhadbhada)

The immersion of Idols/ Tajias in different places at Lower and Upper lake has given rise to settling of soil as well as the chemicals used for the preparation of idols causing adverse effect on the water quality. Bhopal Municipal Corporation under Bhoj Wetland Project has constructed an immersion bay as an alternative arrangement at Prempura. About 35 per cent of the immersion has been successfully diverted to the Prempura bay during the year 1999 and the percentage increased to 38 per cent in the year 2000.

Restoration of Takia Island

Takia' is a small island in the upper lake, its banks have eroded and in the peripheral area, siltation has taken place. Two toe walls have been constructed around the island to prevent siltation.

Afforestation and Creation of Buffer Zones

In order to prevent encroachment, human settlements, movement of the cattles and cultivation etc., buffer zones have been created particularly in the Western, Southern and Northern fringe area of the lake besides this intensive plantation has been carried out in the catchment area of the Upper lake to control and check soil erosion. The species selected in this project are mostly Bio-mass producing and medicinal plants. The plants have been carefully selected not only considering type of the soil, water and climatic conditions of the area but also taking into account of their economic and high medical values. Under this sub-project more than 16.3 lakhs plants have been planted in 936 hectare and land. In the programme under social forestry, about 2.06 lacs of plants have been distributed and planted. The afforestation in this catchment area has proved successful in checking these activities and also reducing the flow of silt from this portion of catchment area into the lake

Prevention of Pollution (Sewage Scheme) and Construction of Garland Drain

The aim of this sub project was to provide an improved sewage system in the areas which drain into upper and lower lake of Bhopal so that sewage and waste water from these areas which are a major cause of pollution in the lakes could be collected and diverted away for treatment. Hygienic environment could be created by giving good sanitation facilities. This sub-project envisages laying of sewerage net work system, construction of a treatment plant at a village near Bhopal and diversion of sewage towards the proposed treatment plant. About 62 kms sewerage pipeline and 23 kms long force mains are being laid in 23 municipal wards of the city which is divided in 3 networks of Gandhinagar, Maholi and Kotra. Four sewerage treatment plants are also being constructed in these areas.

Solid Waste Management

The main objective of this sub-project is to improve the solid waste management and facilities with in the catchment area of upper and lower lake in order to prevent pollution of the lakes by the direct or indirect entry of solid wastes into the lakes. The Solid Waste Management sub-project is being implemented in following 18 municipal wards under the catchment area of the upper and lower lake.

Construction and Development of Lake View Promenade

The Upper lake of Bhopal, on its eastern end is engulfed by the urban areas. The topography is accentuated by comparatively steeper slopes and thus provides for an extremely fast run-off from the catchment. Since the Lake edges are adjunct to the densely populated areas of the city, the tendency of construction and development right upto the edge of the water body has been very strong. Thus on planning grounds, it was imperative that a buffer had to be created between the Lake edge and the Developmental edge.

It was on this principal that the link Road from Retghat and Lalghati was constructed under the project as a buffer between the Human Settlements and the Lake. The effectiveness of this solution and the response from the population in general regarding availability of an easy access to the lake front prompted the replication of similar solution at other fringes.

Protection of Lake Fringes

As is known, the water level in the Upper lake fluctuates regularly with season, from the Full Tank Level of 508.04m at the maximum and an average minimum level of 505m. In the absence of clear demarcation of the FTL of the Upper Lake and the 50 m statutory No Construction Zone boundaries, it is difficult to ascertain these zones, and as a result, statutory control cannot be exercised with vigour.

It is thus proposed that the entire shoreline of the Upper Lake be marked with the Full Tank Level Line as well as the 50 m No Construction Zone line beyond FTL with boundary stones and document them as a survey. This document would serve as the basis for verification of any constructional/developmental work in the fringe area and thus a vigorous control could be exercised. The work is being executed by Bhopal Municipal Corporation which is the statutory body governing constructional and developmental activities in the city.

References

Agarker, M.S. *et al.*, 1994. Biology, conservation and management of Bhoj wetland. *Bionature*.

Ali, S.A., 1981. *Bhopal: Past and Present (A brief history of Bhopal from the hoary past upto the present time)*. Jai Bharti Publishing House, Bhopal, 450 pp.

SAPROF, 1994. *Conservation and Management Project of Lake Bhopal in India*. The overseas economic cooperation fun, Japan (OECF).

Wanganeo, A., 1996–1998. *Impact of Anthropogenic Activities on Bhoj Wetland with Particular Emphasis on Nutrient Dynamics*. Project reports submitted to MOEF, New Delhi.

Section II
Wildlife

Chapter 10

Structure and Composition of Birds in the New Amarambalam Tropical Forests of Kerala, Southern Western Ghats, India

☆ *E.A. Jayson*

Abstract

A total of 2293 birds were seen and altogether 100 taxa of birds were recorded from the whole region. Nilgiri House Swallow (*Hirundo tahitica domicola*), Bluewinged Parakeet (*Psittacula columboides*) and Yellowbrowed Bulbul (*Hypsipetes indicus*) were the most common and dominant species in the evergreen forest where as Rose ringed Parakeet (*Psittacula krameri*), Jungle myna (*Acridotheres fuscus*) and Common Babbler (*Turdoides caudatus*) were the most common and the dominant species in the moist deciduous forest. Species richness was highest in winter months and lowest in May and June. Highest number of birds was recorded in January, February and March. Shannon-Weiner diversity Index showed high value (3.73) which is comparable with other tropical forests of the Western Ghats. Slightly higher bird diversity was recorded in the moist deciduous forests (3.70) than the evergreen forests (3.15). Species richness indices where also higher in moist deciduous forests (R1=11.23) than the evergreen forests (R1=9.88). Highest density of birds was recorded in the moist deciduous forests (775 birds/km^2), followed by Shola (402 birds/ km^2) and the evergreen forests (400 birds/km^2). Only 10 migratory species were confirmed from the area, namely Rufoustailed Flycatcher, Paradise Flycatcher, Great Reed Warbler, Tickell's Leaf Warbler, Plain Leaf Warbler, Greenish Leaf Warbler, Blue Rock Thrush, Forest Wagtail, Yellow Wagtail and Grey Wagtail. Maximum number of species (44) was from insectivorous feeding guild. Presence of eight endemic and threatened species of birds of Western Ghats shows the conservation value of the forest area.

Introduction

Avian community studies are effective tools for monitoring forest ecosystems. In this study, an attempt is made to find the species composition and diversity of birds in the New Amarambalam Reserved Forests of southern Western Ghats. Only a few studies have been carried out in the New Amarambalam forests in the past. Birds of the adjacent Silent Valley National Park have been reported by various workers (Ramakrishnan, 1983; Pramod *et al.*, 1997; Jayson and Mathew, 2000 and Jayson and Mathew, 2000a). Similarly, birds of the nearby Nedumgayam forests also have been listed by the researchers from the Calicut University (Mathew, Pers. Comm.). The objective of this work was to find out diversity of birds in the New Amarambalam forests.

Seasonal changes of tropical forest birds have been reported by several workers from other countries (Morrison *et al.*, 1980; Pyke, 1984). The structure and composition of bird communities are known to vary in different vegetation types (Wiens, 1989). Studies on forest bird communities mainly examined the parameters like structure of forest bird communities (Nilson, 1983), distributions (Howe *et al.*, 1981) and community organisation (Landers and Mac Mohan, 1980). Yorke (1984) and Terborgh *et al.* (1990) described community structure of tropical forest birds. Community studies of birds in the forests of southern Western Ghats are rare. This study attempts to find out the various diversity parameters of the bird community in the tropical evergreen forests of New Amarambalam. The study had many constraints. Due to the logistical reasons, the interior forests were not accessible during the months of monsoon.

Study Area

New Amarambalam Reserved Forest is situated in the Western Ghats biogeographic region of Peninsular India, in Malappuram District of Kerala State. It covers an area of 265.57 km^2 and is located between 11°14′ and 11°24′ N latitude and between 76°11′ and 76°33′ E longitude. The reserved forest is situated North-West of Silent Valley National Park forming a contiguous forest belt in the Kerala part of the Nilgiri Biosphere Reserve. The altitude of the area ranges from 40 to 2254 m long above msl. The Mukurthi National Park of Tamil Nadu State borders New Amarambalam Reserved Forest in the East. Mukurthi is the highest peak (2554 m), which lies at the interstate border between Kerala and Tamil Nadu. New Amarambalam reserved forests occupy the Western slope of Nilgiri Plateau. The upper reaches of this reserve from part of the core area of the Nilgiri Biosphere Reserve along with Silent Valley National Park.

Methods

Occurrence of Birds

Diversity of birds was assessed in representative plots selected in various forest types using point count method (Buckland *et al.*, 1993). Eighty-three points were sampled in the evergreen forests, 80 points in the moist deciduous forest and 8 points were covered in the shola forests. Twenty minutes were spent at each point to record the presence of birds. Species richness and species composition of birds in the area were computed from the census data obtained from the point count method and from field observations. Apart from the systematic sampling from the point counts, whenever a bird was sighted in the forest it was identified and details recorded. The number of species recorded is considered as species richness. The relative dominance of each bird species in various vegetation types was determined by calculating Dominance Index using the following formula.

$$\text{Relative Dominance} = ni \times 100/N$$

where,

- ni: Number of individuals in the species
- N: The total number of individuals of the all species seen during the study period.

Abundance

Total number of birds recorded in each month in the sample plots was calculated using the census data.

Density

Density of birds and individual abundance of selected species were also calculated. The density was estimated using Point Count Method as described in Buckland *et al.* (1993). The density was computed using the software "DISTANCE". Ungrouped data were used for analysis. Flocks were considered as single and only one radial distance to the middle of the flock was measured. A birdcall was considered equivalent to a single individual and was used along with sighting record for the density estimation. The density estimation was done at two levels:

1. Total bird density was calculated for each month pooling data of all species.
2. Total bird density for each vegetation type was calculated pooling data of all species.

Species Richness Indices

Another way of representing species richness is through the species richness indices. These indices provide easily understandable measures of diversity. Species richness as a yardstick of diversity was used in many earlier studies also. Species richness indices like Margalef Index (R1) and Menhinick Index (R2) were calculated for each vegetation type using the formula (Magurran, 1988). The indices take mainly into consideration, S (Number of species) and N (Total number of individuals summed over all species).

Diversity Indices

For the assessment of species diversity in the area Shannon-Weiner Index, Simpson Index and Hill's diversity numbers N1 and N2 were calculated (Ludwig and Reynolds, 1988). Computations were done using the programme SPECDIVERS.BAS developed by Ludwig and Reynolds (1988). Even though species-abundance model provides full description of diversity, they cannot be compared by means of diversity indices. Indices based on the proportional abundance of species provide another approach to measure diversity. They are called heterogeneity indices because they take both evenness and species richness into account (Peet, 1974).

Evenness Indices

Two evenness measures namely, Shannon Evenness and Sheldon Evenness were calculated using the computer programme SPDIVERS.BAS developed by Ludwig and Reynolds (1988).

Results

Species Composition

Species composition and diversity of birds in a forest is related to the vegetation type, altitude, availability of microhabitats and various other factors. One hundred species of birds were recorded from the study area belonging to 13 Orders and 31 Families. Common bird species found in New Amarambalam were Black Bulbul, Black Drongo, Goldenbacked Woodpecker, Jungle Myna and Yellow-browed Bulbul. The population of Passeriformes was highest followed by Coraciiformes and Cuculiformes (Table 10.1). Only 10 migratory species were sighted from the area namely Rufoustailed Flycatcher, Paradise Flycatcher, Great Reed Warbler, Tickell's Leaf Warbler, Plain Leaf Warbler, Greenish Leaf Warbler, Blue Rock Thrush, Forest Wagtail, Yellow Wagtail and Grey Wagtail. Highest species richness was seen in February during summer and then in November (Figure 10.1). Lowest was seen during May and June. Highest number of birds was insectivores (60) followed by frugivores and others.

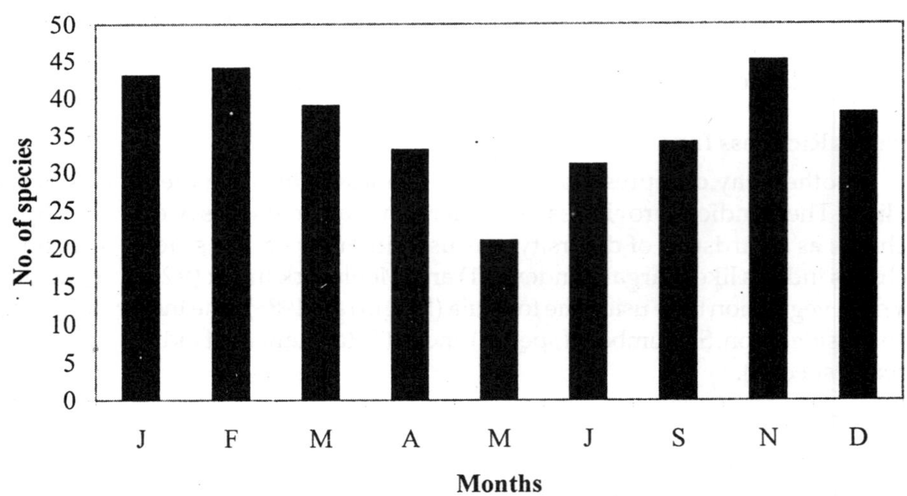

Figure 10.1: Species Richness of Birds in Different Months at New Amarambalam

Table 10.1: Number and Status of Birds Recorded at New Amarambalam

Sl.No.	Order	Status			Feeding Guild						
		R	M	Total	A	I	G	N/F	C	F	O
1.	Pelecaniformes	1	–	1	1	–	–	–	–	–	–
2.	Ciconiiformes	3	–	3	3	–	–	–	–	–	–
3.	Falconiformes	4	–	4	–	–	–	–	4	–	–
4.	Galliformes	4	–	4	–	4	–	–	–	–	–
5.	Columbiformes	5	–	5	–	–	5	–	–	–	–
6.	Psittaciformes	4	–	4	–	–	–	–	–	4	–
7.	Cuculiformes	6	–	6	–	5	–	–	–	–	1
8.	Strigiformes	1	–	1	–	–	–	–	1	–	–
9.	Apodiformes	4	–	4	–	4	–	–	–	–	–
10.	Trogoniformes	1	–	1	–	1	–	–	–	–	–
11.	Coraciiformes	9	–	9	4	2	–	–	–	–	3
12.	Piciformes	2	–	2	–	2	–	–	–	–	–
13.	Passeriformes	46	10	58	–	44	2	7	–	5	–
	Total	90	10	100	8	60	7	7	5	9	4

A: Aquatic feeders; I: Insectivores; G: Granivorous; N/F: Nectar and Frugivorous; C: Carnivorous; F: Frugivorous; O: Omnivorous; R: Resident; M: migrant.

Of the 16 endemic and restricted range species of birds found in the Western Ghats and Kerala, the following eight species were recorded in New Amarambalam.

1. Nilgiri Wood Pigeon (*Columba elphinstonii*)
2. Bluewinged Parakeet (*Psittacula columboides*)
3. Malabar Grey Hornbill (*Tockus griseus*)
4. Southern Tree Pie (*Dendrocitta leucogastra*)
5. Small Sunbird (*Nectarinia minima*)
6. Greyheaded Bulbul (*Pycnonotus priocephalus*)
7. Nigiri Flycatcher (*Eumyias albicaudata*)
8. Wayanad Laughing Thrush (*Garrulax delesserti*)

Among these, the Nilgiri Wood Pigeon is a globally threatened species found only in India. Total number of birds recorded at New Amarambalam in different months is given in Figure 10.2.

Abundance and Density

The mean number of birds seen in each month during the study period is given in Figure 10.2. The number of birds ranged from 85 to 596 at New Amarambalam. A slight reduction in the total number of birds was seen during May and June. Chi square test showed significant difference in total number of birds in various months

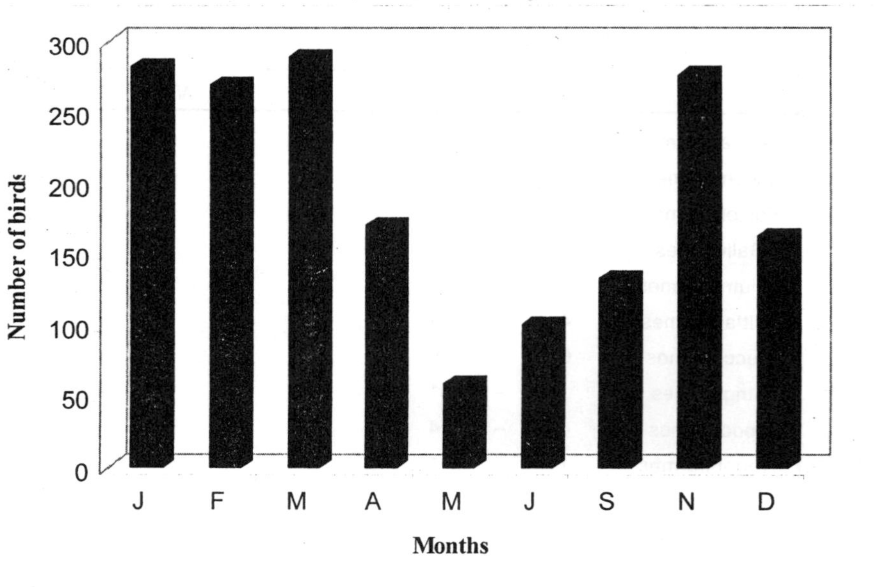

Figure 10.2: Total Number of Birds Recorded at New Amarambalam in Different Months

(χ^2= 318.02, P= 0.05, df=8). Nilgiri House Swallow was the dominant bird species in the evergreen forests where as Little Cormorant and Roseringed Parakeet was dominant in the moist deciduous forest and Nilgiri Flycatcher in the Shola forests. The recording of Little Cormorant as the dominant species should be considered as an exception, because as some point transects covered riversides also the number of Little Cormorants increased in the sampling. Dominance index of selected species of birds from three vegetation types is given in Tables 10.2–10.4.

Table 10.2: Dominance Index of Selected Species of Birds in the Evergreen Forest

Sl.No.	Species	Dominance Index
1.	Nilgiri House Swallow	24.73
2.	Bluewinged Parakeet	9.83
3.	Yellowbrowed Bulbul	7.48
4.	Malabar Whistling Thrush	7.03
5.	Alpine Swift	3.93
6.	Small Leaf Warbler	3.78
7.	Jungle Myna	3.70
8.	Ashy Drongo	2.42
9.	Goldenbacked Woodpecker	2.42
10.	Magpie Robin	2.57

Table 10.3: Dominance Index of Selected Species of Birds in the Moist Deciduous Forest

Sl.No.	Species	Dominance Index
1.	Little Cormorant	9.73
2.	Roseringed Parakeet	7.86
3.	Jungle Myna	7.55
4.	Common Babbler	7.14
5.	Racket-tailed Drongo	4.14
6.	Black Drongo	3.00
7.	Small Leaf Warbler	2.89
8	Bluewinged Parakeet	2.79
9.	Grey Hornbill	2.79
10.	Jungle Crow	2.58

Table 10.4: Dominance Index of Selected Species of Birds in the Shola Forest

Sl.No.	Species	Dominance Index
1.	Nilgiri Flycatcher	12.42
2.	Nilgiri Laughing Thrush	8.28
3.	White throated Ground Thrush	5.91
4.	Nilgiri House Swallow	5.32
5.	Yellow throated Sparrow	4.73
6.	Tickell's Leaf Warbler	4.14
7.	Black Eagle	2.36
8	Black Bird	1.77
9.	House Swallow	1.18
10.	Jungle Myna	1.18

Diversity Indices

Species Diversity

Values of four diversity indices obtained for New Amarambalam is given in Table 10.5. Shannon–Weiner Index showed a value of 3.73 and Simpson's index was 0.04. All the four diversity indices showed high values.

Table 10.5: Bird Diversity in the New Amarambalam Reserve Forest

No. of Species	No. of Individuals	Simpson's Index	Shannon-Weiner Index	Hill's Number N1	Hill's Number N2
100	2265	0.04	3.73	22.65	21.31

Evenness or Equitability Indices

A variety of evenness measures is available. Two of them are based on Shannon-Weiner Index and Simpson's Index. This index measures the evenness of species-abundance, is complimentary to the diversity index concept, and is a measure of how the individuals are appropriated among the species. The ratio of observed diversity to maximum diversity is taken as measure of Evenness (E). Two different evenness measures were calculated. Shannon Evenness E1 was 0.91 and Shannon Evenness E2 was 0.73.

Diversity of Birds

Diversity indices of birds in the three vegetation types were calculated as described earlier. Highest diversity index (H'= 3.70) was obtained for the moist deciduous forest (Table 10.6) followed by evergreen forest and the shola. Highest number of species was recorded from moist deciduous forest followed by evergreen forest and shola.

Table 10.6: Diversity Indices of Birds in Three Vegetation Types

Vegetation Types	*Shannon-Weiner Index*	*Simpson's Index*	*Hill's Number N1*
Moist deciduous forest	3.70	-0.038	40.50
Evergreen forest	3.15	0.09	23.40
Shola forest	2.24	0.01	9.37

Species Richness Indices

Margalef Index and Menhinick Index showed higher value for the moist deciduous vegetation (Table 10.7). This suggests that the bird community in the moist deciduous forest is more diverse. A thematic representation of the birds in different forest types is shown in Figure 10.3.

Table 10.7: Species Richness of Birds in Three Vegetation Types at New Amarambalam

Parameters	*Moist Deciduous Forest*	*Evergreen Forest*	*Shola Forest*
No. of species	78	72	14
Margalef Index R1	11.23	9.88	2.92
Menhinick Index R2	2.53	1.98	1.50

Species-abundance Relationships

Another way of studying diversity in a community is through species-abundance or distribution models. This was introduced by Fischer *et al.* (1943). A species-abundance model utilises all the information gathered in a community and is the most complete mathematical description of the data (Magurran, 1988). Analyses were also carried out to see whether the truncated lognormal model suited the bird

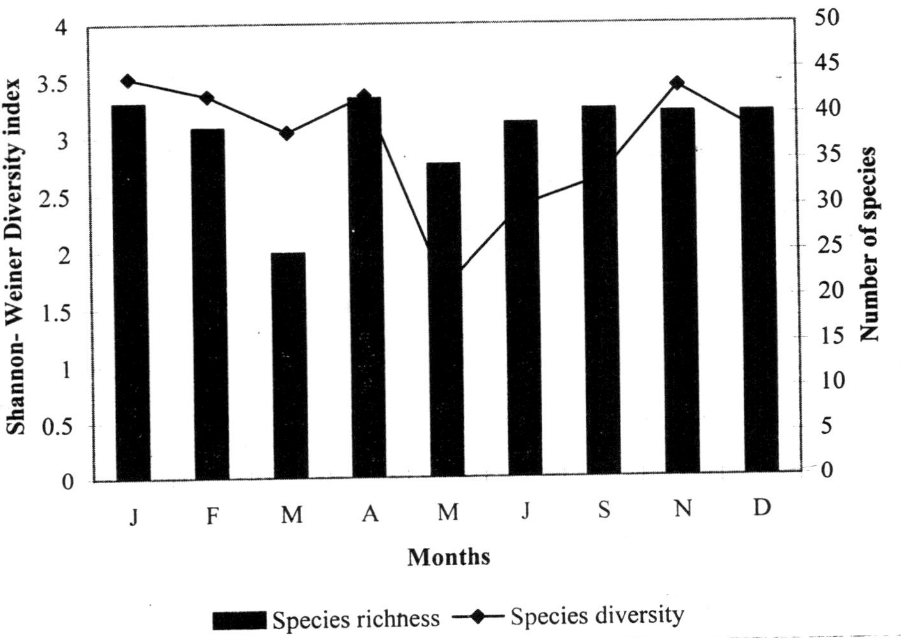

Figure 10.3: Species Richness and Diversity of Birds at New Amarambalam in Different Months

community at the study area. Species-abundance distribution of birds in the evergreen forest areas in semi-log scale is presented in Figure 10.4 and that of moist deciduous forest in Figure 10.5. This distribution indicated the absence of a single dominant species or group of species and the presence of a long series of very rare species.

A truncated lognormal distribution is fitted to the data from the evergreen forest and moist deciduous forest (Tables 10.8 and 10.9) (Pielou, 1975). The observed and expected numbers of species were computed using χ^2 goodness of fit test. The test showed no significant difference between the observed and expected distribution. This indicated that the distribution pattern of species was following truncated log-normal ($\chi^2 = 2.6$; df = 5; p= >0.7) in the evergreen forest (Table 10.11). In moist deciduous forest (4. 12), also the distribution pattern was following a truncated lognormal distribution ($\chi^2 = 6.54$; df=4; p= 0.1)

Seasonal Fluctuations

Species richness and total number of birds varied in different months. Species richness recorded in various months is given in Figure 10.3. An increase in species richness is visible during the months of November, December and January. However, as the rain stopped, new species started to arrive and a maximum of fifty-five species were recorded in the month of January. Reduction in species richness during the monsoon season was observed throughout the study period. This was primarily due

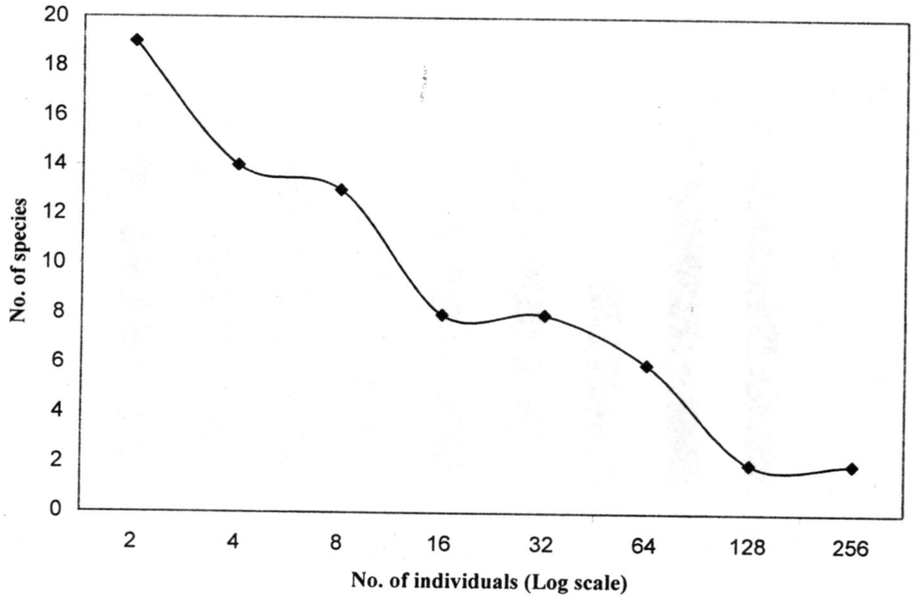

Figure 10.4: Species-Abundance Distribution of Birds in the Evergreen Forest of New Amarambalam

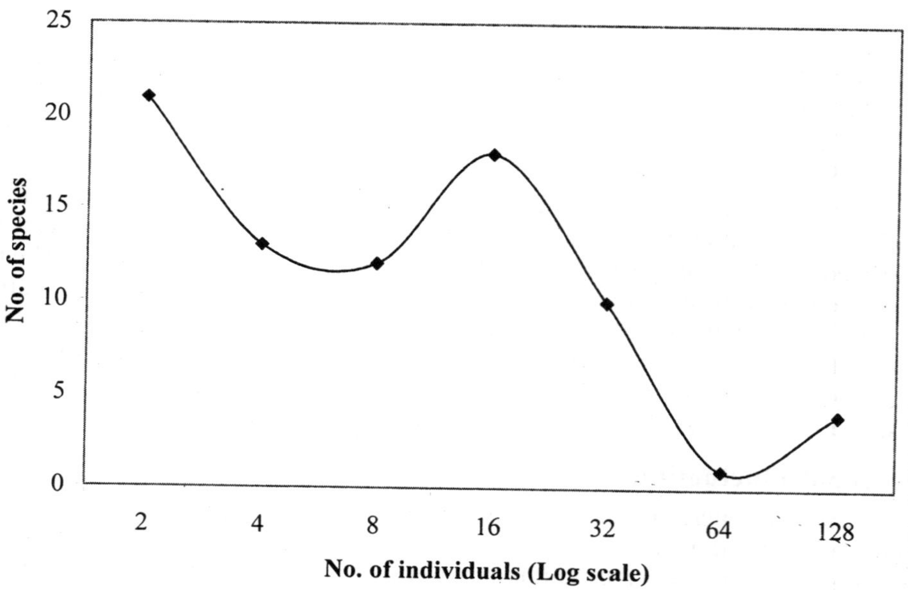

Figure 10.5: Species-abundance Distribution of Birds in the Moist Deciduous Forest of New Amarambalam

to the absence of migratory species such as wagtails and due to the local movement of species like Black bulbul. Similarly, a surge in species richness was recorded during summer, in all the years.

Table 10.8: Truncated Lognormal Distribution in the Evergreen Forests (χ^2 test)

Class	*Upper Boundary*	*Observed*	*Expected*
Behind veil line	0.5	–	4.36
1	2.5	19	19.2
2	4.5	14	11.28
3	8.5	13	12.63
4	16.5	8	11.82
5	32.5	8	8.43
6	64.5	6	5.01
7	128.5	2	2.36
8		2	1.27

Table 10.9: Truncated Lognormal Distribution in the Moist Deciduous Forests (χ^2 test)

Class	*Upper Boundary*	*Observed*	*Expected*
Behind veil line	0.5	–	2.45
1	2.5	21	19.35
2	4.5	12	13.45
3	8.5	13	16.03
4	16.5	18	14.09
5	32.5	10	9.27
6	64.5	1	4.6
7		4	2.34

Density

Highest density of bird population was found in the month of November. Overall density of bird community at New Amarambalam was 510 birds/km^2. Density of birds in different moths were also calculated which is presented in Figure 10.6. Highest density was observed in moist deciduous forest (775 birds/km^2) followed by shola forests (402 birds/km^2). Lowest density was recorded in the evergreen forest (400 bird/km^2). Individual species density of 26 common species of birds was calculated as described in the methods and the results are given in Table 10.10. Highest density was obtained for the species Common babbler followed by other species namely Black bulbul and Small leaf warbler. No comparison is possible as similar data is not available from other areas.

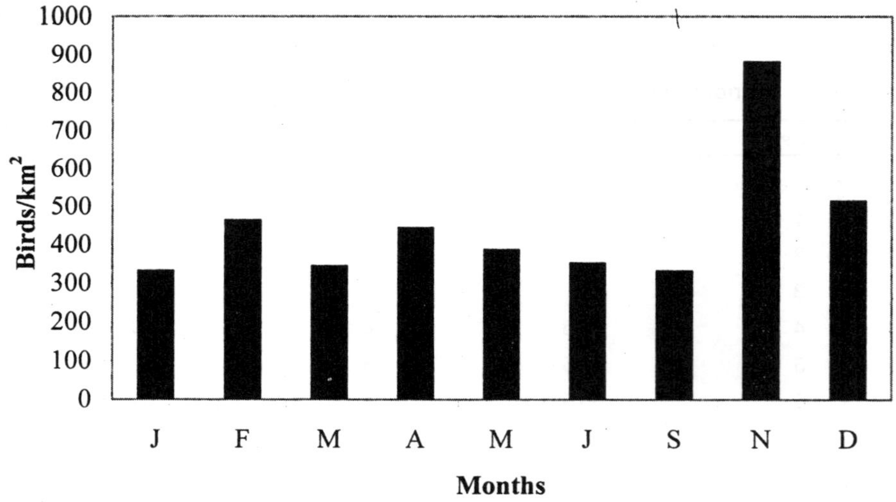

Figure 10.6: Density of Birds Recorded at New Amarambalam in Different Months

Table 10.10: Density of Selected Bird Species at New Amarambalam

Sl.No.	Species of Bird	Density (Birds/Km^2)
1.	Common babbler	691.70
2.	Black bulbul	473.16
3.	Small leaf warbler	469.49
4.	Purple sunbird	402.66
5.	Spotted dove	395.39
6.	Yellow browed bulbul	333.27
7.	Black drongo	308.16
8.	Racket-tailed drongo	265.63
9.	Indian koel	263.88
10.	Nilgiri flycatcher	259.84
11.	Tree pie	220.07
12.	Roseringed parakeet	217.73
13.	Emerald dove	210.40
14.	Grey hornbill	203.75
15.	Goldenbacked woodpecker	197.91
16.	Greyheaded bulbul	164.60
17.	Jungle myna	163.80
18.	Ashy drongo	157.42

Contd...

Table 10.10–Contd...

Sl.No.	Species of Bird	Density (Birds/Km^2)
19.	Grey jungle fowl	131.94
20.	Bluewinged parakeet	131.8
21.	Blossomheaded parakeet	127.32
22.	Magpie robin	126.08
23.	Blackheaded oriole	117.92
24.	Alpine swift	64.96
25.	Malabar whistling thrush	56.61
26.	Small sunbird	53.68

Discussion

Occurrence of Species

Among the 100 species of birds recorded, ten species were migrants. Presence of eight endemic species of birds to the Western Ghats in New Amarambalam showed the importance of the area. Occurrence of Nilgiri Wood Pigeon, which is a globally threatened species, adds to conservation value of the region. A drop in bird abundance during monsoon season with adverse climate (heavy rainfall) as observed in this study was reported by Ramakrishnan (1983) and Jayson (2000) also. Similarly, Morison *et al.* (1980) also reported reduction of birds during non-winter period and their increase during winter. One factor, which may influence the abundance, is detectability. Seasonal differences in detectability are common for most of the bird species (Emlen, 1977). These differences result from changes in weather and habitat structure. Increasing foliage density decreased the visibility of birds. However, in the study area foliage structure was identical in all seasons and only rainfall had some influence on detectability. During monsoon (June to September) a reduction in total number of birds and density was recorded which is reported from other areas in the Western Ghats also. Density obtained for the moist deciduous forest is comparable with other tropical forests. Individual species density of birds showed that highest density was recorded for Common Babbler followed by Black Bulbul. Similar results were reported from Silent Valley also in an earlier study (Jayson, 1994).

Diversity Indices

Shannon-Weiner diversity Index of 3.73 obtained is comparable to the adjacent Silent Valley forest, from where a value of 3.30 was obtained in an earlier study. As reported for the other community parameters, highest index of diversity was obtained for the moist deciduous forests. Species richness indices and diversity indices showed high diversity for New Amarambalam. Species richness of New Amarambalam is comparable to the adjacent Silent Valley forest (Jayson and Mathew, 2000a) (Table 10.11). As reported from other areas of the State, high species richness was recorded for the moist deciduous forest. Lowest species richness was recorded in the month of March, which is not the usual pattern for the Western Ghats.

Table 10.11: Number of Species of Birds Reported from Different Protected Areas in Kerala

Name of Areas	Number of Species	Source
Silent Valley forests (Tropical evergreen and moist deciduous)	212	Jayson (1994)
Periyar Tiger Reserve (Tropical evergreen, grass lands and sholas)	249	Srivasthava *et al.* (1993)
Parambikulam Wildlife Sanctuary (Tropical evergreen, Moist deciduous and teak plantations)	133	Vijayan (1978)
Chinnar Wildlife Sanctuary (Dry deciduous and dry deciduous scrub)	143	Nair *et al.* (1997)
Peechi-Vazhani Wildlife Sanctuary (Moist deciduous and tropical evergreen)	177	Nest (1991)
Waynad Wildlife Sanctuary (Moist deciduous evergreen and plantations)	275	Zacharias (1993)

Species-abundance Relationships

It is reported that the occurrence of a large number of rare species is typical of tropical forests (Lovejoy, 1975). The factors, which control the species richness in an area, are broadly divided into as historical and ecological (Giller, 1984). In New Amarambalam, bird communities in both the evergreen forest and moist deciduous forest follow the truncated lognormal distribution, which is typical for the undisturbed tropical forests.

Comparison with Other Tropical Forests

In the present study, 510 individuals/km^2 were recorded at New Amarambalam. The above values are comparable to a number of similar studies on the composition and abundance of tropical forest birds, which are presently available. Karr (1971) estimated a population density of 1820 pairs of birds/km^2 in Panama's tropical forests. Thiollay (1986) reported a density of 760 pairs/km^2 for a total of 263 species in French Guiana. A long-term study conducted by Brosset (1990) in Gabon on 364 species of birds produced a density of 3,690 individuals/km^2. A study by Bell (1983) on birds of lowland rainforest in New Guinea produced a density of 3450 pairs/km^2. A study by Terborgh *et al.* (1990) on 245 species of Amazonian forest bird community had shown a density of 1910 individuals/km^2.

During this study, the birds were observed 1011 times and a total of 2293 birds were counted. Out of the 100 species located from the whole study area, ten species were migrants and others were residents. New Amarambalam is not a major wintering area of palaeartic migrants and most of the birds were showing only local movements. The migrants that were recorded from here were the Rufoustailed Flycatcher, Paradise Flycatcher, Great Reed Warbler, Tickell's Leaf Warbler, Plain Leaf Warbler, Greenish Leaf Warbler, Blue Rock Thrush, Forest Wagtail, Yellow Wagtail and Grey Wagtail. No wintering waterfowls were recorded from the area. This was due to the lack of wetlands in the study area. Most of the Doves, Pigeons, Parakeets and Black Bulbuls (*Hypsipetes madagascariensis*) were not recorded during rainy season, but were seen returning to the area with the retreat of the rain. From the dominance index, it is clear that barring a few species all others are very rare. In addition to this, local movement of species like Black Bulbul to the study area was observed during summer. Presence of endemic and threatened species shows the conservation value of New Amarambalam forests. It is recommended that the study area may be declared as a protected area.

References

Ali, S. and Ripley, S.D., 1983. *Handbook of the Birds of India and Pakistan.* Oxford University Press, Oxford, 737 p.

Bell, H.L., 1983. A bird community of low land rainforest in New Guinea. 6. Foraging ecology and community structure of the avifauna. *Emu.*, 84(3): 142–159.

Brosset, A., 1990. A long-term study of the rain forest birds in m'Passa (Gabon). In: *Biography and Ecology of Forest Birds,* (Ed.) A. Keast. SPB Academic, The Hague.

Buckland, S.T., Anderson, D.R., Burnham, K.P. and Laake, J.L., 1993. *Distance Sampling: Estimating Abundance of Biological Populations.* Chapman and Hall, London, 445 p.

Emlen, J.T., 1977. Estimating breeding season bird densities from transect counts. *Auk,* 88: 455–468.

Fischer, R.A., Corbet, A.S. and Williams, C.B., 1943. The relationship between the number of species and the number of individuals on a random sample of an animal population. *J. Anim. Ecol.,* 12: 42–58.

Giller, P.S., 1984. *Community Structure and the Niche.* Chapman and Hall, 58 p.

Howe, R.W., Howe, T.D. and Ford, H.A., 1981. Bird distribution on small rain forest remnants in New South Wales. *Australian Wildl. Res.,* 8: 637–652.

Jayson, E.A., 1994. Synecological and behavioural studies on certain species of forest birds. *Ph.D. Thesis,* University of Calicut.

Jayson, E.A. and Mathew, D.N., 2000. Seasonal changes of tropical forest birds in the southern Western Ghats. *J. Bombay Nat. Hist. Soc.,* 97(1): 52–61.

Jayson, E.A. and Mathew, D.N., 2000a. Diversity and species-abundance distribution of birds in the tropical forests of Silent Valley. *J. Bombay Nat. Hist. Soc.,* 97(3): 390–399.

Karr, J.R., 1971. Structure of avian communities in selected Panama and Illinois habitats. *Ecol. Monogr.,* 41: 207–233.

Landers, P.B. and MacMohan, J.A., 1980. Guild and community organisation: Analysis of an Oak Woodland Avifauna in Sonora, Mexico. *Auk,* 97: 351–365.

Lovejoy, T.E., 1975. Bird diversity and abundance in Amazon forest communities. *Living Bird,* 13: 127–191.

Ludwig, J.A. and Reynolds, J.F., 1988. *Statistical Ecology.* John Wiley and Sons.

Magurran, A.E., 1988. *Ecological Diversity and its Measurement.* Croom Helm Ltd., London, 179 p.

Mathew, D.N. *Personal Communication.*

Morrison, M.L., Kimberly, A. and Timossi, I.C., 1980. The structure of a forest bird community during winter and summer. *Wilson Bull.*, 98(2): 214–230.

Nair, P.V., Ramachandran, K.K. and Jayson, E.A., 1993. Distribution of mammals and birds in Chinnar Wildlife Sanctuary, *KFRI Research Report No. 131*, KFRI, Peechi.

Nair, K.K.N., Sharma, J.K., Mathew, George, Ramachandran, K.K., Jayson, E.A., Mohanadas, K., Nandakumar, U.N. and Nair, P. Vijayakumaran, 2002. Studies on the biodiversity of New Amarambalam reserved forests of Nilgiri Biosphere reserve. *KFRI Research Report No. 247*, KFRI, Peechi.

Nest, 1991. *Birds of Peechi–Vazahni Wildlife Sanctuary: A Survey Report*. Nature Education Society, Trichur, 22 p.

Nilson, S.G., 1983. The structure of bird communities of natural forests in Sweden and the grading success of birds nesting in natural cavities. *Ibis*, 125(4): 587–588.

Peet, R.K., 1974. The measurement of species diversity. *Ann. Rev. Ecol. Syst.*, 5: 285–307.

Pielou, E.C., 1975. *Ecological Diversity*. John Wiley & Sons, New York.

Pramod, P., Ranjit Daniels, R.J., Joshi, N.V. and Gadgil, Madhav, 1997. Evaluating bird communities of Western Ghats to plan for a biodiversity friendly development. *Curr. Sci.*, 73: 156–162.

Pyke, G.H., 1984. Seasonal patterns of abundance of insectivorous birds and flying insects. *Emu*, 85(1): 34–39.

Ramakrishnan, P., 1983. Environmental studies on the birds of the Malabar forest. *Ph.D. Dissertation*, University of Calicut.

Srivasthava, K.K., Zacharias, V.J., Bhardwaj, A.K. and Jaffer, P. Mohammed, 1993. Birds of Periyar Tiger Reserve, Kerala, South India. *Indian For.*, 119: 816–827.

Terborgh, J., Robinson, S.K., Parker, T.A. III, Munn, Charles A. and Pierpont, N., 1990. Structure and organisation of an Amazonian forest bird community. *Ecol. Monogr.*, 60: 213–238.

Thiollay, J.M., 1986. Structure comparee du peuplement avien dans trois sites de foret primaire en Guyan. *Revue d'Ecologie la Terre et la Vie*, 41:59–105.

Vijayan, V.S., 1978. Parambikulam Wildlife Sanctuary and its adjacent areas. *J. Bombay Nat. Hist. Soc.*, 75(3): 888–900.

Wiens, J.A., 1989. *The Ecology of Bird Communities, Vol. 1: Foundations and Patterns.* Cambridge University Press, 53 p.

Yorke, C.D., 1984. Avian community structure in two modified Malaysian habitats. *Biol. Conserv.*, 29(4): 345–362.

Zacharias, V.J. and Gaston, A.J., 1993. The birds of Wayanad, southern India. *Forktail*, 8: 11–23.

Chapter 11

An Unusual Breeding Case of the House Sparrow (*Passer domesticus*) at Jammu

☆ *B.L. Kaul and Anil K. Verma*

Abstract

The present study deals with the breeding habits of the house sparrow (*Passer domesticus*) at Jammu (J&K) India. Generally the bird breeds in summer but the present study shows a case of delayed breeding in autumn which is rather unusual in this bird species.

An Unusual Case: The present study concerns a breeding pair that had been re-using the same nest built three years ago inside the 4 inch dia outlet of a modern chimney (Figures 11.4, 11.5) of Sr. Author house at Sarwal, Jammu. The Jammu and Kashmir State extends over 480 km. from the east to west and 640 km from north to south. The valley of Kashmir alone is 15520 sq. km in area. Climatically, Jammu region which is very rich in avifauna is divisible into two groups *i.e.* the plain region and the mountainous region.

The Plain Region (South Shiwaliks)

The outer plains and the outer hills are grouped into tropical and subtropical climate region. This region includes the Kathua and Jammu districts and extends upto Shiwaliks hills in the north. The chief towns of this region Jammu, Kathua, Samba, Hiranagar, Basohli, and Akhnoor observe intense tropical heat in summer. In Jammu, a typical foothill town of outer hill region, the dust storms are common with occasional rains owing to its close vicinity to the plains of Punjab. The nights are charming because the cool wind descends from the Shiwaliks at night time.

The hot air of the summers which steadily blows from west to east is known as "Loo" in local dialect. The summer crops including vegetables are repeatedly irrigated to protect them from withering due to 'Loo' (hot air). The mist and fog is quite common in winters, particularly in the morning hours, which reduces the visibility. Sometimes, for a short period during the day, the intensively less visibility gives a scene of night time, as the vehicular traffic switches on their light for their movement. The frost and the cold winds, which blow from north to south (snow peaks of Chenab valley to the plains), are common in winters.

The Mountainous Region

The middle Himalayas (Pir Panjal and Upper Chenab region), with their rugged topography and slope, have altogether different weather and climatic conditions from that of the plains region of Jammu. The middle Himalayas of Jammu division sprawl between Ravi in east and Poonch in west. Chief towns of this region are Doda, Bhaderwah, Kishtwar, Batote, Ramban, Banihal, Ramsu, Riasi, Rajouri, Poonch and Udhampur. The moisture laden winds of the summer monsoon cause rainfall in the outer plains regions but in middle mountains they become too weak to cause precipitation.

In Jammu, the monthly temperature remains almost above 20°C round the year, thus it enjoys a growing season of full year. The hottest month is the June while January is the coldest month. Sometimes the temperature shoots to 46°C in June. Most of the rainfall comes from July to September so the humidity is also high (69 per cent) during this period. Four months *i.e.* May, June, October and November are generally dry. December to February remains the cold season with a temperature between 13.5 to 17°C in this region of J&K State. The days are often sunny and warm as compared to nights, which are very cold.

Table 11.1: Showing Temperature Regimen and Rainfall at Jammu

Months	*Mean Maximum Temp. °C*	*Mean Minimum Temp. °C*	*Rainfall in mm*
January	19.0	6.7	46
February	20.2	8.4	40
March	23.4	12.0	31
April	32.2	18.1	24
May	38.3	23.0	11
June	38.0	25.3	75
July	34.2	25.1	275
August	35.5	24.8	250
September	33.6	22.4	125
October	31.9	17.5	46
November	26.1	12.3	08
December	22.2	7.7	33

The annual rainfall is about 150 mm. The temperature begins to rise above 17°C in March till it reaches above 40°C in June. The weather becomes mostly dry and very hot. The south-west monsoon arrives in Jammu by first week of July and withdraw by second week of September. The humidity is quite high during monsoon. October and

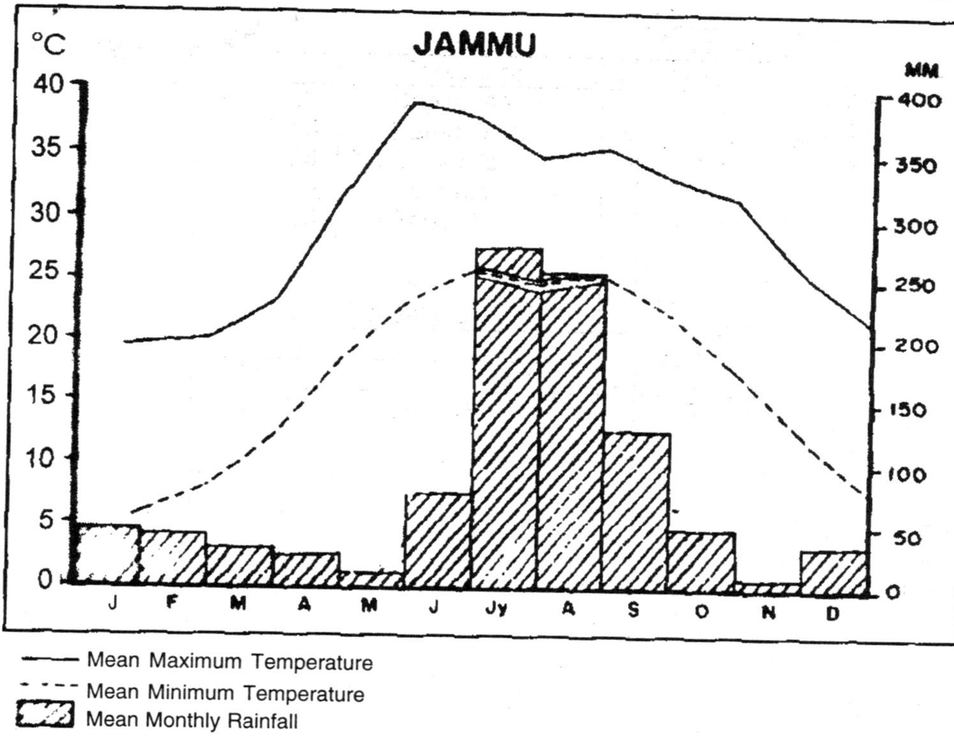

Figure 11.1: Mean Maximum and Minimum Temperature and Mean Monthly Rainfall

November are generally dry with a decrease in temperature to 27°C and 21°C respectively.

The study site had the following physiographical as well as topographical details (Figure 11.1, Table 11.1). They had been breeding in the same nest for the last three years (2006 to 2008) from March to August raising an average of 3 clutches per year.

But this year an unusual breeding activity was observed late in the month of October. The pair refurbished their nest and laid a single egg and successfully hatched it in the second week of November. The single hatchling (Figure 11.3) grew up rather slowly and demanded food even during night may be because insects form the major part of a hatchling's food are not so easily available in November. Finally it grew up into a fledgling and flew away in the second week of December 2008.

General Account

Worldwide there are 12 sub-species of the House Sparrow (*Passer domesticus*). House Sparrows are one of the most common birds found throughout the populated world. Unfortunately, however, it has been observed that their populations are on the decline because new concrete constructions hardly have any provision for nesting sites/holes in them. These are common in agricultural, sub-urban, and urban areas and the only areas they tend to avoid are woodlands, forests, grasslands, deserts, and frozen waste lands. Other than that, a sparse population of House Sparrows generally indicates sparse population of humans. House Sparrows have no established migratory pattern *i.e.* they are year round residents. It. is believed that 90 per cent of adults stay within a 1.25 mile radius during nesting season. Flocks of juveniles and non-breeding adults will move 4 to 5 miles from nesting sites to seasonal feeding areas.

Food and Feeding Habits

The House Sparrows mostly feed on grains, weed seeds; insects and other arthropods during breeding season. They forage on the ground for seeds and may pierce flowers to get at nectar. 60 per cent livestock feed in fields, at waste feed, or from animal dung (wheat, oats, cracked corn, sorghum); 18 per cent cereals (grains from field or storage); 17 per cent weed seeds (major species: ragweed, crabgrass, bristle grass, knotweed); and 4 per cent insects. Urban birds are known to feed on food waste (breed, restaurant waste, etc.).

Nestlings diet is about 68 per cent insects, and 30 per cent livestock feed on average. For young (1–3 days old), invertebrates make up 90 per cent of diet decreasing to 49 per cent by 7 days. Insects feed include what is abundant, including *alfalfa* weevils, bark beetle larvae, periodic *cicadas*, Dung beetles, and *Melanopus* grasshoppers.

Song Cheep

Its song is a single cheep (up to 4 types). Males cheep continuously, 60–100 times/hour, males calling more often. Females do not cheep much (4 calls/hour) unless they are trying to attract a mate. They tend to sing less during cold or rainy weather and short winter days.

Nesting Activity

Males aggressively defend their chosen nest site, spending nearly 60 per cent of their perching time at nest boxes during reproduction. They attack adults of other species, peck eggs and remove of nestlings from nest boxes. Occasionally male House Sparrows attack other male house sparrows and female house sparrows attack other females. Females usurping other house sparrow nests regularly commit infanticide.

House sparrows being *monogamous*, appeared to be more closely bonded to a nest site than a mate *i.e.*, if a mate is lost, they remain at the nest site trying to attract another female. Females prefer males with a larger bib size; males with larger bibs are more active sexually.

House sparrows can and often use any old nest suitable for other birds. Females prefer hole-type nests over tree nests, but there is no imprinting on the parental nest site (Figures 11.4 and 11.5). However, smaller birds may be able to enter smaller holes.

Nestlings may prematurely fledge 10 days after hatching. The nest may contain fleas, blow fly and/or mites (*Pellonyssus reedi*).

Incubation

Usually begins with the penultimate (next-to-last) egg of a clutch, and is by both the male and female, lasting 10–13 days. Both the male and female was observed to sit on the eggs. In the initial period female sits for a longer duration than the male, but later it is the male which sits or eggs for longer time. Male sessions average 9 minutes, females 11 minutes. Eggs hatch 1 day after the last egg was laid, over about a 1.3 days period.

Development

Hatchling are red, fading to pink or light gray after 6–10 hours. The mouth is red, and the rectal flange is yellow. At 6 days, feather sheaths on ventral and dorsal tracts and wing coverts split. Feathers begin to fan at 7–8 days. The female primarily broods nestlings, with time decreasing as they grow older. Both parents feed (15–19 visits/hour) young by regurgitation; males feed about 40 per cent of the time.

Fledging

Usually 14–17 days after hatching. Youngs are capable of more or less sustained flight upon fledging. Young stay with adult male for a few days, then gather with other young into foraging and roosting flocks. They are independent (feeding themselves) 7–10 day after leaving the nest.

Dispersal

Birds generally remain near breeding colonies. Shortly after fledging, perhaps 50–75 per cent of juveniles may wander 1–2 km. to new feeding areas. Breeding individuals generally stay within this same range.

Number of Broods

House Sparrows may raise 2–4 average of 3 clutches each breeding season (March to August). Nests may be reused, with a re-nest or subsequent brood typically beginning 8 days after nest failure or after young leave nest.

References

Ali, Salim, 1996. *The Book of Indian Birds.* Bombay Natural History Society.

Ara, Jamal, 1970. *Watching Birds.* National Book Trust, New Delhi.

Dhindsa, M.S., 1984. Status of avian fauna in Punjab and its management. In: *Status of Wildlife in Punjab.* The Indian Ecological Society, Ludhiana, India.

James, Fisher, 1946. *Watching Birds.* Pelican Books.

John, Stidworthy, 1983. *Birds: Spotting and Studying.* Macdonald and Co., London.

Joseph, J. Hicky, 1943. *A Guide to Bird Watching.* Oxford University Press.

Laeeq, Futehally and Ali, Salim, 1959. *About Indian Birds.* Blackie, Bombay.

Richard, Grimmett, Roberts, Tom and Inskipp, Tim, 2008. *Birds of Pakistan.* Christopher Helm, London and Yale University Press, New Heaven.

Chapter 12

Avifauna of Jammu and Kashmir

☆ *O.P. Sharma*

Abstract

Jammu and Kashmir is rich in sub-tropical, temperature and alpine avifauna. The fascinating avifauna includes Bee eaters, Egrets, Herons, Snow Cocks, Quails, Partridges, Pheasants, Ducks, Swifts, Hill pigeons, Doves, Falcons, Crows, Choughs, Dippers, Redstarts, Whistling thrushes, Woodpeckers, Barbets, Kingfishers, Cuckoos, Wheatears, Warblers, Larks, Wagtails, Accentors, Snow finches, Mountain finches, Rose finches and many other species.

Introduction

Birds are fascinating and enchanting feathered friends of mankind. Their fastness, flight and colourful feathers are amazing. Although evolved from more ferocious dinosaurs, yet they are elegant and graceful. They help mankind in a number of ways. They pollinate flowers, disperse seeds, enhance seed germination, devour insect pests, form crucial links in the food chains and decorate our green surroundings with their presence. With their chirping and melodious songs they enliven our surroundings. They are facing threat due to loss of their habitat as a result of human interference. It calls for restoration of their habitat.

Common Birds

Acridotheres tristis is our common myna called Hor in Kashmiri Language. The name 'Hari Parvat' is derived from this bird. It is also called Saarika and Goddess Sharika temple has legend linked to it. *Acridotheres ginginianus* is our Bank myna which has colonized all crowded environments in our cities. It is called Saanch in Dogri. *Acridotheres fuscus* is the Jungle Myna. Allied genus *Sturnus* is for starling locally called Tilyar (*Sturnus vulgaris*). Rosy Pastor (*Sturnus roseus*) and Pied Myna

(*Sturnus contra*) are other birds seen in Jammu plains. Hill Myna, *Gracula religiosa* is the state bird of Jharkhand. It can be made to imitate human speech the way parakeets do.

Accentors are high altitude sparrow like passerine birds belonging to Genus *Prunella. Prunella vulgaris* is one of our mountain medicinal herb locally called Kalveuth but *Prunella collaris* is our accentor. *Anser indicus* is our Bar headed geese which migrates to our Gharana Wetland during winter months. Two black stripes on its head form its distinguishing feature. It is called Rajhans in Hindi. Babblers, Barbets, Bee eaters are quite common in Jammu region. Babblers are noisy birds. *Turdoides striatus* is our Jungle Babbler and *T. caudatus* is our common Babbler, usually seen foraging in a group of seven and therefore has got its local name as Sat-Bhai. Barbets spend their winter in silence but become active on the onset of spring season.

Bee eaters are green birds with longish tails and are scientifically identified as merops orientalis. In Kashmir, *Merops apiaster* is quite common bird with a chest nut head. They dig out tunnels in the soil and lay eggs inside those tunnels. Barbets (*Megalaima zeylanica*) are called Kudroo and Bee eaters as Toti in Dogri. Blue throated Barbet (*Megalaima asiatica*) and Copper-smith Barket (*Megalaima haemacephala*) are also seen in Jammu forests. Great Barket (*Megalaima virens*) locally called Traihoo is seen in the hills.

Cormorants are crow like water birds found of eating fish. Great Cormorant, *Phalacrocorax carbo* visits Gharana in Jammu and Wullur lake in Kashmir. Indian Comorant, *Phalacrocorax fuscicollis* is also common in Jammu. Cuckoos are our familiar birds. *Cuculus micropterus* is Indian Cuckoo called Pivoke Bhejo in Dogri. Dogra legend associated with this bird relates the story of a married girl who wanted to visit mother's house for eating yellow raspberries but was not allowed by in-laws and died without fulfilling her wish. Same girl now wanders in the form of familiar Indian Cuckoo.

Doves are peace loving birds. One of Dove, *Streptopelia orientalis* locally called Kalmoonha is almost threatened due to indiscriminate shooting by the hunters. This Rufous Turtle Dove favours feast of fruits of Monkey face tree (*Mallotus philippinensis*). Drongos are black coloured bold birds. One of the Drongo, *Dircurus leucophaes* called Gankaat in Dogri, is bold enough to eat honey from beehive directly. Racket tailed Drongos are fascinating as they have racket like tail extensions but are seen in central India.

Eagles are preying birds. Palla's Fish Eagle (*Haliaeetuc leucoryphus*) is globally threatened and is in vulnerable category. Egrets are our wetland birds but cattle egrets (*Bubulcus ibis*) love grasslands and remain in association of cattle.

Finches, Faintails, Fire throats, Flycatchers are small sized fascinating birds of our hills. Flower peekers are found in Jammu plains. Black Partridge, *Francolinus francolinus* locally called Kala Teetar is common in Jammu and is the state bird of Haryana. Grey partridge is also quite common in Akhnoor-Sunderbani area. Floricans are endangered grassland birds of India. Hoatzin, the national bird of Guyana is not seen in India.

Indian Roller, *Coracias benghalensis* locally called Leelari is quite sacred bird in the Dogra culture. In the adjoining village of Bhamiyal, Villagers initiate Mundan (Tonsure) ceremony only with the touch its auspicious feather. European Roller, *Coracias garrulous* is common in Kashmir.

Jacana is the familiar Lily-trotter bird of Kashmir wetlands. It is called Gundkav in Kashmir and Jalmor in Dogri. Its long legs and unwebbed feet make it comfortable to trott on the leaves of Water lilies. Its long tail has earned it a common name as Pheasant tailed Jacana (*Hydrophasianus chirurgus*).

Kaleej pheasant, *Lophura leucomelana* is our familiar Kolsa of hills which is indiscriminately shot by the hunters and has now entered in the threatened category. Kingfishers are fish eating birds of wetlands. Kultonch is the Kashmiri name of *Alcedo atthis* whose Dogri equivalent is Machhmar.

Lapwings are wading birds whereas Larks are grassland birds. Laughing thrushes are familiar mountain birds. A thrush locally called pargulli in Dogri and Sheen-e-Pin in Kashmiri is *Garrulax lineatus*. Its English name is streaked laughing Thrush. Magpies are most common in temperate mountains. Yellow Billed Blue Magpie (*Urocissa flavirostris*) locally called Bansaar or Kushaar is most common in our hill resorts like Kud, Patnitop, Bhaderwah, Batote and Sannasar but Red Billed Blue Magpie, *Urocissa erythrorhycha* is restricted to Bani-Basholli area inside J&K. Minivets, gaudy passerine birds of our coniferous forests are crimson red with a mix of black in the wings. Monal pheasant, *Lophophorus impejanus* is another magnificent bird which is the state bird of Uttrakhand. It is found in coniferous forests throughout J&K but its population is on the decline.

Nutcrackers scientifically called *Nucifraga caryocatactes* are feathered friends of our Fir forests. Night jars are nocturnal birds which are locally called as Shapaaki and scientifically named as *Caprimulgus asiaticus*. *Nectarinia asiatica* is brilliantly coloured purple sun-bird, which is a frequent visitor to our flowering parks and gardens.

Orioles are yellow coloured birds of woodlands, male is brilliant golden yellow and female is dull but streaked. It is called Peelkar in Dogri language. Owls (*Bubo bubo*, *Athene brama*) are night birds which keep rodent population under control.

Paradise fly catcher is praiseworthy partner of nature. Male has black crested head and white coloured very long tail streamers. Female is grayish brown and duller in contrast to male bird. This bird is called Poonchiri in Dogri, Doodhraj in Hindi and *Terpsiphone paradisi* in scientific parlance. Peacock pheasant is found in our North East but is encountered in the North Indian Delhi Zoo. It is *Polyplectron bicalcaratum* and is quite different from other pheasants of our area. *Pica pica* is Black Billed Magpie commonly seen in Leh and Kargil districts of Ladakh region. It is named as Kataangputit in Ladakhi. Like other magpies and treepies, it is also a close relative of crows of the family Corvidae.

Quails are common in our sub-tropical Kandi belt. They inhabit crops and grassland areas. Black "anchor" mark on throat and streaky body pattern is diagnostic feature. Tail is quite stumpy in these birds. Scientifically they are called *Coturnix coturnix*.

Rockchats and Robins are common birds of our city surroundings. Rockchat is *Cercomela fusca* and Robin is *Saxicoloides fulicata*. Oriental Magpie Robin (*Copsychus saularis*) is common in our orchards and shrubberies. It sings very sweetly and its melodies attract our attention in the early morning hours. It is the national bird of Bangladesh. Dogri name of this bird is Guaal Piddi.

Snowfinches (*Montifribgilla*) are confined to our alpine meadows, Sunbirds to our flowering gardens, Shovellers and Snipes to our rivers and wetlands, Sandgrouse to our grasslands. Swallows are very active birds usually seen near freshwaters and human habitations.

Tailor birds (*Orthotomus sutorius*) are expert in stitching leaves for their nests. They are frequent visitors to our lawns, parks and gardens. River Terns visit our Gharana wetlands. Treepies locally called Laangardumbas are crow cousins and are familiar in our city surroundings. *Treron sphenura* is our preu which is fondly hunted by hunters and has become almost threatened in our Kandi belt. This bird is frugivorous and is fond of feasting on figs. It is called Green Pigeon in English.

Upupa epops is our Hudhud and quite common in our towns and cities. Its black tipped fan like crest is diagnostic feature and always attracts our attention.

Vultures (*Gyps benghalensis, G. indicus*) once common in our landscape have almost disappeared at an alarming rate in last ten years, thanks to urbanization, use of agrochemicals, felling of old fig trees and over use of veterinary medicine Diclofenac.

Vanellus indicus is our common Tateehri, very noisy during moonlit midnights, uttering its familiar note–'Did ye do it, Pity to do it', in the open flat grounds near water courses. Its English name is Red Wattled Lapwing, Northern Lapwing, *Venellus vanellus* is migratory bird and seen in Billawar area of Jammu. Its black crest and dark green upper parts are quite characteristic.

Wagtails and Warblers are quite familiar to us. Wagtails are usually seen during winter months. They constantly wag their tails up and down and are called Mamolas or Dhubbi in Dogri and Khanjan in Hindi. Warblers (*Ploceus benghalensis*) are yellow headed sparrow like birds having expertise in weaving a retort shaped nest usually seen hanging on Acacia trees in and around Kandi belt of Jammu. Woodpeckers (*Dinopium benghalense*) are seen in forests and are insectivorous. Whistling thrush (*Myophonus caeruleus*) is our Gurkol or Bhatt Bhatel, dark blue with yellow beak and white spots scattered in front. It inhabits mountain streams and rivulets.

Yellow Bittern (*Ixobrychus sinensis*), as well as Little Bittern (*I. minutus*) is seen in our wetlands. *Zosterops palpebrosus* is our familiar white eye, usually seen foraging in our fruit orchards. Its eye ring is white and throat bright yellow. It is quite active in searching its food among foliage and fruits.

Birds of Wetlands, Agriculture Fields, Forests and Grasslands

Following birds are found in our wetlands, agricultural fields, forests, grasslands and mountains.

Spot-billed Duck (*Anas poecilorhyncha*)

58–60 cm sized water bird with a yellow tipped bill and darker crown.

Mallard (*Anas platyrhynchos*)

50–65 cm sized green headed duck found in wetlands.

Eurasian Wigeon (*Anas penelope*)

45–50 cm water bird with a yellow forehead and forecrown, chestnut head and pink coloured breast.

Gadwall (*Anas strepera*)

40–42 cm sized water bird seen in wetlands.

Common Teal (*Anas crecca*)

34–38 cm sized chestnut headed water bird.

Tufted Duck (*Aythya fuligula*)

40–46 cm sized wetland bird found in Hokersar and Wullar lake of Kashmir.

Common Pochard (*Aythya ferina*)

42–48 cm sized water bird with chestnut head and pale grey back.

Red Crested Pochard (*Rhodonessa rufina*)

52–56 cm sized water bird with rusty orange head, black neck and breast.

Bar Headed Geese (*Anser indicus*)

Wetland bird with two black stripes on the head. It breeds in Ladakh and visits Kashmir and Jammu wetlands during winter months.

Grey Heron (*Ardea cinerea*)

90–98 cm sized water bird with long curved neck and white head with black head plumes.

Purple Heron (*Ardea purpurea*)

78–90 cm long necked marsh bird which is fond of eating fish.

Indian Large Cormorant (*Phalacrocorax carbo*)

Large sized fish eating water bird. It is found in Lower Kishanganga valley and around Wullar lake. Also visits Gharana wetland in Jammu.

Indian Cormorant (*Phalacrocorax fuscicollis*)

62 cm slimmer long tailed water bird.

Black Stork (*Ciconia nigra*)

Long necked and red legged glossy black marsh bird found in Jammu wetlands.

Woolly Necked Stork (*Ciconia episcopus*)

75–92 cm sized blackish stork with woolly white neck.

Painted Stork (*Mycteria leucocephala*)

94–100 cm long necked and long legged marsh bird which occasionally visits Gharana wetland (Guldev Raj, 2005).

Black Stork

White Throated Kingfisher

Monal Pheasant (*Lophophorus impejanus*)

Painted Stork

Common Hoopoe

Common Parakeet (*Psittacula krameri*)

Figure 12.1

Common Mallard (*Anas platyrhynchos*)

Bar Headed Geese (*Anser indicus*)

Purple Heron

Large Cormorant

Figure 12.2

Common Hoopoe (*Upupa epops*)

Black striped chocolate brown bird with the graceful crest and long curved bill. It is called Satut in Kashmiri and Tarkhan or Sukdudu in Dogri.

Himalayan Snow Cock (*Tetraogallus himalayensis*)

72 cm sized bird with striped neck and white breast contrasting with dark grey underparts. Resident 4250–5500 m in the rocky slopes and alpine meadows. It is locally called Bilji in Shina language, Galaundh in Dogri and Him Teetar in Hindi.

Snow Partridge (*Lerwa lerwa*)

38 cm sized bird with fine vermiculated upperparts and chestnut streaking on underparts. Resident 4000–5000 m. in alpine mountains, it is locally called Vikir.

Chukor (*Alectoris chukar*)

38 cm sized bird with red bill and red legs. It is national bird of Pakistan. It is found in Gilgit, Gurez, Bhaderwah, Kishtwar and Leh mountains.

Himalayan Monal (*Lophophorus impejanus*)

70 cm sized purple coloured crested bird with iridescent green patches and cinnamon-brown tail. It is locally called lest and lenth in Shina language, Neel in Dogri and Monaal in Hindi.

Kashmir Koklas (*Pucrasia macrolopha*)

58–64 cm sized green headed pheasant, underparts chestnut. Breeds in Gilgit and Ladakh. Found in Kishanganga velley, Kashmir, Bhaderwah and Sud-Mahandev mountains.

Brown Dipper (*Cinclus pallasii*)

High altitude brown coloured bird seen dipping in icy cold mountain streams. It is called Galkar in Kashmiri, Baitingiri in Shina, Chubba in Dogri and Chubia in Hindi.

Ruddy Shelduck (*Tadorna ferruginea*)

60–65 cm rusty orange bird which breeds in Ladakh wetlands and also visits Hokersar wetland in Kashmir. It is called Sako in Kashmiri, Maru and Nguru in Ladakhi and Surkhaab in Hindi.

Brown Fronted Woodpecker (*Dendrocopos auriceps*)

20 cm sized white barred blackish bird with well defined streaking on underparts. It is called Kulthuthar in Shina, Chirka in Dogri and Kathphorba in Hindi.

Scaly-belled Woodpecker (*Picus squamatus*)

35 cm sized forest bird with red eyes, black eye stripe and unstreaked throat. It is locally called Kulthakthak in Pahadi and Harit Kashtkoot in Hindi.

Common Hoopoe (*Upupa epops*)

30 cm sized orange buff crested bird with black and white barred hind parts. It is called Dudbadshah in Pahadi, Satut in Kashmiri and Huthutse and Phukshukshali in Ladakhi.

European Roller (*Coracias garrulus*)

30 cm sized bird with turquoise head and underparts contrasting with rufous cinnamon back. It is called Kujuk in Ladakhi and Neelkrash in Kashmiri.

White Throated Kingfisher (*Halcyon smyrensis*)

Medium sized chocolate brown bird with bluish lower back and tail. It is called Machhmar in Dogri, Neeldadil in Pogli and Kultonch in Kashmiri.

Common Kingfisher (*Alcedo atthis*)

16 cm sized greenish blue bird with orange ear-coverts and underparts. Common in Wullar lake area. It is called Ngyza in Ladakhi and Kultonch in Kashmiri.

Crested Kingfisher (*Megaceryle lugubris*)

40 cm sized riverside or lakeside bird with black bill and barred upperparts. It is called tont and Kultonch in Kashmiri. It breeds in Kishanganga and Kashmir valley.

European Bee-eater (*Merops apiaster*)

25–35 cm sized greenish bird with a chestnut brown head, pale-yellow throat and a black gorget. It is called Rodbubri in Kashmiri.

Large Hawk Cuckoo (*Hierococcyx sparverioides*)

38 cm sized brown coloured bird with strongly barred underparts. Seen in and around Wullar lake in Kashmir. It is called Phuffu by locals.

Eurasian Cuckoo (*Cuculus canorus*)

32–34 cm sized forest bird with barred whiter underparts and yellow eyes. It is called Hoorkuk and Kuku in Kashmiri. It is found in Kishanganga and Kashmir valley.

Indian Cuckoo (*Cuculus micropterus*)

33 cm sized grayish brown bird with broad barring on underparts. It is called Piyoke Bhejo in Dogri and Kaphal Pako in Hindi.

Common Swift (*Apus apus*)

16 cm sized sooty brown bird with white throat and prominent forked tail. It is locally called Shamachai in Shina.

Long Eared Owl (*Asio otus*)

35 cm sized nocturnal bird found in coniferous forests of J&K.

Eurasian Eagle Owl (*Bubo bubo*)

56–66 cm sized owl with upright ear-tufts and dark brown mottled upperparts. It is called Ghughnu in Gojri and Dogri.

Snow Pigeon (*Columba leuconota*)

34 cm sized slaty grey headed pigeon with creamy white underparts. It is called Chatkotur in Kashmiri.

Speckled Wood Pigeon (*Columba hodgsonii*)

38 cm sized mountain pigeon. It is found in Sonamarg area mainly in temperate forests.

Oriental Turtle Dove (*Streptopelia orientalis*)

32 cm sized stocky bird with rufous scaled scapulars and vinaceous pink underparts. It is locally called Kanooli in Gurez and Kalmoonha in Jammu.

Plum Headed Parakeet (*Psittacula cyanocephala*)

36 cm long plum headed green bird with a white tipped blue green tail.

Rose-ringed Parakeet (*Psittacula krameri*)

42 cm long tailed red billed green coloured bird commonly seen in and around Jammu region.

Slaty Headed Parakeet (*Psittacula himalayana*)

40 cm long grey headed dark green bird with yellow tipped tail. Found in well wooded areas.

Greater Coucal (*Centropus sinensis*)

48 cm sized crow like bird with bright chestnut wings and black underwing coverts. It is common in R.S. Pura area of Jammu.

Water Rail (*Rallus aquaticus*)

24–28 cm sized stumpy tailed bird with pink legs and streaked upperparts. It is found in marshes and wet fields.

White Breasted Water Hen (*Amauronis phoenicurus*)

32 cm sized white breasted marsh bird with rufous cinnamon undertail coverts.

Purple Swamphen (*Porphyrio porphyrio*)

45–50 cm dark purple marsh bird with a red bill and red legs.

Tibetan Sandgrouse (*Syrrhaptes tibetanus*)

48 cm sized pin-tailed plump bird with finely barred breast and white belly. It is called Kakeling in Ladakhi and vyangpaad in Hindi.

Solitary Snipe (*Gallinago solitaria*)

30 cm sized dull coloured marsh bird with a streaked body pattern and prominent white spotting.

Ruff and Reeve (*Philomachus pugnax*)

26–32 cm sized bird with ruff in the neck. Reported from Baltal in Sonamarg area of Kashmir.

Black Tailed Godwit (*Limosa limosa*)

36–44 cm sized long beaked and rufous orange necked bird.

Ibis Bill (*Ibidorhyncha struthersii*)

38–40 cm sized black faced bird with a down curved dark red bill. Seen in Pahalgam and Dachigam National Park of Kashmir.

Common Sandpiper (*Actitis hypoleucas*)

19–20 cm wetland bird. Locally common in Kishanganga valley of Gurez.

Green Sandpiper (*Tringa ochropus*)

21–24 cm wading bird with greenish legs and darker upperparts.

Common Greenshank (*Tringa nebularia*)

30–34 cm long legged wading bird, upperparts grey, underparts whitish.

Lesser Sand Plover (*Charadrius mongolus*)

20 cm sized long legged riverside or lakeside bird which breeds in high mountains. It is called Tittru in Gojri.

Brown-headed Gull (*Larus brunnicephalus*)

42 cm sized black headed bird with pale yellow bill and iris. Breeds in high altitude lakes of Ladakh. It is called Neyazor in Ladakhi.

Osprey (*Pandion haliaetus*)

55–58 cm sized bird of prey, head whitish upperparts black brown, underparts white.

Pariah Kite (*Milvus migrans*)

55–70 cm sized Kite with shallowly forked tail. It breeds upto 3000 m.

Lammergeier (*Gypaetus barbatus*)

100–115 cm sized bird of prey with jet black upperparts and rufous buff underparts. It is locally called Paanteel in Shina and Jatayu in Dogri.

Booted Eagle (*Hieraetus pennatus*)

45–52 cm sized eagle with long square ended tail. It is seen in Gilgit and Gurez area.

Cinereous Vulture (*Aegypius monachus*)

100–110 cm sized vulture. Recorded from Kishanganga valley.

Himalayan Griffon (*Gyps himalayensis*)

115–125 cm sized vulture with pinkish legs and buff white wing coverts. It is locally called Bring in Shina, Gridh or Girj in Dogri.

Eurasian Marsh Harrier (*Circus aeruginosus*)

48–58 cm sized broad winged stocky bird which glides in the sky. It is called Upshi in Shina language and Pancheel in Hindi.

Shikra (*Accipiter badius*)

30–36 cm sized bird with fine brownish orange barring on underparts.

Golden Eagle (*Aquila chrysaetos*)

75–88 cm sized broad winged bird with gold coloured crown and nape. It is called Khakai in Shina, Garud in Dogri and Shunth in Pogli.

Common Kestrel (*Falco tinnunculus*)

32–35 cm sized rufous brown bird with distinct streaking of black. It is called Khermutiya in Hindi.

Lesser Kestrel (*Falco naumanni*)

30–32 cm sized slimmer bird with rufous unmarked upperparts and whitish claws. It is globally threatened and is in vulnerable category in J&K.

Peregrine Falcon (*Falco peregrinus*)

38–48 cm sized stocky build bird of prey, upperparts slaty-grey and underparts whitish. It is national bird of United Arab Emirates and Angola.

Eurasian Jay (*Garrulus glandarius*)

32–36 cm reddish brown bird with black moustachial stripe and blue barred wings. It is called Ban-Bakkra in Gojri.

Yellow Billed Blue Magpie (*Urocissa flavirostris*)

60–66 cm long mountain-bird with bluish upperparts and white underparts. It is called Latraj in Kashmiri, Gushantal in Padri, Bansaar in Dogri and Guchkichi in Kanthiyali.

Black-billed Magpie (*Pica pica*)

45–50 cm sized black and white bird with metallic green-purple tail. It is found in Kargil and is called Khataah and Kataang Putit in Ladakhi and Ladakhan Laangardumbi in Dogri.

Red Billed Chough (*Pyrrhocorax pyrrhocorax*)

36–40 cm long red billed high altitude bird which resembles crow but forages upto 6000m. It is locally called Kiyuni or Kangal in Shina and Chunka in Ladakhi.

Yellow Billed Chough (*Pyrrhocorax graculus*)

38 cm sized yellow billed bird of highland areas. It is common in Leh and Kargil mountains.

Eurasian Jackdaw (*Corvus monedula*)

24–38 cm grey napped and white eyed crow like bird of Kashmir. It is called Chor Kaua in Hindi.

Large Billed Crow (*Corvus macrorhynchos*)

45–60 cm sized jet black woodland crow.

Common Raven (*Corvus corax*)

58–68 cm sized crow with prominent throat hackles.

Eurasian Golden Oriole (*Oriolus oriolus*)

25 cm yellow coloured bird with flesh coloured bill and black wings. It is called Poshnul in Kashmiri, Peelkad in Dogri and Peelak in Hindi.

Long Tailed Minivet (*Pericrocotus ethologus*)

20 cm sized forest bird with blackish upperparts and dark red underparts. It is called Rangchiri and Bhangtit in Kanthiali, Bozulmini in Kashmiri and Gulabchasham in Hindi.

Ashy Drongo (*Dicrurus leucophaeus*)

28 cm sized slaty bird with red eyes and ashy grey underparts. It is called Gankaat, Gankhaav in Kishtwari and Shyam Angarak in Hindi.

White Throated Dipper (*Cinclus cinclus*)

20 cm sized riverside bird of mountain streams. Throat and breast are pure white whereas back has distinct dark scaling. It is called Dungal and Galkaar in Kashmiri and Chubiya in Hindi.

Brown Dipper (*Cinclus pallasii*)

20 cm sized coffee brown bird of mountain streams, tail is stumpy. It is locally called Baiteengri in Shina and is quite common in and around Dawar.

Blue Rock Thrush (*Monticola solitarius*)

20 cm sized indigo blue bird of rocky mountains.

Blue Whistling Thrush (*Myophonus caeruleus*)

32 cm sized glistening blue bird with black legs, yellow bill and white spotting infront. Common along mountain streams. It is locally called Kagulli and Kaangal in Shina, Bhatt Bhatel and Gudkol in Dogri.

Eurasian Black Bird (*Turdus merula*)

25–28 cm sized blackish brown bird of juniper scrub vegetation. It is called Koo Kastur in Kashmiri and Kasturika in Hindi.

Kashmir Flycatcher (*Ficedula subrubra*)

12 cm sized grey coloured bird with red underparts. It is called Kashmir Chutki in Hindi.

Verditer Flycatcher (*Eumyias thalassina*)

15 cm sized greenish blue bird with black lares.

Himalayan Paradise Flycatcher (*Terpsiphone paradise*)

20 cm sized black headed white bird with a distinct crest and long tail streamers. It is called Poonchiri in Dogri, Talwaru in Pahadi Sultan Bulbul in Gojri and Doodh raj in Hindi.

White Winged Redstart (*Phoenicurus erythrogaster*)

18 cm stocky built bird with white cap and white patch on the wings. Underparts red-brown. It is called Sentik in Ladakhi and Thirthira in Hindi.

White Capped Redstart (*Chaimarrornis leucocephalus*)

18 cm sized riverside bird with snow white head, black wings and chestnut brown underparts. It is called Dadchai in Shina and Gadchidoli in Bhaderwahi.

Spotted Fork Tail (*Enicurus maculates*)

25 cm white forheaded, black breasted and fortailed mountain bird. It is found in Bani, Sarthal and Bhaderwah area of Jammu and is called Dhubban in Dogri.

Little Forktail (*Enicurus scouleri*)

12 cm sized mountain bird with a white head, blackish upperparts and buff white underparts. It is called Shakhellot in Kashmiri and Dsang in Dogri.

Kashmir Nut Hatch (*Sitta cashmirensis*)

12 cm sized chestnut bellied bird of Kashmir forests.

Wall Creeper (*Tichodroma muraria*)

16 cm sized bird with a down curved bill and crimson wing coverts. It is called Lamb diddar in Kashmiri and Bheetsarpi in Hindi.

Fire Capped Tit (*Cephalopyrus flammiceps*)

10 cm sized greenish bird with yellow underparts and scarlet orange fore-crown. It is called Pudkanu in Dogri.

Winter Wren (*Troglodytes troglodytes*)

9–10 cm sized stubby tailed bird with dark barred brownish upperparts. It is called Beeshu in Shina language and Chikti in Hindi.

White-Throated Tit (*Aegithalos niveogularis*)

11 cm sized high altitude bird with white forehead, throat and cinnamon underparts. It is called Pudkano in Gojri.

Himalayan Bulbul (*Pycnontus leucogenys*)

20 cm brown crested bird with a black throat, yellow vent and white cheeks. It is called Vilvichur in Kashmiri.

Red Vented Bulbul (*Pycnontus cafer*)

20 cm long black headed and crested bird with scaly pattern on wings and red vent. It is called Peenja in Dogri and Bulbul in Pahadi.

Black Bulbul (*Hypsipetes madagascariensis*)

15 cm sized bird with red bill and red legs. It is called Darkaal in Dogri.

White Browed Tit Warbler (*Leptopoecile sophiae*)

10 cm sized bird with white supercilium, rufous crown and purple lilac plumage.

Long Billed Bush Warbler (*Bradypterus major*)

13 cm sized high altitude bird with whitish underparts and spotted throat.

Lesser White Throat (*Sylvia curruca*)

13 cm sized bird with brownish grey upperparts and grey crown with darker ear coverts. It is called Chatkika in Hindi.

Mountain Chiffchaff (*Phylloscopus sindianus*)

11 cm sized olive green bird. It is called Koojni in Hindi.

Tickell's Leaf Warbler (*Phylloscopus affinis*)

11 cm sized greenish brown bird with bright yellow supercilium. It is called Phudki or Koojni in Hindi.

Buff-barred Warbler (*Phylloscopus pulcher*)

10 cm sized sub alpine bird with buff orange wing bars and yellowish supercilium. It is called Madornu in Dogri and Peetbhra Tarukoojni in Hindi.

Greenish Warbler (*Phylloscopus trochiloides*)

10 cm sized greenish bird with whitish grey underparts. It is called Tarukoojni in Hindi.

Tytler's Leaf Warbler (*Phylloscopus tytleri*)

10 cm sized olive green bird with no wing bars but with a long fine supercilium. It is called viri-tiriv in Kashmiri.

Whistler's Warbler (*Seicerus whistleri*)

10 cm sized high altitude bird with yellow underparts and dark green upperparts. Seen in Wangat and Sindh valley.

Gold Crest (*Regulus regulus*)

9 cm sized bird with yellow centred crown and buff white underparts. It is the National Bird of Luxumberg. It is common in Kashmir.

Streaked Laughing Thrush (*Garrulax lineatus*)

20 cm sized grey brown bird with fine streaks and grey panel in the wings. It is locally called Ushkoor in Shina, Sheen-e-Pipin in Kashmiri and Pargulli in Dogri.

Rufous Sibia (*Heterophasia capistrata*)

20 cm sized black capped rufous bird with black grey bands on rufous tail. It is called Teedi-meedi in Dogri and Krishan Sheersh Shrivad in Hindi.

Horned Lark (*Eremophila alpestris*)

18 cm sized bird with a black breast band and black and white patterned head. It is called Ukpothkir in Ladakhi.

White Wagtail (*Motacilla alba*)

18 cm sized bird with black and white plumage and longish tail. It is called Mamola in Dogri.

Citrine Wagtail (*Motacilla citreola*)

18 cm sized yellow faced bird with yellow underparts and dark grey upperparts. Common in Dawar area of Gurez. It is called Peeli Mamoli in Dogri.

Alpine Accentor (*Prunella collaris*)

15–16 cm sized grayish brown bird with black barring on throat and black band across wing coverts. Frequents Gangbal pastures. It is called Tirdu in Dogri.

Robin Accentor (*Prunella rubeculoides*)

16 cm sized grey coloured bird with a rusty orange band across breast and whitish belly. It is called Tsilder in Ladakhi.

Plain Mountain Finch (*Leucosticte nemoricola*)

15 cm sized bird with a boldly streaked upperparts and pale braces.

European Goldfinch (*Carduelis carduelis*)

13–15 cm sized red faced mountain bird with black and yellow wings. Frequently seen at Kanzalwan and Radwan. It is called Sehaara in Kashmiri and Sonpari Tooti in Hindi.

Spectacled Finch (*Callacanthus burtoni*)

18 cm sized bird with blackish head having red orange eye spectacles. Seen at Achhura and Chorwan in Gurez.

Common Rose finch (*Carpodacus erythrinus*)

14–15 cm sized red fronted bird with streaked upperparts and double wing bar. Common in Pahalgam area. It is called Lolchai in Shina and Gulabi Tooti in Hindi.

Orange Bullfinch (*Pyrrhula aurantica*)

14 cm sized orange buff bird with black tail and wings. It is called Sontsar in Kashmiri. It is seen in Gulmarg and Dachigam area of Kashmir. It is called Sontsar in Kashmiri and Kaalpuchh in Hindi.

Rock Bunting (*Emberiza cia*)

16 cm grey headed black striped mountain bird.

White Capped Bunting (*Emberiza stewarti*)

15 cm long bird with a chestnut breast band and a black supercilium. It is called Van-tsar in Kashmiri.

References

Ali, Salim, 1996. *The Book of Indian Birds*. Bombay Natural History Society, Bombay.

Ara, Jamal, 1970. *Watching Birds*. National Book Trust, India.

Campbell, W.D., 1959. *Bird Watching as a Hobby*. Stanley Paul, London.

Dindsa, M.S., 1984. Status of Avian Fauna in Punjab and its Management. *The Ecological Society*, Ludhiana, India.

James, Fisher, 1946. *Watching Birds*. Pelican Books.

Joseph, Hicky J., 1943. *A Guide to Bird Watching*. Oxford University Press.

Koul, S.C., 1957. *Birds of Kashmir*. Normal Press Publication, Srinagar.

Ripley, S.D., 1961. *A Synopsis of the Birds of India and Pakistan*. Bombay Natural History Society, Bombay.

Stuart, Smith, 1945. *How to Study Birds*. Collins.

Chapter 13

Responses of Thyroid, Parathyroid, Calcitonin Producing C Cells and Adrenal Cortex of *Rattus norvegicus* to Sublethal Heroin Administration

☆ *S.R. Barai, S.A. Suryawanshi and A.K. Pandey*

Abstract

Intramuscular (im) administration of the sublethal dose (16.4 mg/kg body weight/ day; 0.75 LD_{50}) of heroin elicited a progressive increase in plasma Ca level during the first seven days, thereafter the level declined ($P<0.001$) on day 15 and 30. However, plasma Pi level of the heroin-treated rats registered a progressive increase with the peak value ($P<0.001$) on day 30. Plasma Na level of the treated rats showed progressive decline ($P<0.01$) at 24 hours with the minimum value (126.53±2.68 meEq/l) on day 30 whereas plasma K level recorded progressive increase during entire period of the treatment with peak (8.78±0.23 meEq/l) on day 30. Sublethal heroin administration for 30 days inflicted degenerative changes in the cytoplasm and nuclei of the follicular epithelial cells. There were reduction in the size of follicle and follicular epithelial height suggesting hypoactivity in the gland. The treatment also induced degenerative changes in parathyroid gland as evident by Cytoplamic vacuolization, presence of more pyknotic nuclei and occurrence of patchy areas among the chief cells. Degenerative changes were also noticed in critae of Mitochondria, Golgi complex and Eendoplasmic reticulum. There were decrease in the chromatin material in the nucleus and loss of hormone granules in the cytoplasm. Oxyphil cells of the heroin treated rat depicted dilation of endoplasmic reticulum and mitochondria with damaged cristae. Sublethal heroin administration in the rat for 30 days induced dilation in endoplasmic reticulum of the

C cells. Though sublethal heroin administration for 30 days elicited cytoplasmic vacuolation in all the three zones of adrenal cortex, much of cytological alterations were observed in zona glomerulosa and zona fasciculata cells. In zona glomerulosa cells degenerative changes in the organelles were more pronounced as evident by the loss of typical cristae in the mitochondria and hormone granules were rarely seen in these cells. Though rough endoplasmic reticula were scanty, many lipid granules encountered in zona glomerulosa cells of the treated rat.

Keywords: *Heroin, Thyroid, Parathyroid, C Cells, Adrenal cortex, Plasma electrolytes, Rattus norvegicus.*

Introduction

Heroin (Brown Sugar) abuse is a burning problem of the society (Martin, 1984; Neri-Serneri and Modesti, 1991; Griffiths *et al.*, 1994; Kringsholm *et al.*, 1994; Sporer, 1999). The drug (diacetylmorphine) is metabolized into 6-acetylmorphine and subsequently to morphine in the human body (Sawynok, 1986; Goldberger *et al.*, 1994; Jenkines *et al.*, 1994; Cami and Farrie, 2003). Nephrotoxic and hepatotoxic effects of the drug have been documented in mammals (Charuvastra *et al.*, 1980; Weller *et al.*, 1984; Gomez-Lechon *et al.*, 1987; Campistol *et al.*, 1988; Volochine *et al.*, 1988; Kringsholm and Christoffersen, 1989; Barai *et al.*, 2004a). Haematological as well as biochemical profiles of serum of dog, rabbit and rat to the exogenous drug administration have also been recorded (Hussain and Kumar, 1988; El-Daly, 1994; Sharma *et al.*, 2001; Barai *et al.*, 2004b). Though there exist few reports suggesting that the chronic heroin addiction/administration modulates the pituitary-thyroid as well as hypothalamus-pituitary-adrenal (HPA) functions and calcitonin secretion in mammals (Glass *et al.*, 1973; Brambilla *et al.*, 1980; Tagaliaro *et al.*, 1984; Kringsholm and Christoffersen, 1989; Sarnyai *et al.*, 2001; Laorden *et al.*, 2002) but the observations are highly conflicting (George *et al.*, 2005). An attempt has, therefore, been made to record the changes occurring in thyroid, parathyroid, calcitonin-producing C cells and adrenal cortex as well as plasma calcium (Ca), inorganic phosphate (Pi), sodium (Na) and potassium (K) levels of *Rattus norvegicus* in response to sublethal heroin administration.

Materials and Methods

Healthy male rats (*Rattus norvegicus*) weighing 150-200 gm were procured from Bombay Municipal Corporation, Mumbai. They were acclimatized under the ambient laboratory conditions (temparature 28±2°C; photoperiod 14L:10 D) for 10 days, fed *ad libitum* on rat feed (Lipton, Bangalore) and clean water was provided for drinking. 50 rats were randomly selected and divided into two equal groups–experimental and control. The experimental group rats were given intramuscular injection of 0.75 LD_{50} dose (16.4 mg/kg body weight) of heroin (the drug was initially dissolved in small quantity of alcohol and the desired dose was prepared in physiological saline) while the control rats received equal volume (0.2 ml/kg body weight) of the physiological saline. Animals from both the groups were dissected on day 1, 7, 15 and 30 of the treatment. Blood samples were collected in sterilized glass syringe from post-caval vein of rats under mild ether anaesthesia and plasma was separated by centrifugation at 3,500 rpm. Plasma Ca and Pi levels were estimated according to methods of Trinder (1960) and Fiske and Subbarow (1925), respectively while plasma Na and K levels were estimated by the flame photometry method (Wootton, 1974).

Thyroid, parathyroid and adrenal glands were surgically removed and fixed immediately in Bouin's solution for light microscopic studies. After routine processing, sections were cut at 6 μm and stained with hematoxylin-eosin (H&E) and lead-hematoxylin (PbH) (Solcia *et al.*, 1969). For electron microscopic observations, the tissues were fixed in 3 per cent glutaraldehyde maintained at 4°C. They were washed thoroughly with 0.1N cacodylate buffer to remove traces of glutaraldehyde and kept in 1 per cent osmium tetraoxide for 2 hours at 4°C. They were dehydrated through ascending series of alcohol, cleared in propylene oxide and transferred to the mixture of equal parts of propylene oxide and araldite solution for one hour to facilitate infiltration. They were kept overnight at room temperature in araldite solution A and then in the araldite solution B for 1 hour at room temperature.

For preparation of tissue block, the glands were embedded in plastic capsule KDB filled with araldite solution B. They were kept at 60°C for 48 hours for polymerization and hardening. The blocks were removed from the capsule and trimmed with a surgical blade under stereomicroscope. Semithin sections (1μm) were cut using ultramicrotome, spread on glass slides and fixed by gentle heating. The sections were stained with toluidine blue and examined under the light microscope. Ultrathin sections (600-800A°) were cut from the selected area with glass knife and mounted on 400 mesh copper grids. The tissues were double stained with 10 per cent alcoholic uranyl acetate for 20 minutes and with Reynold's lead citrate for 10 minutes. Sections were scanned under Jeol-100 electron microscope.

Results and Discussion

Plasma Electrolytes

Alterations in plasma Ca and Pi levels of *Rattus norvegicus* in response to sublethal heroin administration have been summarized in Table 13.1. Plasma Ca level of control rat ranged between 9.53±0.32–9.88± 0.22 mg/100 ml while Pi concentration fluctuated between 4.55±0.18–4.71±0.24 mg/100 ml. Sublethal heroin administration elicited a progressive increase in plasma Ca level during the first seven days, thereafter the level declined to 7.43±0.82 mg/dl by day 30. However, plasma Pi level of the heroin-treated rats registered a progressive increase with the peak ($P<0.001$) on day 30. Pedrazzoni *et al.* (1993) also recorded increased blood ionized and urinary Ca in heroin addicts.

Effects of sublethal heroin treatment on plasma Na and K levels of *Rattus norvegicus* have been summarized in Table 13.2. Plasma Na and K levels of the control rats fluctuated between 153.14±2.88–157.23±2.16 meEq/l and 5.04±0.32–5.63±0.41 meEq/l, respectively. Plasma Na level of the treated rats recorded progressive decline ($P<0.01$) at 24 hours with the minimum value (126.53±2.68 meEq/l) on day 30 whereas plasma potassium level registered a progressive increase during entire period of the treatment with peak value (8.78±0.23 meEq/l) on day 30. Earlier observations in rats have documented the normal range of Plasma Na between 137-146 mEq/l and that of K to the tune of 3-5 mEq/l (Wootton, 1974; Chester Jones and Henderson, 1980; Gorbman *et al.*, 1983). Though mild (but insignificant) hyponatremia was recorded, plasma K level was dangerously high (hyperkalemia) in a patient with heroin abuse (Pearce and Cox, 1980).

Table 13.1: Effects of Heroin Administration on Plasma Calcium and Inorganic Phosphate Levels (mg/100 ml) of *Rattus norvegicus*

Group	*Duration*			
	24 Hours	*7 Days*	*15 Days*	*30 Days*
		Plasma calcium		
Control	9.88±0.22	9.64±0.16	9.69±0.57	9.53±0.32
Heroin ($0.50\ LD_{50}$)	10.82±0.31[a] (+15)	12.16±0.43[c] (+30)	9.02±0.12 (−9)	7.84±0.38[b] (−19)
Heroin ($0.75\ LD_{50}$)	10.93±0.55[a] (+11)	12.71±0.28[c] (+32)	9.24±0.35 (−5)	7.43±0.42[c] (−22)
		Plasma inorganic phosphate		
Control	4.62±0.21	4.68±0.32	4.55±0.18	4.71±0.24
Heroin ($0.50\ LD_{50}$)	5.31±0.28 (+9)	5.68±0.56[a] (+20)	6.25±0.21[c] (+28)	7.36±0.89[c] (+63)
Heroin ($0.75\ LD_{50}$)	5.24±0.36 (+13)	5.83±0.14[b] (+25)	6.35±0.35[b] (+40)	7.14±0.32[c] (+52)

Values are mean±SD of 5 specimens. Values in parenthesis indicate per cent increase (+) or decrease (–) over control. Significant responses: [a]P<0.05; [b]P<0.01; [c]P<0.001.

Table 13.2: Effects of Sublethal Heroin Administration on Plasma Sodium and Potassium Levels (mEq/l) of *Rattus norvegicus*

Group	*Duration*			
	24 Hours	*7 Days*	*15 Days*	*30 Days*
		Plasma sodium level		
Control	156.57±2.30	155.89±2.51	153.14±2.88	157.23±2.16
Heroin	149.64±2.76[a] (−4)	141.38±3.49[b] (−9)	135.47±3.91[b] (−12)	126.53±2.68[b] (−20)
		Plasma potassium level		
Control	5.04±0.92	5.19±0.32	5.63±0.81	5.12±0.63
Heroin	5.72±0.24 (+14)	6.88±0.67[a] (+33)	7.94±0.31[b] (+41)	8.78±0.46[b] (+72)

Values are mean±SD of 5 animals. Values in parentheses indicate per cent increase (+) or decrease (–) over control. Significant responses: [a]p<0.01; [b]p<0.001.

Thyroid Gland

Thyroid gland of the control *Rattus norvegicus* comprised follicles surrounding the lumen and blood vessels traversing in the interfollicular spaces. The follicles (height 5.98±0.34 μm) were lined by columnar epithelial cells and lumia filled by eosinophilic colloidal material. At places, peripheral vacuolization of colloidal materials were also seen in the control rats (Figure 13.1). Sublethal heroin administration for 30 days inflicted severe degenerative changes in follicular epithelial cells. The follicular epithelial height (4.82±0.42 μm) registered a significant ($P<0.001$) reduction as compared to the control and the cells became cuboidal in shape. Some of the epithelial cells displayed vacuolation and pyknosis in the epithelial cells. The colloid appeared homogenous but contained numerous degenerated/desquamated epithelial cells (Figure 13.2). Thyroid gland of *Rattus norvegicus* resembles to those described for other mammalian species (Fujita, 1986; Ekholmes, 1990). Brambilla *et al.* (1980) studied the effects of chronic heroin addiction (lasting from 8 months to 4 years) on pituitary-thyroid function in man. They found no difference in basal levels of thyroid stimulating hormone (TSH), thyroxine (T_4) and triiodothyronine (T_3) between the addicts and controls. However, in the present investigation, sublethal heroin administration inflicted severe degenerative changes in the gland as the epithelial cells showed vacuolation and pyknosis. In case of fetal intravenous administration of morphine and heroin, thyroid gland exhibited nodular stroma in the follicles (Kringsholm and Christoffersen, 1989). Recently, degenerative changes in the thyroid gland of albino rats were also observed in response to sublethal (oral; 0.5 g/kg body weight) administration of Marijuana (Yadav *et al.*, 2004).

Parathyroid Chief Cells

Parathyroid glands made their first phylogenetic appearance in tetrapods and their number varies from two to four in different class of vertebrates (Roth and Schiller, 1976; Clark *et al.*, 1986; Pandey, 1991, 1992). There are instances of the occurrence of functional accessory parathyroids too (Dacke, 1979; Pandey, 1985; Clark *et al.*, 1986; Suryawanshi *et al.*, 1997; Dhande *et al.*, 2003). While four parathyroids have been reported in bat, dog, goat and man, two glands are present in mouse, mole, shrew, hedgehog, pig, koala and seal (Roth and Schiller, 1976; Clark *et al.*, 1986). In *Rattus norvegicus*, only one pair of parathyroid glands occasionally embedded in thyroid tissue was observed. It consisted mainly of chief cells arranged in elongated and branching cords, separated by connective tissue stroma, capillaries and sinusoids (Figure 13.3). The chief cells were bound by unit membranes, clear desmosomes and terminal bars joining the plasma membranes. The nucleus was large, spherical or oval structure containing many small granules which were more concentrated towards the periphery. The cytoplasm contained prominent rough endoplasmic reticulum, Golgi apparatus was composed of straight or curved stacks or membranes with small vesicles and granules. The mitochondria were distributed throughout the cytoplasm. Only a few electron dense secretory granules could be seen in the cytoplasm. Besides these, lipid, glycogen and lysosomal bodies were also noticed in the chief cells (Figures 13.5 and 13.6). Similar ultrastructures of the chief cells have also been recorded in mammals including man and bats (Capen and Rowland, 1968a, b; Roth

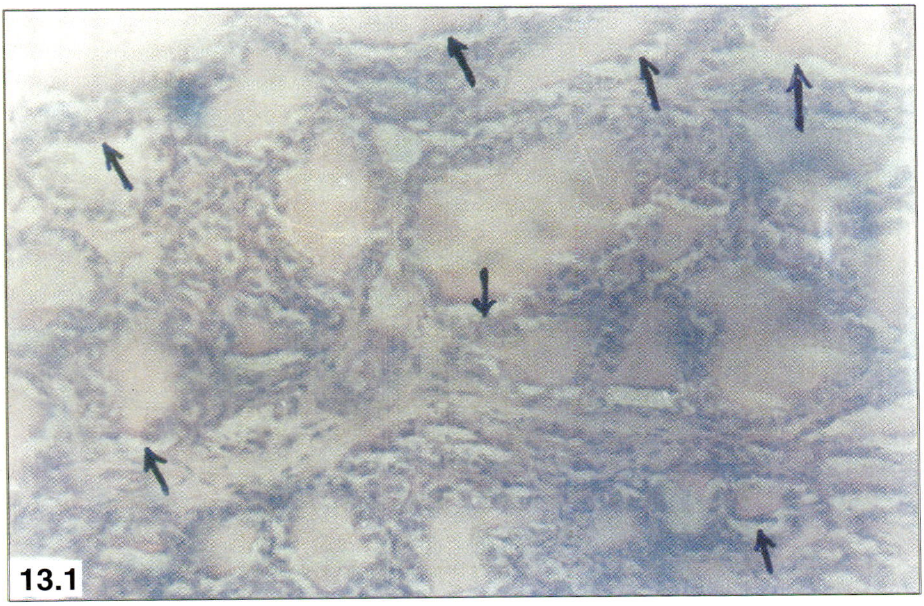

Figure 13.1: Thyroid Gland of the Control *Rattus norvegicus* Showing Large Follicles Lined by Columnar Epithelial Cells and Filled with Colloid Material. Mark the vacuolation in the colloid at places (arrow). H&E. x 240.

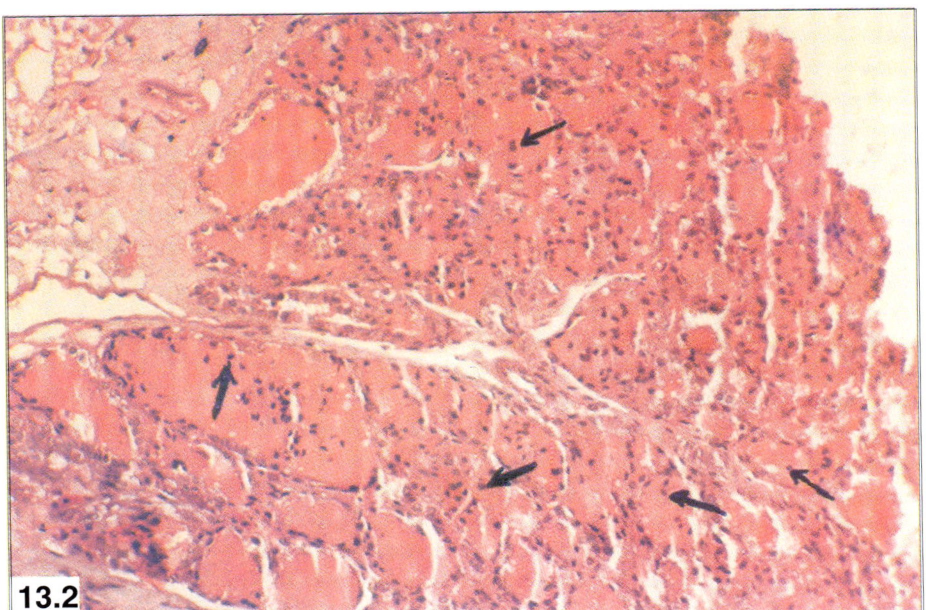

Figure 13.2: Thyroid Gland of the Rat Treated with Heroin for 30 Days Exhibiting Small Follicles Lined by Cuboidal Epithelial Cells and Lumina Filled with Homogenous Colloid Material. Numerous desquamated epithelial cells are seen in the lumina (arrow). H&E. x 240.

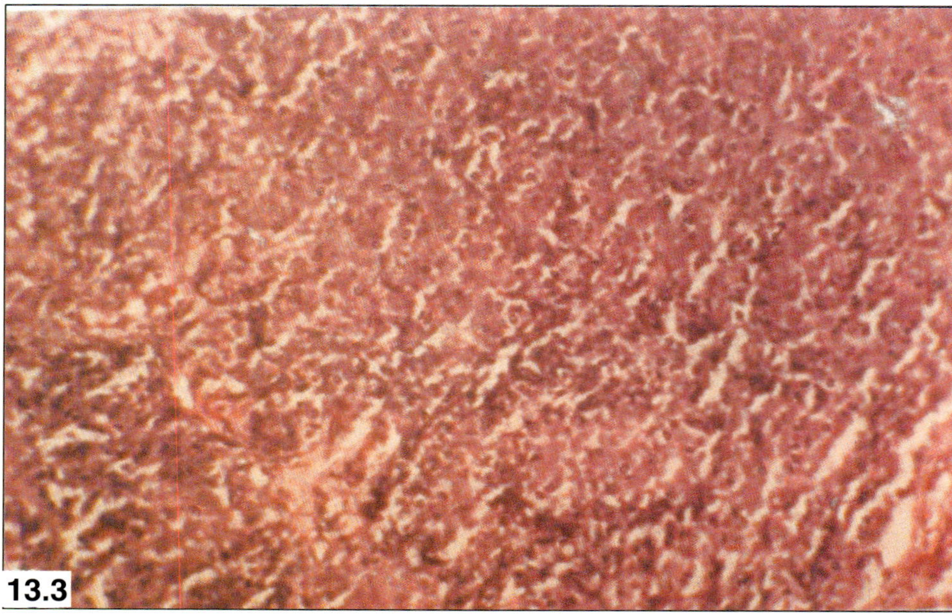

13.3

Figure 13.3: Parathyroid Gland of Control *Rattus norvegicus* Depicting Distribution of Chief Cells. H&E. x 100.

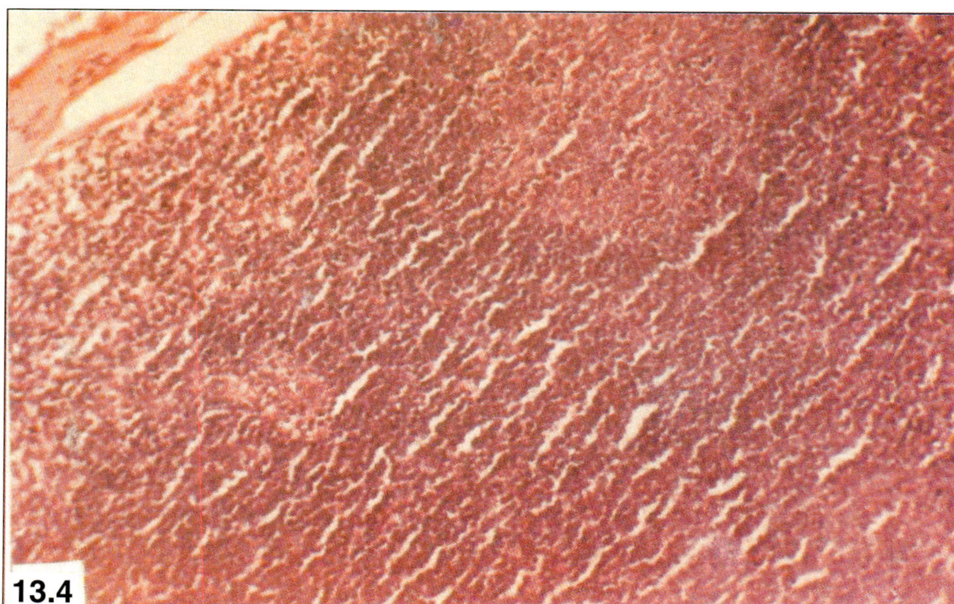

13.4

Figure 13.4: Parathyroid Gland of Rat on Day 30 of Heroin Injection Showing Disintegrated and Patchy Areas of Chief Cells. H&E. x 100.

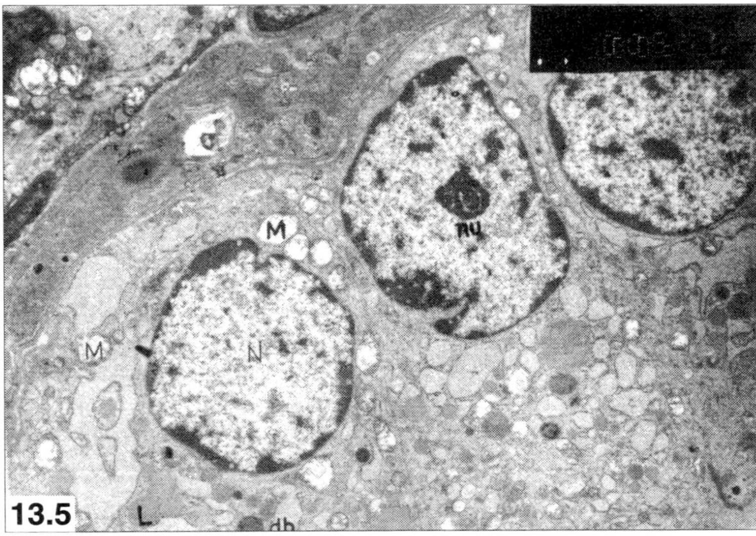

Figure 13.5: Chief Cells of Control Rat Exhibiting Nucleus (N), Nucleolus (nu), Mitochondria (M), Lipid Droplet (L), Dense Body (db) and Rough Endoplasmic Reticulum. x 3,000.

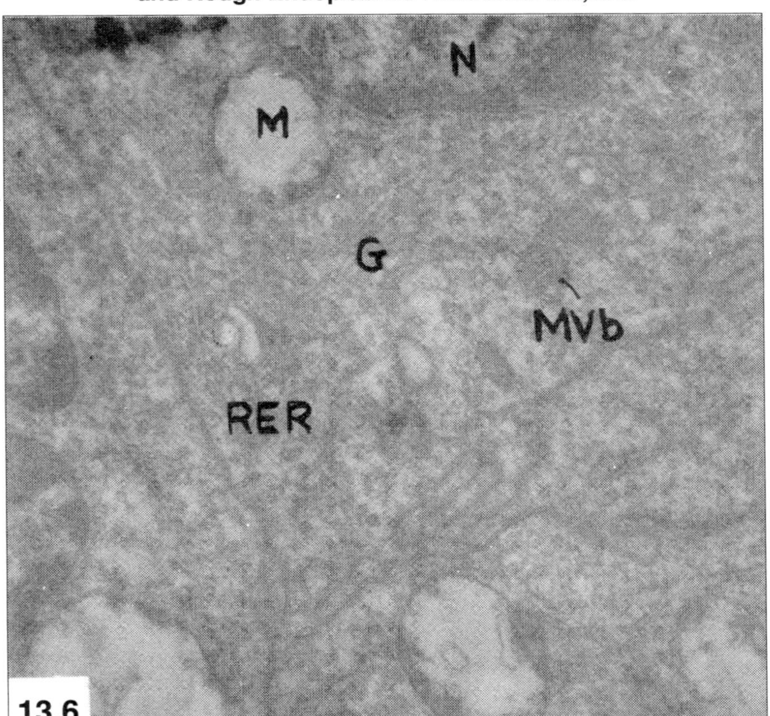

Figure 13.6: Chief Cell of Control Rat Showing Prominent Nucleus (N), Rough Endoplasmic Reticulum (RER), Golgi Complex (G), Mitochondria (M) and Multivesicular Body (MVb). x 15,000.

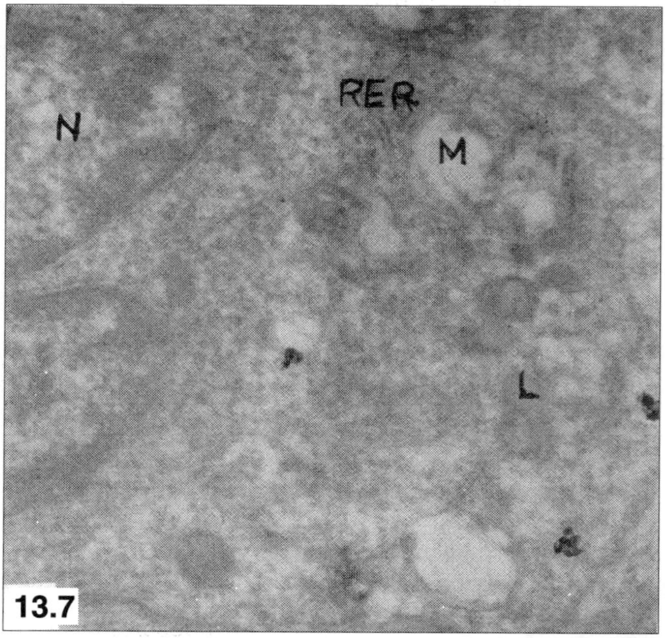

Figure 13.7: Chief Cell of Rat Treated with Heroin for 30 Days Exhibiting Part of Nucleus (N), Rough Endoplasmic Reticulum (RER), Lipid Droplet (L) and Disintegrated Mitochondria (M). x 15,000.

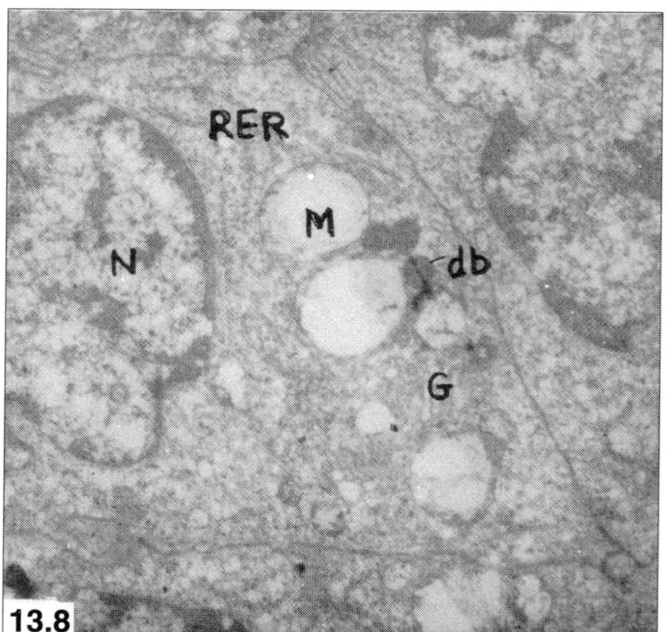

Figure 13.8: Oxyphil Cell of Control *Rattus norvegicus* Depicting Nucleus (N), Dense Body (db), Mitochondria (M) and Golgi Complex (G). x 8,000.

and Schiller, 1976; Clark *et al.*, 1986). Since PTH is secreted continuously in the blood circulation from the gland (Gorbman *et al.*, 1983; Clark *et al.*, 1986; Dacke, 2000), we found less number of secretory granules in the chief cells. Though there are reports on the existence of light and dark chief cells in mammals (Roth and Schiller, 1976; Clark *et al.*, 1986), we could not find such variations in the chief cells of parathyroid gland of *Rattus norvegicus*. It is assumed that light and dark variants correspondingly represent the inactive and active phases of the chief cells (Clark *et al.*, 1986). Sublethal heroin administration for 30 days induced degenerative changes in the parathyroid gland as evident by cytoplamic vacuolization, presence of more pycnotic nuclei and occurrence of patchy areas among the chief cells (Figure 13.4). Degenerative changes were also noticed in cristae of mitochondria, Golgi complex and endoplasmic reticula. There were decrease in the chromatin material in the nucleus and loss of hormone granules in cytoplasm of the chief cells (Figure 13.7).

Oxyphil Cells

Though function of oxyphil cells is not clearly defined, they are found only in mammals and their number increase with age (senility) (Roth and Schiller, 1976; Setoguti, 1977; Clark *et al.*, 1986). We observed a few oxyphil cells, polygonal in shape and larger than the chief cells in parathyroid gland of *Rattus norvegicus*. Their nuclei were smaller, irregular and denser than those of the chief cells (Figure 13.8). The abundant cytoplasmic area of the oxyphil cells was filled with numerous large mitochondria. The endoplasmic reticula, Golgi apparatus and secretory granules were poorly developed in these cells. Oxyphil cells of the heroin treated rats depicted dilation of endoplasmic reticula and damaged mitochondrial cristae on day 30 (Figure 13.9).

C Cells

Calcitonin-producing C cells of the rat were unevenly distributed in thyroid follicular cells. They were larger in size with more electronlucent cytoplasm as compared to those of thyroid follicular cells. Ultrastructurally, they possessed conspicuous endoplasmic reticulum, prominent Golgi apparatus, numerous mitochondria (both circular and elongated types) and dark electron dense secretory granules in their cytoplasm (Figure 13.10). In a few cells, desmosomes and terminal bars were also encountered. Similar structures of C cells have also been recorded for other mammalian species including bats (Kameda, 1976; Nunez and Gershon, 1978; Robertson, 1986). C cells of *Rattus norvegicus* were rich in mitochondria and secretory granules. In fact, Pearse and Carvalheira (1967) distinguished these cells from the follicular epithelial cell for the first time based on these characteristics. Presence of large number of secretory granules in the cytoplasm of C cells suggests accumulation of the hormone (calcitonin) under the normal physiological condition (Nunez and Gershon, 1978; Robertson, 1986). Since these cells get stimulated under hypercalcemic challenge, they are source of calcitonin (Pearse and Carvalheira, 1967; Calvert, 1975; Kameda, 1976; Nunez and Gershon, 1978; Robertson, 1986). Though dark and light parafollicular cells have been reported in electron micrographs of bat, tree shrew, dog, mouse and guinea pig (Nunez and Gershon, 1978), we could not find such variants of C cells in thyroid gland of *Rattus norvegicus*. Sublethal heroin administration in the rat for 30 days induced dilation in endoplasmic reticulum and loss of secretory granules of the C cells (Figure 13.11).

Parathyroid glands appeared first only in tetrapods (Clark *et al.*, 1986; Pandey, 1991, 1992), probably to protect against the development of hypocalcemia and to maintain skeletal integrity in terrestrial animals (Wendalaar Bonga and Pang, 1991). Parathyroid hormone (PTH) is a predominant hypercalcemic and hypophosphatemic factor which controls the plasma Ca and Pi metabolism of mammals in concert with calcitonin (CT) and 1,25-dihydroxyvitamin D_3 (active metabolite of vitamin D_3). These hormones exert their control in an integrated manner through three main processes–(i) the balance between the rate of deposition and mobilization of Ca and Pi in the bone (reservoir), (ii) the urinary excretion of Ca and Pi and (iii) the absorption of Ca and Pi from the gastrointestinal tract (Tayler, 1984; Pang and Schreibman, 1989; Wendelaar Bonga and Pang, 1991; Aurbach *et al.*, 1992; *et al.*, 1996; Dacke, 2000). There exist limited observations on the effects of opioid drugs on calcium regulating hormones suggesting that the circulating levels of calcitonin may be higher in heroin addicts (Tagliaro *et al.*, 1984, 1985; Spagnolli *et al.*, 1987, 1988), however, Tagliaro *et al.* (1992) observed increase only in "calcitonin-like" material, but not immunoreactive calcitonin, in such subjects. The opioids may play a role in regulation of calcitonin secretion in rat (Gozariu *et al.*, 1985). In the present study, sublethal heroin administration in rat for 30 days elicited degenerative changes in chief and oxyphil cells of the parathyroid gland as well as calcitonin-producing C cells. Though oxyphil cells made their first phyletic appearance in parathyroid gland of mammals, its function remains obscure except an increase in number with advancing age (Roth and Schiller, 1976; Setoguti, 1977; Clark *et al.*, 1986). The observed hypocalcaemia and hyperphosphataemia in the rat due to prolonged sublethal heroin administration appears to be due to degenerative changes in the chief cells which are the source of PTH in mammals. Pedrazzoni *et al.* (1993) also recorded lower PTH (1-84) levels in the patients with a history of chronic heroin abuse.

Adrenal Cortex

Adrenal gland of *Rattus norvegicus* were paired encapsulated structures lying in anterior region of each kidney. The gland was made up of two separate secretory tissues–the adrenal medulla, comprising chromaffin cells, located in the centre while the surrounding adrenal cortex produces steroid hormones. It has been reported that chromaffin cells initially arise from the neighbouring paraganglion cells of the neural crest complex and migrate to lie adjacent to the adrenocortical cell groups which are mesodermal in origin (Chester Jones, 1976; Balment *et al.*, 1980; Chester Jones and Phillips, 1986). Like other mammals (Long, 1975; Idelman, 1978; Chester Jones and Phillips, 1986), the adrenal cortex of *Rattus norvegicus* was divisible into three zones–zona glomerulosa, zona fasciculata and zona reticularis (Figure 13.12). The glomerulosa was a thin zone consisted of 4-6 layers of ovoid or columnar cells immediately inside the outer connective tissue capsule. The nuclei varied in shape from sausage to oval with one or two nucleoli and cytoplasm contained large number of lipid droplets (liposomes) lying between nucleus and side of the cell that abut the capillary (Figure 13.14). Mitochondria were rod or thread-like, smaller in size and

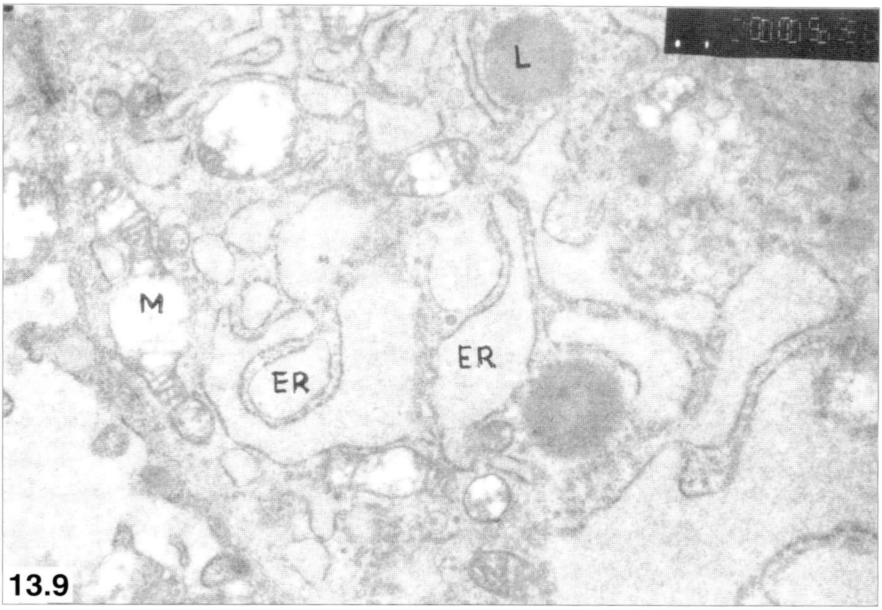

Figure 13.9: Oxyphil Cell of Rat Treated with Heroin for 30 Days Showing Mitochondria with Damaged Cristae (M), Dilated Endoplasmic Reticulum (ER) and Lipid Body (L). x 10,000.

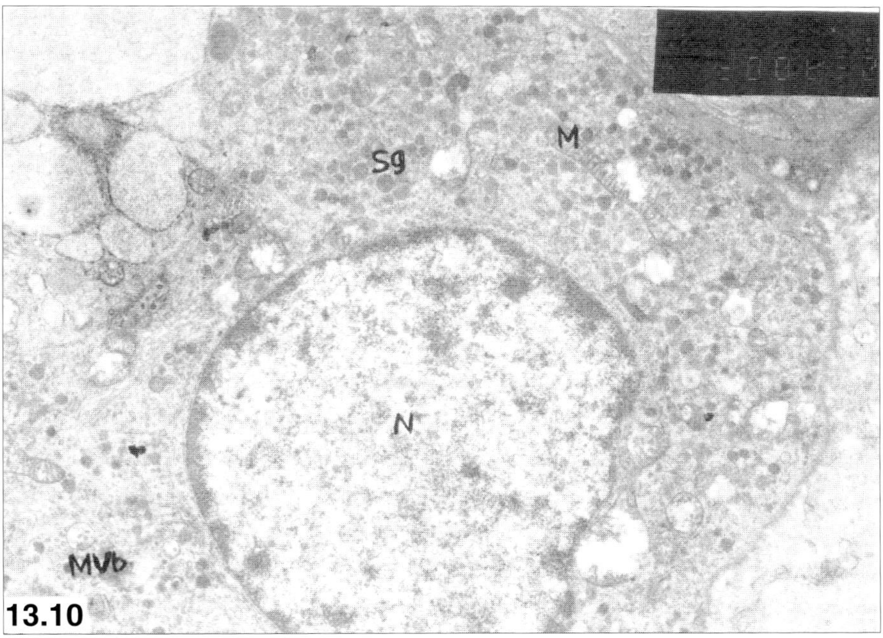

Figure 13.10: C Cell of Control *Rattus norvegicus* Exhibiting Prominent (N), Mitochondria (M), Multi-vesicular Body (MVb) and Large Number of Secretory Granules (Sg). x 6,000.

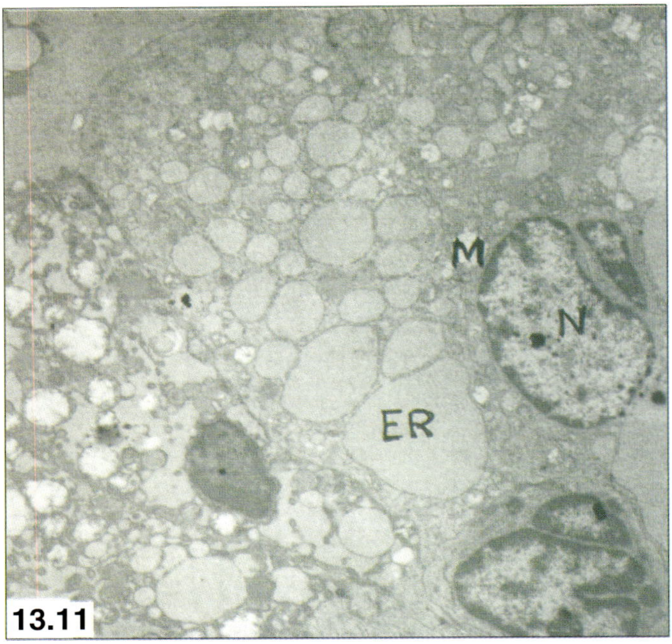

Figure 13.11: C Cell of Rat Treated with Heroin for 30 Days Depicting Nucleus (N), Mitochondria (M), Dilated Endoplasmic Reticulum (ER) and Loss of Secretory Granules. x 4,000.

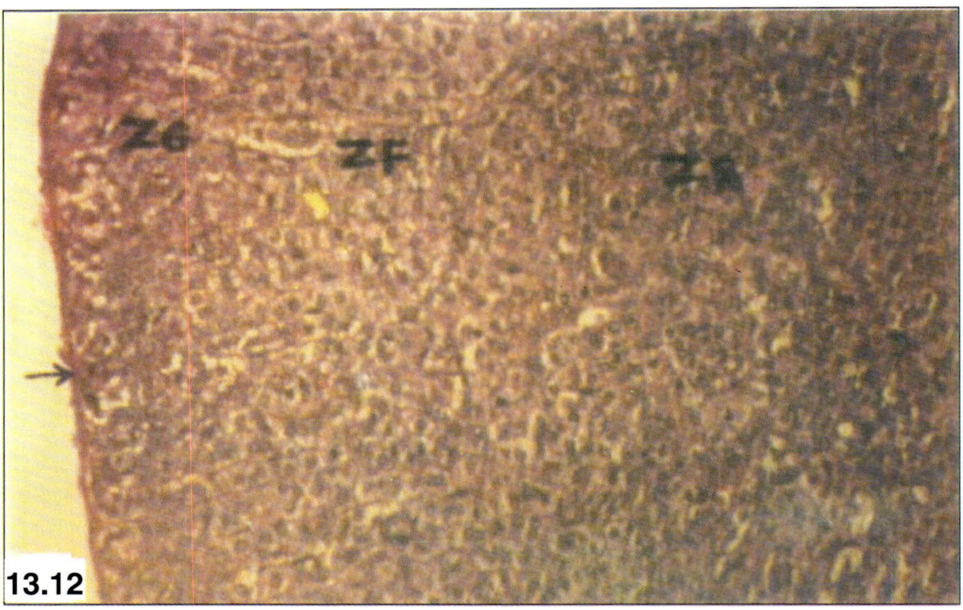

Figure 13.12: Adrenal Cortex of Normal Rat Showing Outer Connective Tissue Capsule (Arrow), *Zona glomerulosa* (ZG), *Zona fasciculata* (ZF) and *Zona reticularis* (ZR). H&E. x 100.

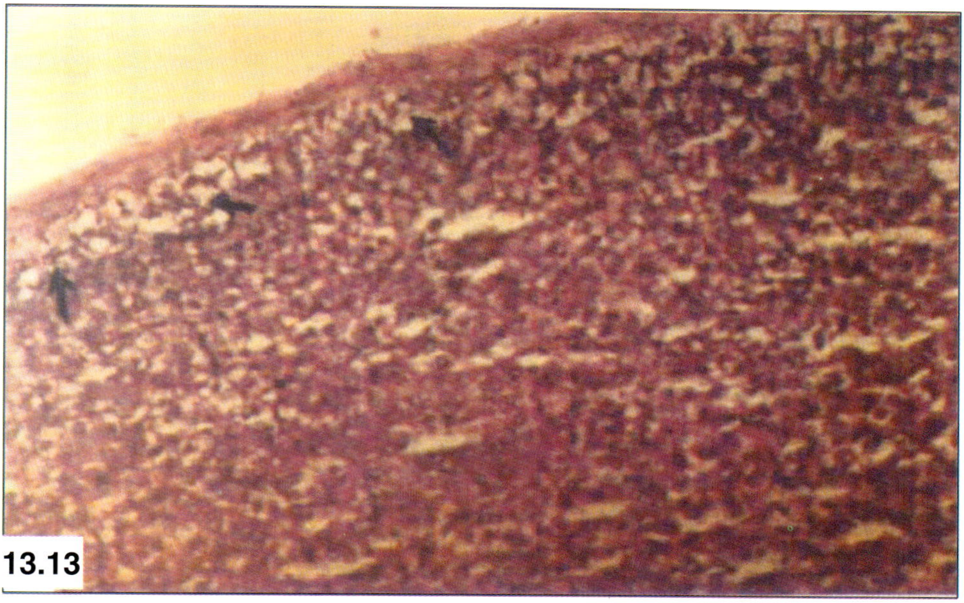

Figure 13.13: Adrenal Cortex of Rat on Day 30 of Heroin Administration Depicting Degenerative Changes in Cells of all the Three Zones. Mark the excessive degeneration in *Zona glomerulosa* cells (arrow). H&E. x 100.

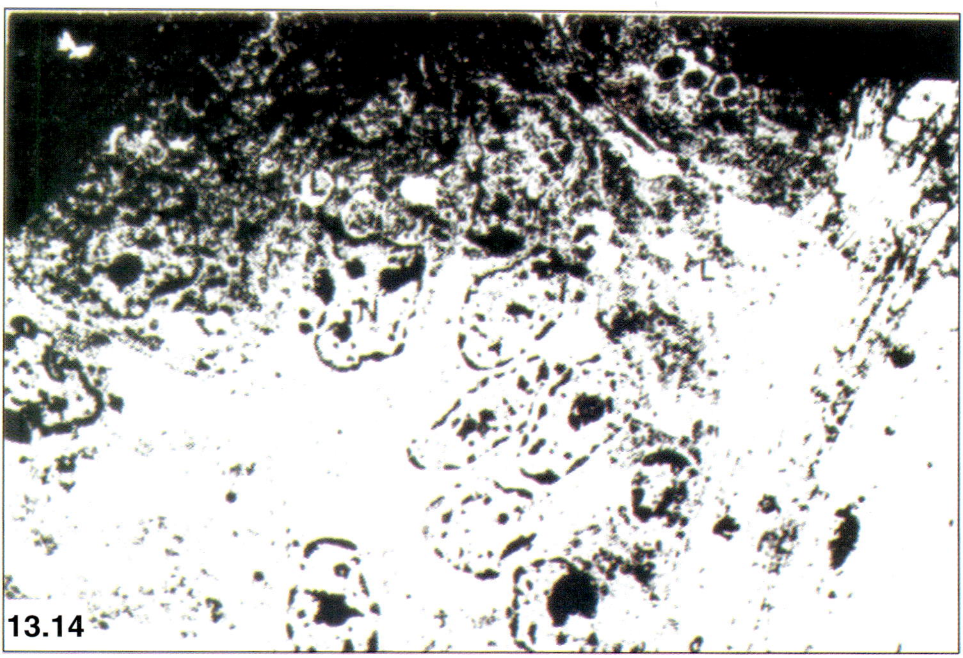

Figure 13.14: *Zona glomerulosa* Cells of Control *Rattus norvegicus* Exhibiting Abundant Nuclei (N) and Numerous Lipid Droplets (L). x 2,000.

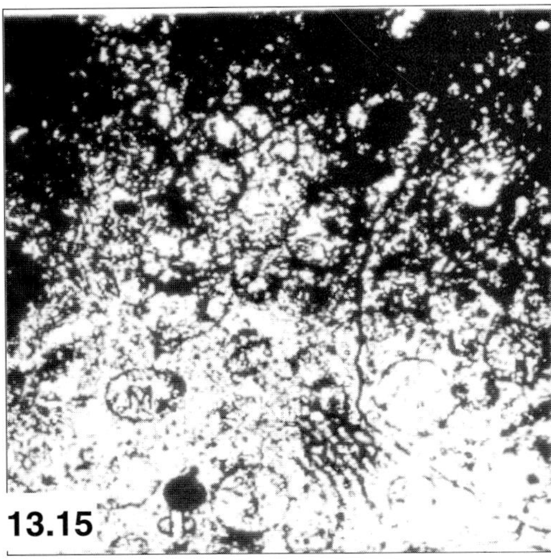

Figure 13.15: *Zona fasciculata* Cell of Control Rat Showing Mitochondria (M), Rough Endoplasmic Reticulum (RER) and Dense Body (db). X 10,000.

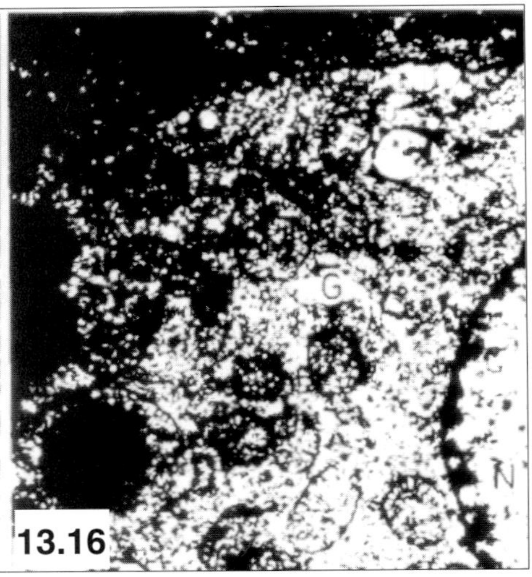

Figure 31.16: *Zona fasciculata* Cell of Rat Depicting Part of Nucleus (N), Numerous Mitochondria (M), Golgi Complex (G) and Lipid Droplets (L). x 10,000.

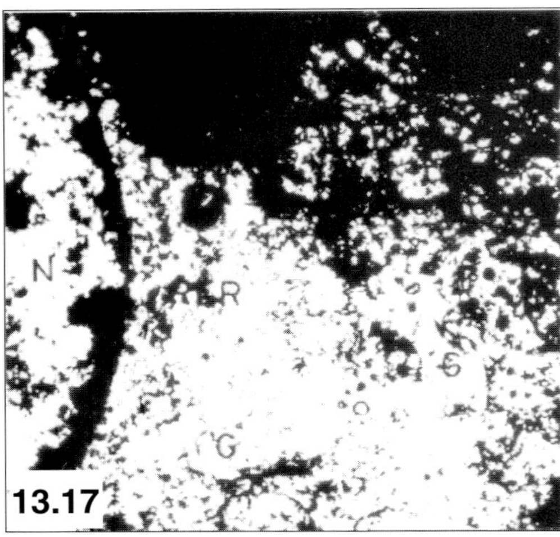

Figure 13.17: *Zona reticularis* Cell of *Rattus norvegicus* Exhibiting Part of Nucleus (N), Lipid Droplet (L), Rough Endoplasmic Reticulum (RER) and Golgi Complex (G). x 15,000.

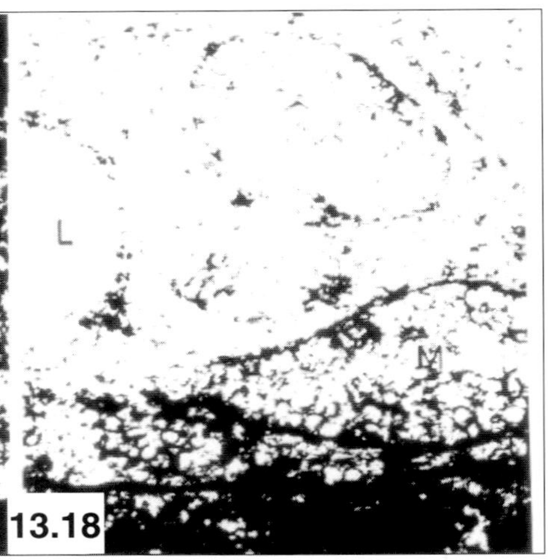

Figure 13.18: *Zona glomerulosa* Cell of Rat Treated with Heroin for 30 Days Showing Many Large Mitochondria (M) with Damaged Cristae, Desmosome (D) and Lipid Droplet (L). x 25,000.

their number was quite large. Branching and anastomosing tubules of smooth-surfaced endoplasmic reticula were also seen. Golgi complex was well developed lying adjacent to nucleus without articular orientation. Similar structures of the zona glomerulosa cells have also been recorded in other mammals (Lentz, 1971; Chester

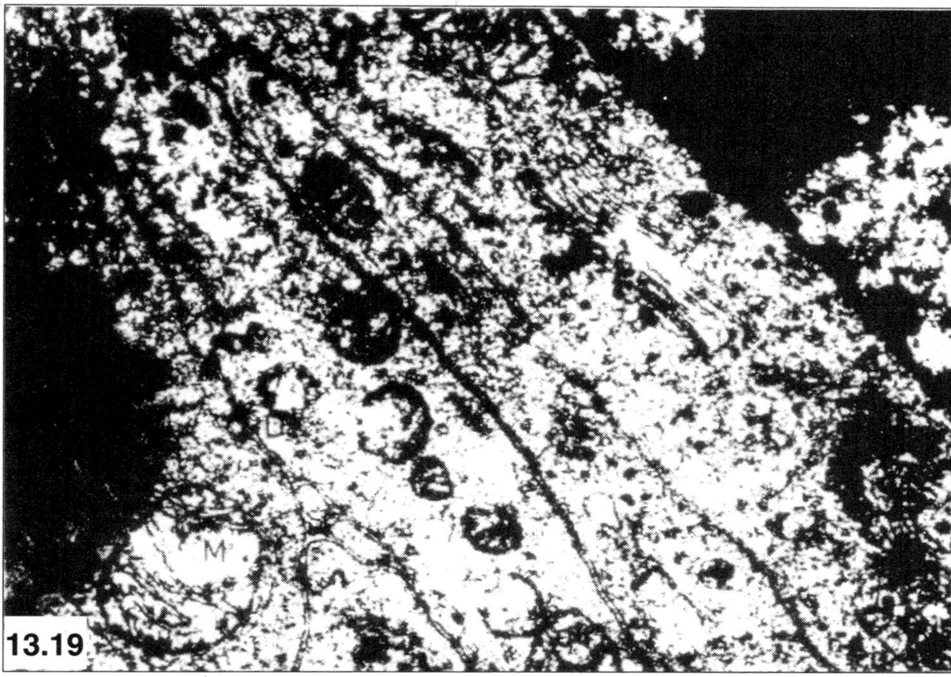

13.19

Figure 13.19: *Zona glomerulosa* Cell of Rat on Day 30 of Heroin Administration Showing Many Large Mitochondria (M) with Damaged Cristae, Desmosome (D) and Lipid Droplet (L). Mark the degenerative changes in Golgi complex (G). x 25,000

Jones and Henderson, 1980; Chester Jones and Phillips, 1986). These cells are responsible for secretion of aldosterone (the mineralocorticoid) which is involved in regulation of serum electrolytes (Chester Jones and Henderson, 1980).

The zona fasciculata of the rat was the widest zone of adrenal cortex. It comprised radial cords of polyhedral cells which were considerably larger, had round central nucleus with dense chromatin towards periphery, one or two nucleoli and cytoplasm packed with lipid droplets (Figures 13.15 and 13.16). Though branching and anastomosing tubules of smooth-surfaced endoplasmic reticula were extensive, the rough-surfaced endoplasmic reticulum was more abundant in this zone as compared to cells of zona glomerulosa The mitochondria were large, generally oval to spherical and Golgi apparatus well developed (Figures 13.15 and 13.16). Lysosomes and lipofusin pigments were also encountered. The zona fasciculata cells secrete glucocorticoids (cortisol and corticosterone) which are involved in protein as well as carbohydrate metabolism (Chester Jones and Henderson, 1980).

The innermost *zona reticularis*, bordering the adrenal medulla of the rat, consisted of anastomosing networks of reticular cell cords. The cells were smaller than those of the rest of the cortex with less lipid content. They possessed smooth-surfaced endoplasmic reticula, stacks or whorls of rough-surfaced lamellae, cytoplasmic ribosomes, Golgi apparatus, lipid droplets and lysosomes. However, mitochondria were more elongated and lipofusin pigment granules were numerous in the cytoplasm (Figure 13.17). Similar ultrastructures of the zona reticularis cells have also been observed in other mammalian species (Lentz, 1971; Chester Jones and Henderson, 1980). These cells are responsible for the secretion of sex steroids, particularly androgens (Chester Jones and Phillips, 1986).

Though sublethal heroin administration for 30 days elicited varying degrees of vacuolization in endocrine cells of all the three zones, cells of zona glomerulosa exhibited severe degenerative changes in cytoplasm and nuclei as compared to those of zona glomerulosa and zona fasciculata (Figure 13.13). Ultrastructurally, degenerative changes were observed in various cell organelles. Mitochondria lost the typical cristae and hormone granules were rarely seen (Figure 13.18). Though rough endoplasmic reticula were scanty, degenerative changes were also observed in these structures as well as in Golgi complex. Many lipid granules and vacuolization were also observed in zona glomerulosa as well as zona fasciculata cells of the treated rat (Figure 13.19).

Several endocrine dysfunctions such as abnormal adrenal metabolism and insufficiency, abnormalities in circadian rhythm of corticosteroid secretion, increased levels of thyroxine (T_4) and triiodothyronine (T_3), thyroxine-binding globulin (TBG) and reduced T_3 level, abnormalities in insulin-glucose metabolism, reduction in testosterone level and abnormal FSH and LH levels have been recorded in the patients with the history of chronic opiate misuse (Glass *et al.*, 1973; Pechnick, 1993; Spangel, 1999; Sarnyai *et al.*, 2001; George *et al.*, 2005; Al-Gommer *et al.*, 2007). There exist reports that opioids affect the adrenal cortex function by involving hypothalamus-pituitary-adrenal (HPA) axis (Pechnick, 1993; Spangel, 1999; Sarnyai *et al.*, 2001; Laorden *et al.*, 2002; George *et al.*, 2005; Blesener *et al.*, 2006). Though sublethal heroin administration for 30 days elicited cytoplasmic vacuolization in cells of all the three zones of adrenal cortex of *Rattus norvegicus*, degenerative changes were more marked in zona glomerulosa cells. Our study demonstrates that sublethal heroin administration induces alterations in plasma Na and K metabolism by affecting the zona glomerulosa cells of adrenal cortex.

Acknowledgements

One of us (SRB) is grateful to the Council of Scientific and Industrial Research (CSIR), New Delhi for the award of Fellowship to carry out this work. We are obliged to Hon'ble Justice Mrs. K.K. Baam, the then High Court Judge, Bombay for permitting us to work on heroin and to Mr. Rahul Rai Sur, the then Deputy Commissioner of Police, Narcotics Cell, Greater Mumbai for the procurement of the drug. We are thankful to Mr. Vijay Kate and Mr. Dilip Kanaskar of the Electron Microscopy Division, Jaslok Hospital, Mumbai for their assistance in this study.

References

Al–Gommer, O., George, S. and Haque, S., 2007. Sexual dysfunctions in male opiate users: a comparative study of heroin, methadone and buprenorphine. *Addict. Disord. Their Treat.*, 6: 137–143.

Aurbach, G.D., Marx, S.J. and Spiegel, A.M., 1992. Parathyroid hormone, calcitonin, and the calciferols. In: *Williams Handbook of Endocrinology* (Eds.) Wilson, J.D. and Foster, D.W. W. B. Saunders, Philadelphia, pp. 397–1476.

Balment, R.J., Henderson, I.W. and Chester Jones, I., 1980. The adrenal cortex and its homologues in vertebrates: evolutionary considerations. In: *General Endocrinology and Clinical Endocrinology of the Adrenal Cortex. Vol. 3* (Eds.) Chester Jones, I. and Henderson, I.W. Academic Press, New York, pp. 525–562.

Barai, S.R., Suryawanshi, S.A. and Pandey, A.K., 2004a. Effect of sublethal heroin administration on liver and kidney of *Rattus norvegicus. J. Natcon.*, 16: 149–155

Barai, S.R., Suryawanshi, S.A. and Pandey, A.K., 2004b. Effect of sublethal heroin administration on acid phosphatase and alkanine phosphatase activities of *Rattus norvegicus. J Exp. Zool. India*, 7: 329–332.

Blesener, N., Albrecht, S., Schwager, A., Wecbecker, K., Litchermann, D. and Lingmuller, D., 2006. Plasma testosterone and sexual function in men receiving buprenorphine maintenance for opioid dependence. *J. Clin. Endocrinol. Metab.*, 90: 203–206 (2005).

Brambilla, F., Nobile, P., Zanoboni, A., Muciaccia, W. and Meroni, P.L., 1980. Effects of chronic heroin addiction on pituitary–thyroid function in man. *J. Endocrinol. Invest.*, 3: 251–255.

Calvert, R., 1975. Structure of rat ultimobranchial bodies after birth. *Anat. Rec.*, 181: 561–580.

Cami, J. and M. Farrie, 2003. Drug addiction. *N. Eng. J. Med.*, 349: 975–986.

Campistol, J.M., Montoliu, J., Soler–Amigo, J., Darnell, A. and Revert, L., 1988. Renal amyloidosis with nephrotic syndrome in a Spanish subcutaneous heroin abuser. *Nephrol. Dial. Transplant.*, 3: 471–473.

Capen, C.C. and Rowland, G.N., 1968a. The ultrastructure of the parathyroid glands of young cats. *Anat. Rec.*, 162: 327.

Capen, C.C. and Rowland, G.N., 1968b. Ultrastructural evaluation of parathyroid glands of young cats with experimental hyperparathyroidism. *Z. Zellforsch. Mikrosk.– Anat.*, 90: 495–507.

Chester Jones, I., 1976. Evolutionary aspects of the adrenal cortex and its homologues. The Dale Lecture. *J. Endocrinol.*, 71: 1P–31P.

Chester Jones, I. and Henderson, I.W., 1980. *General Endocrinology and Clinical Endocrinology of the Adrenal Cortex. Vol. 3.* Academic Press, New York.

Chester Jones, I. and Phillips, J.G., 1986. The adrenal and interrenal glands. In: *Vertebrate Endocrinology: Fundamental and Biomedical Implications. Vol. 1. Morphological Considerations* (Eds.) Pang, P.K.T. and Schreibman, M.P. Academic Press, San Diego, pp. 319–350.

Clark, N.B., Kaul, K. and Roth, S.I., 1986. The parathyroid glands. In: *Vertebrate Endocrinology: Fundamentals and Biomedical Implications. Vol. 1. Morphological*

Considerations (Eds.) Pang, P.K.T. and Schreibman, M.P. Academic Press, San Diego, pp. 207–234.

Dacke C G., 1979. *Calcium Regulation in Submammalian Vertebrates*. Academic Press, London.

Dacke, C.G., Danks, J., Caple, I. and Flik, G., 1996. *The Comparative Endocrinology of Calcium Regulation*. Journal of Endocrinology Press, Bristol, United Kingdom.

Dacke, C.G., 2000. The parathyroids, calcitonin and vitamin D. In: *Avian Physiology* (Ed.) Sturkie, P.D. Academic Press, San Diego and London, pp. 473–488.

Dhande, R.R., Suryawanshi, S.A. and Pandey, A.K., 2003. Seasonal changes in plasma calcium and inorganic phosphate levels in relation to ultimobranchial gland of the grey quail, *Coturnix cotrunix coturnix* Linnaeus. *Proc. Zool. Soc.* (*Calcutta*), 56: 19–27.

Ekholm, R., 1990. Biosynthesis of thyroid hormones. *Int. Rev Cytol.*, 120: 243–288.

El–Daly, E.S., 1994. Effect of morphine and stadol on lipid content in liver of rat. *Life Sci.*, 55: 1419–1426.

Fiske, C.H. and Subbarow, V., 1925. The colorimeteric determination of phosphorus. *J. Biol. Chem.*, 66: 375–400.

Fujita, H., 1986. Functional morphology of the thyroid. *Int. Rev. Cytol.*, 113: 145–185.

George, S., Murali, V. and Pullickal, R., 2005. Review of neuroendocrine correlates of chronic opiate misuse: dysfunctions and pathophysiological mechanisms. *Addict. Disord. Their Treat.*, 4: 99–109.

Glass, L., Rajegowda, B.K., Mukherjee, T.K., Roth, M.M. and Evans, H.C., 1973. Effect of heroin on corticosteroid production of pregnant addicts and their fetuses. *Am. J. Obstet. Gynecol.*, 117: 416–418.

Goldberger, B.A., Cone, E.J., Grant, T.M., Caplan, Y.H., Levine, B.S. and Smialek, J.E., 1994. Disposition of heroin and its metabolites in heroin–related deaths. *J. Anal. Toxicol.*, 18: 22–28.

Gomez–Lechon, M.J., Ponsoda, X., Jover, R., Pabra, R., Trullenque, R. and Castell, J.V., 1987. Hepatotoxicity of the opioids morphine, heroin, mepridine and methadone to cultured human hepatocytes. *Mol. Toxicol.*, 1: 453–463.

Gorbman, A., Dickhoff, W.W., Vigna, S.R., Clark, N.B. and Ralph, C.L., 1983. *Comparative Endocrinology*. John Wiley and Sons, New York.

Gozariu, L., Orbai, P., Safta, L., Cuparencu, B., Barabas, E. and Gozariu, M., 1985. The effect of some enkephalins on calcitonin secretion. *Endokrinologie*, 23: 201–204.

Griffiths, P., Gossop, M., Powis, B. and Strong, J., 1994 Transition in patterns of heroin administration–a study of heroin chasers and heroin injectors. *Addiction*, 83: 301–309.

Hussain. K. and Kumar, A., 1988. Physiological, haematological, biochemical and clinical effects of epidural morphine in dogs. *Indian Vet. J.*, 65: 491–495.

Idelman, S., 1978. The structure of the mammalian adrenal cortex. In: *General Endocrinology and Clinical Endocrinology of the Adrenal Cortex. Vol. 2* (Eds.) Chester Jones, I. and Henderson, I.W. Academic Press, New York, pp. 1–99.

Jenkines, A.J., Keenan, R.M., Henningfield, J.E. and Cone, E.J., 1994. Pharmacokinetics and pharmacodynamics of smoked heroin. *J. Anal. Toxicol.*, 18: 317–330.

Kameda, Y., 1976. Fine structural and endocrinological aspects of thyroid parafollicular cells. In: *Chromaffin, Enterochromaffin and Related Cells* (Eds.) Coupland, R.E. and Fujita, T. Elsevier Sci. Pub., Amsterdam, pp. 155–170.

Kringsholm, B. and Christoffersen, P., 1989. Morphological findings in fatal drug addiction. An investigation of injection marks, endocrine organs and kidneys. *Forensic Sci. Int.*, 40: 15–24.

Kringsholm, B., Kaa, E., Steentoft, A., Worm, K. and Simonsen, K.W., 1994. Deaths among drug addicts in Denmark in 1877–1991. *Forensic Sci. Int.*, 67: 185–195.

Laorden, M.L., Castells, M.T. and Milanes, M.V., 2002. Effect of morphine and morphine withdrawal on brainstem neurons innervating hypothalamic nuclei that controls the pituitary–adrenocortical axis in rat. *Br. J. Pharmacol.*, 136: 67–75.

Lentz, T.L., 1971. *Cell Fine Structure*. W.B. Saunders, Philadelphia and Pennsylvania.

Long, J.A., 1975. Zonation of the mammalian adrenal cortex. In: *Handbook of Physiology. Vol. VI* (Eds.) Blaschko, H., Sayers, G. and Smith, A.D. American Physiological Society, Washington, pp. 13–24.

Martin, W.R., 1984. Pharmacology of opioids. *Pharmacol. Rev.*, 35: 282–323.

Neri–Serneri, G.G. and Modesti, P.A., 1991. Medical complications connected with the use of drugs. *Ann. Ital. Med. Int.*, 6: 313–324.

Nunez, E.A. and Gershon, M.D., 1978. Cytophysiology of parafollicular cells. *Intern. Rev. Cytol.*, 52: 1–80.

Pandey, A.K., 1985. Occurrence of accessory parathyroid and connective tissue septae in parathyroid gland of *Bufo melanostictus* (Schneider). *J. Adv. Zool.*, 6: 109–111.

Pandey, A.K., 1991. Endocrinology of calcium metabolism in reptiles: a comparative aspect in lower vertebrates. *Biol. Struct. Morphogen. (Paris)*, 3: 159–176.

Pandey, A.K., 1992. Endocrinology of calcium metabolism in amphibians, with emphasis on the evolution of hypercalcemic regulation in tetrapods. *Biol. Struct. Morphogen. (Paris)*, 4: 102–1256.

Pang, P.K.T. and M.P. Schreibman, 1989. *Vertebrate Endocrinology: Fundamentals and Biomedical Implications, Vol. 3: Regulation of Calcium and Phosphate.* Academic Press, New York and San Diego.

Pearce, C.J. and Cox, J.G.C., 1980. Heroin and hyperkalemia. *Lancet*, ii: 923.

Pearse, A.G.E. and Carvalheira, A.F., 1967. Cytochemical evidence for an ultimobranchial origin of rodent thyroid C cells. *Nature*, 214: 929–930.

Pechnick, R.N., 1993. Effects of opioids on the hypothalamo–pituitary–adrenal axis. *Ann. Rev. Pharmacol. Toxicol.*, 33: 353–882.

Pedrazzoni, M., Vescovi, P.P., Maninetti, L., Michelini, M., Zaniboni, G., Pioli, G., Costi, D., Alfanso, F.S. and Passeri, M., 1993. Effect of chronin heroin abuse on bone and mineral metabolism. *Acta Endocrinol.* (*Kbh*), 129: 42–45.

Robertson, D.R., 1986. The ultimobranchial body. In: *Vertebrate Endocrinology: Fundamentals and Biomedical Implications. Vol.1. Morphological Considerations* (Eds.) Pang, P.K.T. and Schreibman, M.P. Academic Press, New York and San Diego, pp. 235–259.

Roth, S.I. and Schiller, A.L., 1976. Comparative anatomy of parathyroid glands. In: *Endocrinology. Vol. 7* (Eds.) Greep, R.O., Estwood, E.B. and Aurbach, G.D. American Physiological Society, Washington, pp. 281–311.

Sarnyai, Z., Shahan, Y. and Heinrichs, S.C., 2001. The role of corticotropin–releasing factor in drug addiction. *Pharmacol. Rev.*, 53: 209–244.

Sawynok, J., 1986. The therapeutic use of heroin: a review of the pharmacological literature. *Can. J. Physiol. Pharmacol.*, 64: 1–6.

Setoguti, T., 1977. Electron microscopic studies of the parathyroid glands of senile dogs. *Am. J. Anat.*, 148: 65–84.

Sharma, S., Mohite, V., Rangoonwala, S.P. and Suryawanshi, S.A., 2001. Heroin induced changes in the lipid profiles in rabbit, *Lepus cuniculus. Biochem. Cell. Arch.*, 1: 109–113.

Solcia, E., Capella, C. and Vassallo, C., 1969. Lead–haematoxylin as a stain for endocrine cells. *Histochemie*, 20: 116–126.

Spagnolli, W., Torboli, P., Mattarei, M., De Venuto, G., Morcolla, A. and Miori, R., 1987. Calcitonin and protactin serum levels in heroin addicts: study on a methodone treated group. *Drug Alcohol Depend.*, 20: 143–148.

Spagnolli, W., De Venuto, G., Mattarei, M., Dal Plaz, A., Merz, O. and Miori, R., 1988. Immunoheterogeneity of serum calcitonin in heroin addicts. *Drug Alcohol Depend.*, 22: 165–167.

Spanagel, R., 1999. The hypothalamic–pituitary–adrenocortical system: a biological substrate of vulnerability to drug addiction. *Basic Clin. Sci. Subst. Rel. Disord.*, 168: 1–6.

Sporer, K.A., 1999. Acute heroin overdose. *Ann. Intern Med.*, 130: 584–590.

Suryawanshi, S.A., Dhande, R.R. and Pandey, A.K., 1997. Effect of parathyroidectomy on plasma calcium and inorganic phosphate levels of the grey quail, *Coturnix coturnix coturnix* Linnaeus. *Nat. Acad. Sci. Letters*, 20: 14–18.

Tagliaro, F., Capra, F., Dorizzi, R., Luisetto, G., Accordini, A., Renda, E. and Parolin, A., 1984. High serum calcitonin levels in heroin addicts. *J. Endocrinol. Invest.*, 7: 331–333.

Tagliaro, F., Dorizzi, R., Lafisca, S., Maschio, S. and Marigo, M., 1985. Calcitonin serum levels in heroin addicts: effects of methadone and clonidine detoxication treatment. *Drug Alcohol Depend.*, 16: 181–183.

Tagliaro, F., Dorizzi, R., Ghielmi, S., Comberti, E., Alvera, P., Manzato, E. and Marigo, M., 1992. Immunoreactive "calcitonin–like" material in heroin addicts varying reactivity with different antibodies. *Intern. J. Legal Med.*, 104: 309–312 (1992).

Taylor, C.W., 1984. Calcium regulation in vertebrates: an overview. *Comp. Biochem. Physiol.*, 82A: 249–255.

Trinder, P., 1960. Colorimetric microdetermination of calcium in serum. *Analysts*, 85: 889–894.

Volochine, M.L., Rondeau, E., Viron, B., Mougenot, B., Beaufils, H., Pourriat, J.L. and Chauveau, P., 1988. Renal disease associated with heroin abuse. *Nephrologie*, 9: 217–221.

Weller, I.V., Cohen, D., Sierralta, A., Mitcheson, M., Ross, M.G., Mantano, L., Scheuer, P. and Thomas, H.C., 1984. Clinical, biochemical, serological, histological and ultrastructural features of liver disease in drug abusers. *Gut*, 25: 417–423.

Wendelaar Bonga, S. E. and Pang, P.K.T., 1991. Control of calcium regulating hormones in vertebrates: parathyroid hormone, calcitonin, prolactin and stanniocalcin. *Int. Rev. Cytol.*, 128: 139–213.

Wootton, I.D.P., 1974. *Microanalysis in Medical Biochemistry. 5th Edn.* Churchill Livingston, Edinburgh and London.

Yadav, S.D., Barai, S.R. and Pandey, A.K., 2004. Marijuana (Bhang) induced alterations in thyroid gland of albino rat. In: *Sixth Indian Agricultural Scientists and Farmers Congress* (21–22 February 2004). Bioved Research Society, Allahabad, p. 113.

Chapter 14

Biodiversity Resources of Bhitarkanika Mangrove Ecosystem: Significance and Threat

☆ *Sudhakar Kar*

Abstract

Bhitarkanika is a deltaic area formed by the alluvial deposits of river Brahmani and Baitarani and is a Wildlife Sanctuary as per provisions of the Wildlife (Protection) Act, 1972. The core area of Bhitarkanika Wildlife Sanctuary has been notified as a National Park (145 km^2) during 1998. It has also been declared as a Ramsar site *i.e.* Wetland of International Importance.

Bhitarkanika mangrove ecosystem supports the rich biodiversity including mangroves and associates (71 species), largest population of Estuarine Crocodiles, largest Indian Lizards (Water monitor), poisonous and non-poisonous snakes like King cobra and Python, varieties of resident and migratory birds (217 species) and number of mammalian species (Spotted deer, Sambar, Fishing cat, Otter, Dolphin, etc.).

It is to be noted that, the coastal sandy beaches of Bhitarkanika *i.e.* the Gahirmatha is the world's largest Olive Ridley Sea turtle rookery.

This study highlights the ecological, economic and threat aspects of both the wetlands and also conservation measures need to be taken to preserve these two unique wetlands/ Ramsar sites of Orissa as well as biodiversity conservation for a sustainable society.

Introduction

Orissa is a maritime state located in the eastern coast of Indian peninsula and having a coastline of 480 km. Eighteen Sanctuaries, a National Park and one proposed

National Park constitute the Protected Area network of Orissa that covers 6611.12 sq.km of land area which is 4.25 per cent of the geographical area and 11.37 per cent of the forest area of the state of Orissa. Bhitarkanika (672.00 sq.k.m) has been designated as a Wildlife Sanctuary in 1975 as per the Wildlife (Protection) Act, the core of which has been notified as a National Park (145.00 sq.km) in 1998. Bhitarkanika has also been designated as one of the Ramsar sites (Wetlands of International Importance) in the country on 19-08-2002 which is second such site the state.

Bhitarkanika ecosystem is a hotspot of rich biological diversity. With 71 species of mangroves and mangrove associates, the area supports largest population of estuarine crocodiles (1498 as per Jan., 2008 census) in the country, largest Indian Lizards, varieties of resident and migratory birds (271 species) and a number of rare and endangered mammalian species. In comparison to the national status, the composition of vertebrate fauna of Bhitarkanika site represents 8 per cent mammals, 17.70 per cent birds, 9.40 per cent reptiles and 2.5 per cent amphibians. The Gahirmatha sea beach (now a marine sanctuary), bordering the Bhitarkanika sanctuary attracts over half a million of Olive Ridley sea turtles for mass nesting / egg laying (World's largest rookery) during the winter months (January to April).

This ecosystem supports a range of interconnected food webs within it to maintain the balance of nature. This ecosystem is complex and is intricately mixed with each other. The livelihood patterns of the villagers surrounding Bhitarkanika directly or indirectly influences the very existence and survival of the flora and fauna in this ecosystem. Fishing is the main stay of the villagers. There is much anthropogenic and biotic pressure on the biodiversity. The present study highlights the ecological significance and threat aspects of this ecosystem and also emphasizes the conservation measures need to be taken to preserve the rich biodiversity.

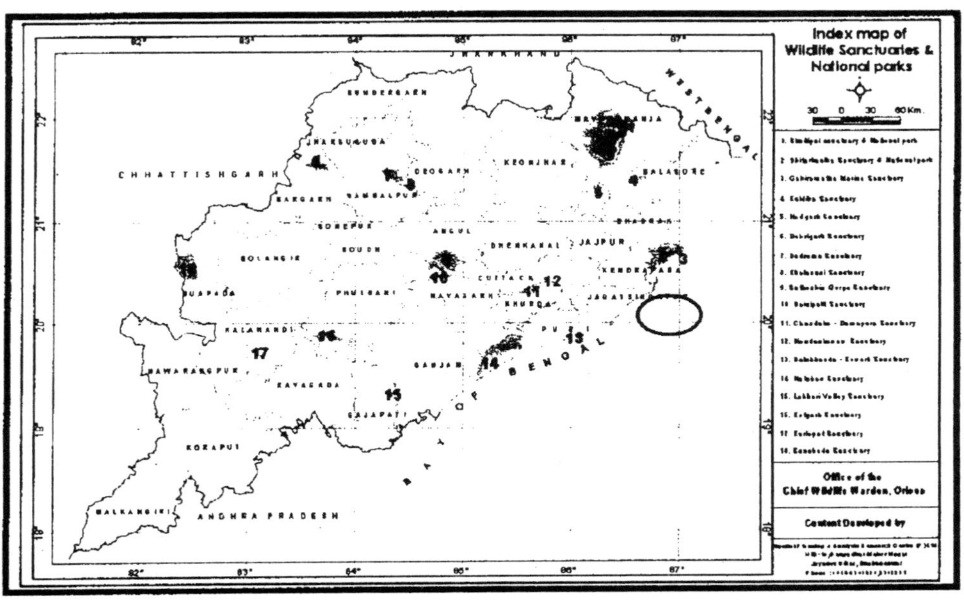

Figure 14.1: Wildlife Sanctuaries and National Parks of Orissa

Bhitarkanika is endowed with a very complex and dynamic ecosystem and is highly fragile in nature. The essential factors for maintenance of such ecosystem is regular influx of fresh water from adjoining land and tidal inflow from the sea. Any change in the regime of either factor is likely to effect a corresponding change in the mangrove ecosystem.

Biodiversity Resources

The wetland supports one of the largest mangrove ecosystems after Sundarbans, Gujarat and Andhra Pradesh in the Indian mainland. It has more than 300 plant species, which include mangroves, mangrove associates and non mangroves. The floral diversity of Bhitarkanika wetland is known to be largest in India and second largest after Papua New Guinea in the world. Considering the genetic diversity of the wetland and its importance, the mangrove steering committee of Govt. of India have established its National Mangrove Genetic Resource Conservation Centre in one of the islands of this wetland *i.e.* Kalibhanjadia island.

Endemism and Biological Uniqueness

Endemism in Bhitarkanika is not fully explored. Yet, it is expected to be there particularly in sectors like mangrove flora and benthic fauna, soil fauna as well as aquatic flora and fauna. Among the three species of Sundari trees (*Heritiera* sp.) available in the area, *Heritiera kanikensis* or Kanika Sundari is endemic to Bhitarkanika.

Special Features

The resilient mangroves serve the protective functions to a greater extent. It protects the hinterland against cyclonic storms during cyclones, super cyclones, tidal surges and other natural catastrophes acting as an effective shelterbelt. In the unprecedented super cyclone of October 1999, the mangroves has withstood the onslaught of cyclonic wind and saved the life and property of millions of people.

Mangrove wetlands perform a variety of productive as well as protective functions. This mangrove wetland in particular is a repository of biological diversity in terms of flora and fauna.

This ecosystem harbours the largest number of saltwater crocodile population in the Indian sub-continent. Other reptilian fauna include Monitor lizard, Indian python, King cobra and varieties of other snake species. It also harbours a number of endangered animals like Fishing cat, Leopard cat, Dolphins and Porpoises.

Bhitarkanika's famous Gahirmatha coast finds a prominent place in the turtle map of the world because of the distinction of having one of world's largest nesting and breeding congregation of Olive Ridley Sea turtles. Mangrove wetlands including mudflats provide ideal feeding, perching and nesting facilities to a variety of resident and migratory waterfowl.

Ecosystem Value/Significance

Mangroves have been considered as "land builders". It is believed that the roots of mangroves secrete a substance, which modifies the coarse particles into fine ones and help in soil formation. Network of mangrove roots provide firm anchorage to the

Figure 14.2: Bhitarkanika River with Luxuriant Mangrove Vegetation

Figure 14.3: Rhizophora, One of the Dominant Mangrove Species

Figure 14.4: A Large Male (20 feet +) Saltwater Crocodile

banks of tidal rivers, creeks and also the coast line. It effectively arrests river bank and coastal erosion and ultimately helps in controlling flood damages. It also exercises a moderating influence on the cyclonic wind and storm surges. In the past, serve cyclones and tidal surges of the coastal Kendrapara district; particularly the Rajnagar area, is known to have been effectively controlled due to the presence of thick mangrove vegetation in the zone of Bhitarkanika and the adjoining Mahanadi deltaic area.

Ecosystem Functions

Mangrove areas support a range of interconnected food webs, which directly sustain the fisheries. Algae and detritus sustain shrimps and prawns, which provide

Figure 14.5: Mass Nesting (Arribada) of Olive Ridley Sea Turtles

a food source for species such as Bhekti (*Lates* sp.), Cat fishes etc. Fish and prawns spend most of their adult life at sea and return to the mangrove areas and vice versa to spawn. Mud skippers, a typical fish reside around and in the mangroves. These fishes are able to survive short periods of aerial exposure, skip around on the water and mud and build chimney like burrows.

Economic Value

Bhitarkanika mangrove wetland is one of the most productive ecosystems. It adds to the coastal fishery production. The rivers and creeks in the wetland are a major source of variety of indigenous fish. The sheltered waters of mangroves provide nursery ground for commercially harvested prawns and shrimps. Several fish species come to the estuary for breeding. Fishing is the mainstay of the villagers those who do not have any landed property. In addition, the local people depend on the mangrove vegetation for collection of honey, wax and medicinal plants. Around 50 quintal of honey is available per year in Bhitarkanika forests.

Socio-Cultural Value

The wetland has a good number of ancient monuments like palace of ex-zamindar, Shiva temple inside Bhitarkanika forest block, Jagannath temple at Righagarh and Keradagarh, Panchubarahi goddess temple at Satabhaya and others such small temples which are culturally significant to the inhabitants.

Scientific Value

The wetland is endowed with a variety of habitats and microhabitats to shelter wide ranging aquatic, terrestrial and avifauna. The animals and birds associated with the mangrove and wetland can be broadly categorized into two groups namely invertebrate and vertebrate. Vertebrate fauna include a variety of fishes, amphibians, birds, reptiles and mammals.

The Saltwater crocodile "rear and rehabilitation" operation is a success story in Bhitarkanika and the crocodile population in the Bhitarkanika river system has been gradually built up. The captive reared young crocodiles have been released in the creeks and estuaries and above 2200 crocodiles have been released in phases since 1977. Some of the released crocodiles have bred successfully in the wild and above 65 clutches of eggs have been located, which is 10 times more in comparison to 1975-76.

Ecotourism Potentiality

Bhitarkanika has become an identified tourist destination in Orissa and is a paradise for nature lovers, conservationists, and biologists. However, the Ecotourism potentiality is yet to be fully explored. Some infrastructure are presently available in places like Chandbali, Dangmal, Dhamara, Habalikhati, Gupti and Ekakula for catering the need of tourists which are being developed and upgraded.

Threats

Aquaculture

In and around the site, a large chunk of the agricultural land adjacent to rivers and creeks have been converted to prawn farms. Even number of people from outside the area have purchased private land along the coast as well as along the creeks and converted the same in to aquaculture farms. They are discharging the untreated effluents from the farm to nearby rivers and creeks and thereby affecting the aquatic fauna and the mangroves.

Fishing

Fishing in the rivers and creeks by the surrounding local people is posing several adverse factors, the major being obstruction of migratory routs of fishes and blocking of free movement of crocodiles. Sometime, fishing by the local people leads to virtual closure of creeks, thereby the tidal inundation is hampered to a considerable extent.

Grazing of Cattle

An estimated 70,000 cattle depend on the forest and meadow located therein for grazing during cropping season. This puts pressure on mangrove vegetation especially *Avicennia* species.

Conservation Measures

To wean the poachers away from poaching, a massive awareness programme has been undertaken. The efforts are supplemented with the establishment of anti-poaching camps at strategic points. To encourage ecotourism, training camps for eco-guides and boat-man associations are being organised.

Habitat development inside the sanctuary is being done with funds received from MoEF of Govt. of India. These measures include raising up of plantations, digging and renovation of creeks and digging of ponds.

Management Strategy

The State Forest and Environment Department have taken several measures for conservation and management of this unique ecosystem and its rich biodiversity, with the support of the Ministry of E&F, Govt. of India. These measures include:

1. Building of Data base
2. Protection of salt water crocodiles and sea turtles
3. Protection of migratory waterfowl and other species prone to poaching for meat
4. Weed control
5. Restoration of the feeding and roosting habitat of water fowls
6. Creation of awareness about the values and functions of mangroves and wetland
7. Research and development activities
8. Community participation
9. Capacity building
10. Institutional strengthening
11. Promotion of eco-tourism

Conclusion

A long term multidisciplinary study on the value and resource of Bhitarkanika area along with formulation of broad based management plans are the key parameters to preserve and sustain this unique ecosystem.

Acknowledgments

I am thankful to the Principal Chief Conservator of Forests (Wildlife) and Chief Wildlife Warden, Orissa for encouragement and support.

References

Behura, B.K., 1999. *Bhitarkanika: The Wonderland of Orissa*. Nature and Wildlife Conservation Society of Orissa, Bhubaneswar, pp. 1–63.

Bustard, H.R., 1976. World's largest sea turtle rookery. *Tiger Paper*, 3(3): 25.

Dash, M.C. and Kar, C.S., 1989. *The Turtle Paradise Gahirmatha: An Ecological Analysis and Conservation Strategy*. M/s Interprint Publishers, New Delhi.

Daniel, J.C. and Hussain, S.A., 1975. A record saltwater crocodile, *Crocodylus porosus* (based on their study in Bhitarkanika during 1973). *J. Bombay Nat. Hist. Soc.*, 71(2): 309–312.

Das, G.P., 1951. *Forest Rules and Regulations of Kanika Raj*. Dewan of Kanika Raj, pp. 1–14.

Hussain, S.A., Mohapatra K.K. and Ali, S., 1984. Avifaunal profile of Chilika Lake: A case for conservation. *Bom. Nat. Hist. Soc. Technical Report*, 4: 46.

Kar, S.K. and Bustard, H.R., 1982. Occurrence partial albinisan in a wild population of Saltwater crocodile (*Crocodylus porosus* Schneider) in Bhitarkanika Wildlife Sanctuary, Orissa, India. *Brit. J. Herpetol.*, 6: 220–221.

Kar, S.K. and Bustard, H.R., 1991. Status of saltwater Crocodiles, *Crocodylus porosus* Schneider in Bhitarkanika Wildlife Sanctuary, Orissa, *J. Bombay Nat. Hist. Soc.*, 86(2): 141–150.

Kar, S.K. and Patnaik, S.K., 1999. *Status, Conservation and Future of Saltwater Crocodile in Orissa: Envis (Wildlife and Protected Areas)*. Wildlife Institute of India, Dehradun, 2(1): 24–28.

Kar, S.K., 1984. Conservation future of the Saltwater crocodile (*Crocodylus porosus* Schneider) in India. In: *Crocodiles Proc. of the 6th Working Meeting of the Crocodile Specialist Group of the IUCN/SSC*. Victoria Fall, Zimbabwe and St. Lucia Estuary, South Africa, 19–30 September, pp. 29–32.

Kar, S.K., 1991. Checklist of birds in Bhitarkanika Wildlife Sanctuary, Orissa, India. *Newsletter for Bird Watchers*, 31(11 and 12): 3–6.

Mohanty, S.C., Kar, C.S., Kar, S.K. and Singh, L.A.K., 2004. *Wild Orissa*. Wildlife Organisation, Forest Department, Govt. of Orissa, Bhubaneswar, pp. 1–81.

Mohanty, S.C., Singh, L.A.K., Kar, S.K., Kar, C.S., and Nair, M.V., 2006. *Nesting Animals of Orissa*. Wildlife Organisation, Forest Department, Govt. of Orissa, Bhubaneswar, pp.1–60.

Singh, L.A.K., Kar, S.K., Kar, C.S., Nair, M.V., Mishra, S.K. and Mohanty, S.C., 2007. *Wildlands of Orissa*. Wildlife Organisation, Forest Department, Govt. of Orissa, Bhubaneswar, pp. 1–68.

Chapter 15

Granulosa: Oocyte Interactions during Folliculogenesis in Mammals

☆ *R.K. Sharma and M.B. Sharma*

Introduction

The female germ cells in the ovaries of mammals differentiate and mature in close morphological association with surrounding somatic cells which are generally designated as nurse, follicle, nutrient, supporting or granulosa cells. These somatic cells show great diversity in their origin, proliferation, development, distribution, three-dimensional organization and functions during folliculogenesis. The association between germ cells and somatic granulosa cells persists throughout growth, differentiation, maturation and fertilization of the oocyte (Buccione *et al.*, 1990; Sharma and Sawhney, 1999). This association is crucial to both the growth and differentiation of the oocyte as well as the granulosa cells that ensures the ultimate success of oogenesis. The granulosa cells regulate oocyte growth, meiotic maturation and help in acquisition of competence for fertilization and embryogenesis. Oocyte on the other hand influences the development and function of granulosa cells. It is becoming evident that the development and differentiation of both somatic and germ cell components of the mammalian ovarian follicles are intimately associated and interdependent. The development concerning the intercellular associations between somatic and germ cells are reviewed in the article and vistas in the perspective development of this important area are pointed out, understanding of which in near future will enhance their applications in medicine, health and animal productivity.

Somatic Cell-Germ Cell Interaction during Development

The proximal region of the epiblast, adjacent to the extraembryonic ectoderm, the site of future amnion, is the progenitor of primordial germ cells in mouse (McLaren, 1983; Lawson, 1994). Within, the epiblast population that gives rise to germ cells, neither germline nor somatic lineages exists prior to germ line allocation. During the migration of germ cells to primitive gonad apart from fibronectin and TGF-β_1, gene products of *white-spotting* (W) and *steel* (Sl) from somatic cells influence the germ cells. Leukemia inhibitory factor, and basic fibroblast growth factor also have an impact on germ cells. All these signals from the surrounding somatic cells ensure the survival, control proliferation at specific stage and guide them to the developing gonad (McLaren, 1994).

The germ cells enter prophase of meiosis and pass through pachytene and get arrested at diplotene before or after birth. This meiotic arrest may lost for months and is regulated by the surrounding somatic cells (Eicher and Washburn, 1986). The decision as whether to enter meiosis or mitotic arrest is unrelated to the chromosomal constitution of the germ cells (Nagamine *et al.*, 1987; Palmer and Burgoyne, 1991; Zamboni and Upadhyay, 1983). The ectopic oocytes generally fail to grow because of failure of somatic cell interaction; lack of organizer follicle cells and absence of specific factors (Eicher and Washburn, 1986; McLaren, 1991). The germ cells in the male adrenal and some other mesonephric region enter meiosis before birth suggests that mouse germ cells enter meiosis spontaneously about 2 weeks after fertilization irrespective of their chromosomal constitution unless prevented by some short-range diffusible molecules emanating from somatic components of testis (Byskov and Hoyer, 1994). These germ cells develop as oocytes. It is evident that the somatic cells function to direct the development of germ cells to male or femaleness. Thus phenotypic sex of the germ cells depends on the sex of the gonad in which they are located (Eicher and Washburn, 1986).

Follicle or pregranulosa cells are mesonephric in origin (Guraya, 1998) form very thin and long processes partially or completely encompassing the oogonia. A large number of loose mesenchymal tissue cells and blood vessels separate the individual clusters consisting of germ cells and pregrenulosa cells (Makabe *et al.*, 1991). Onset of meiosis and breaking up of ovarian cords to form primordial follicles start simultaneously. The number of germ cells per cluster and their number vary in the same cluster as well. The germ cells are distinguishable from pregranulosa cells by their larger size and various cytological characteristics (Makabe *et al.*, 1991). The mouse and rabbit ovaries contain tightly packed oocytes in clusters and show synchronous stages of differentiation and more often they are held to one another by intercellular bridges. The pre-follicular cells continue to surround the germ cells that are near to being oocytes, even when intercellular bridges are eliminated and fragmentation of the nests occurs, these cells presumably play an active role in this fragmentation. In these foilicles desmosomes and small gap junctions can be observed among the germ cells and surrounding granulosa/follicular cells (Byskov, 1986). It is yet to be established as which specific pre follicular cell derived factor regulate this phenomenon and the molecular mechanism thereof need to be studied.

Pregranulosa cells form a close association with the germ cells leaving an intercellular space of less than 30 μm (Makabe *et al.*, 1991; Gondos, 1970). Plasma membranes of adjacent somatic cells form multiple interdigitating folds by the development of short, weak, cytoplasmic projections. Sometimes elongated projections of pregranulosa cells form indentation of the plasma membranes of the germ cells and may develop contact with nuclear envelope (Guraya, 2000). Long, thin cytoplasmic processes of pregranulosa cells containing cell organelles penetrate between adjacent germ cells, thus induct their complete separation (Makabe *et al.*, 1991). Pregranulosa cells possess well-developed cell organelles specific of a protein synthesizing cell which show changes in structure, number and distribution with the initiation of meiosis.

After initiation of meiosis during late fetal or early postnatal life, oocytes become arrested at prophase of the first meiotic division. Before initiation of meiosis the germ cells are called oogonia and thereafter they are referred as oocytes (Eicher and Washburn, 1986). The dictyate stage can last from a few weeks to several years depending on the species. Dictyate oocytes become enclosed in single layer of somatic pregranulosa cells and this unit is referred as a primordial follicle. In majority of the mammalian species most have attained this level of follicle formation when a female young one is borne. Groups of primordial oocytes enter a growth phase at species-specific time during which they remain arrested at diplotene stage. It is still controversial whether it is a genetically programmed event or requires an exogenous signal. As the germ cells acquire growth they become competent to resume meiosis in response to endogenous surge of LH or upon atretogenic degeneration of follicle (Buccione *et al.*, 1990). Concomitant with the initiation of oocyte growth the surrounding somatic cells start multiplication and flattered granulosa cells became cubiodal enclosing the oocytes in several layers. In fetal gonads, an antiapoptotic effect of KIT-KIT ligand interactions on primordial germ cells, oogonia and oocytes has been demonstrated. In the postnatal ovaries, the initiation of follicular growth from primordial pool and progression beyond the primary follicle stage appear to involve KIT-KIT ligand interactions (Zhao *et al.*, 2000). The initial phase of follicle growth and preantral follicle development seems to be independent of gonadotrophic hormones but lot many growth factors are required (Sharma and Guraya, 1990); multi laminar granulosa cells subsequently acquire extracellular space which expands and antrum is formed in response to endogenous hormonal interplay. Seemingly alike looking granulosa cells are divisible into two groups the mural cells lying at the periphery and cumulus cells that enclose oocyte. Each population and its subsequent groups perform very specific functions (Anderson and Albertini, 1976). The factors regulating this structural and functional portioning of granulosa cells are still to be worked out. The molecular mechanism of this process of differentiation would be rewarding.

Role of Granulosa Cells in Oocyte Growth

Oocytes are coupled to the surrounding granulose cells throughout folliculogenesis by specific membrane specialization known as gap junctions (Buccione *et al.*, 1990; Brower and Schultz, 1982). Oocyte growth requires gap junctional communication with the granulosa cells and fail to grow if this communication is disrupted (Eppig, 1979; Eppig, 1994; Herlands and Schultz, 1984). These gap junctions function to transfer nutrients metabolites and low molecular weight growth factors for normal oocyte development (Colonna *et al.*, 1989; Lawrence *et al.*, 1978). The highly specialized membrane connections mediate the transfer of

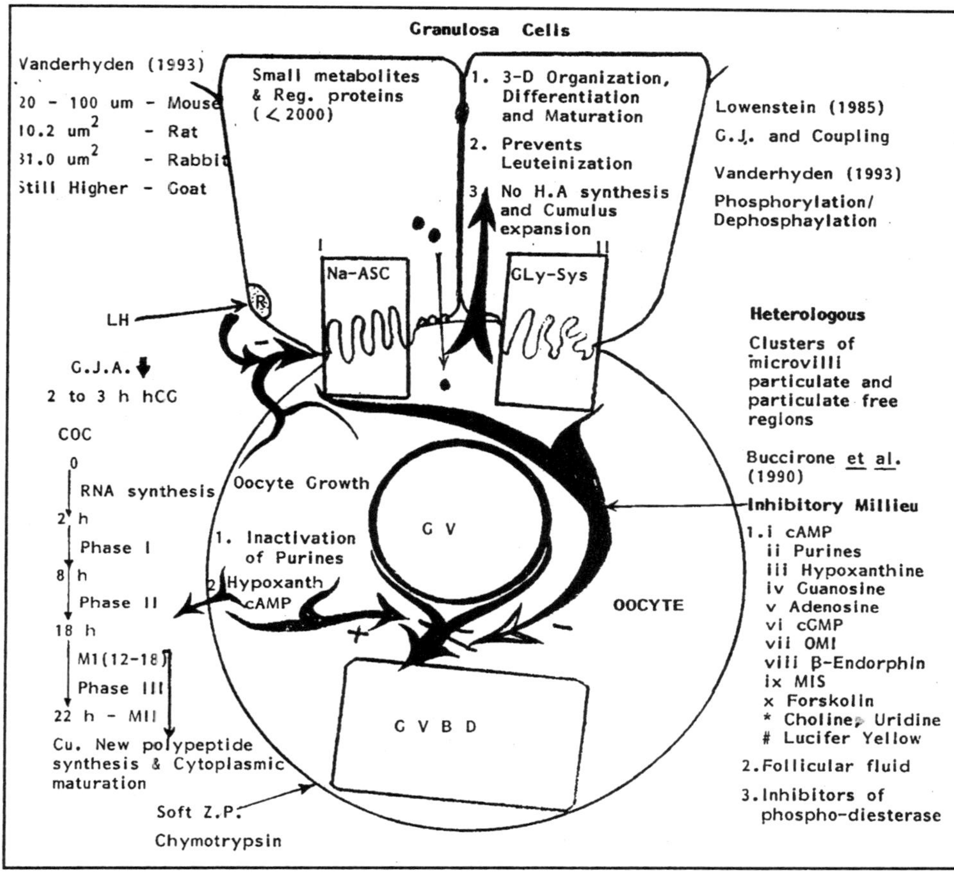

Figure 15.1: Granulosa Oocyte Interaction

small (<2000MW) metabolites and regulatory growth factors from one cell to another (Bachvanrova *et al.*, 1980). Granulosa cells, mural and cumulus and the oocyte, therefore, form a gap junction-mediated syncytium.

During folliculogenesis, the oocyte diameter increases from 15–20 μm to 100–120 μm in a short period of time. For attaining this increase gap-junctional association between granulosa cells and oocyte is essential and this coupling must be maintained or reestablished for oocyte growth *in vitro* (Colonna, 1983). These junctional complexes are involved in the transport of small metabolites, energy substrates, nucleotides, and amino acids from granulosa cells into the oocyte (Eppig, 1979; Cross and Brinster, 1974; Haghigha and Winkle, 1990; Moor *et al.*, 1980; Colonna *et al.*, 1983). The oolemma is equipped with two exchange operated systems (1) the Na^+ dependent ASC system for the transport of amino acids having short linear or polar chains of *alanine, serine* or *cysteine* and (2) glycine-dependent system for the exchange of amino acids large aliphatic or aromatic side chains for transport of glycine (Moor *et al.*, 1980; Winkle, 1988; Herlands and Schultz, 1984). *In vitro* studies have demonstrated the reestablishment of intercellular communication between granulosa cells and germ cells and growth oocytes of such cultures were better than communication-incompetent cell lines (Herlands and Schultz, 1984; Colonna *et al.*, 1989). In fact, it is only cell granulosa cells not other somatic cell types that regulate the level of synthesis of proteins and the pattern of protein phosphorylation in growing oocytes thus directly regulating their metabolism Canipari *et al.*, 1984).

Plasma membrane of the adjacent granulosa cells are closely applied leaving an intercellular distance of less than 300A°. The membrane of the adjacent cells are thrown into multiple folds to form short, narrow cytoplasmic projections entering into the neighbouring granulosa cells as well as into the oocyte. Granulosa cells play a crucial role in supplying nutrients during oocyte growth. Even though oocyte growth is strictly dependent on intercellular communication with the surrounding granulosa cells, it appears that acquisition of meiotic competence is independent of both oocyte growth and gap junctional association with granulosa cells (Skinner and Parrott, 1994), suggesting that this might be an autonomous oocyte programme. Among different functions performed by granulosa cells (McGaughey *et al.*, 1990), gap junctions are described as the sites of molecular and ionic exchange and metabolic coupling functions between the cells (Zamboni, 1974), which serve to coordinate cell functions in many tissues. Regulation of gap junctions is an important area for future investigations.

Cellular associations are represented by two types of junctional devices. Tight junctions and spot and belt desmosomes serve to seal the intercellular spaces and maintain intercellular environment. Belt desmosomes contain actin myofilaments in the form of a band that girdles the inner surface of the cell membrane. Spot desmosomes are localized circular areas of mechanical attachment of keratin tonofilaments. It the early antral follicles, the apposed cell membrane in the areas of contact associate by means of desmosomes (Anderson, 1979; Amsterdam and Lindner, 1979a; Amsterdam and Lindner 1979b; Zamboni). As the follicle develops, the number and size of individual junctions increase and a large variety of shapes are seen (Albertini and Fawcett, 1975; Bjevsing, 1978; Albertini, 1980; Espay and Stutts, 1972).

Gap junctions or nexuses are highly differentiated portions of plasma membrane between adjoining granulosa cells and appear as minute attachment sites. The long zones of contact (abutment nexuses) and spherical inclusions (annular nexuses) in rabbit follicle have also been reported (Bjevsing, 1978). One important feature of granulosa cell gap junction is the presence of intracellular vesicles of gap junction. These circular sectioned profiles were earlier designated as annular gap junctions and are formed from deep invagination of membrane localised gap junctions which are pinched off during internalization (Albertini and Fawcett, 1975; Albertini, 1980; Espay and Stutts, 1972; Merk *et al.*, 1973). The segments of granulose nexuses frequently invaginate into either of the adjoining cells and become isolated cytoplasmic entities. The changes in the abundance and size of surface nexuses as well as internalized nexuses in growing and mature ovarian follicles during the ovulatory process have been recorded (Albertini and Anderson, 1974). Mature follicles have greater number of surface nexuses and also more annular gap junctions or interiorized nexuses as compared to immature follicles (Amsterdam and Lindner, 1979a). The biological significance of the internalization (phagocytosis) of certain gap junctions is still not known. This may represent a mechanism for clearing the surface of "used" receptors, provided there is reformation of the gap junctions (Coons and Espey, 1979; Guraya, 1985).

Gap junctions are the channels for cell-to-cell communication and are made up of protein belonging to connexin family (Park-Sarge and Mayo). It is believed that LH induced loss of gap junction is the trigger for resumption of oocyte meiotic maturation (Park-Sarge and Mayo). Sutovsky *et al.* (1993) employed various methods of immuno-fluorescence, immuno-electron microscopy, micro injections of fluorescent dyes like Lucifer yellow, quantitative freeze fracture analysis and *in vitro* culture experiments of cattle oocyte cumulus complexes. Their result indicated that connexin-43 positive gap junctions present between the corona radiata cells and oolemma, disappears after 6 hours of culture, as they found that cumulus cells were unable to transfer Lucifer yellow to the oocyte. Synthesis of another gap junction protein connexin-32 starts in the oocyte cytoplasm after 6 hours of culture. The cytoskeletal structure of gap junctions opens a new world for investigation which is not yet fully explored. The gap junction channels are formed of two hemichannels (connexons). Each connexon is composed of 6 identical sub-units or connexins that form the central pore. They found the expression of four connexin genes (Cx-26, Cx-303, Cx-32 and Cx-43) in porcine ovaries (Sutovsky *et al.*, 1993). Till date only fragmentary information is available and a systematic studies on its molecular aspects need to be undertaken.

The gap junctions of granulosa cells are believed to be responsible for many biochemical events associated with growth, maturation and ovulation. Autoradiographic analysis have shown that only the thecal cells and mural granulosa cells possess LH receptors (Itahana *et al.*, 1996; Amsterdam *et al.*, 1975) and the gap junctions between granulosa cells and between cumulus cells and oocyte serve to propagate an LH-initiated signal towards the interior or the follicle, as the cumulus cells do not possess these receptors (Espay and Stutts, 1972; Itahana *et al.*, 1996; Oxbeny and Greenwald, 1982; Amsterdam *et al.* 1976).

The function and structure of gap junctions are regulated by numerous factors including hormone growth factors and intracellular regulators Only Cyclic AMP is recognized as an important intracellular regulator of gap junction in several tissues (Amsterdam *et al.*, 1979a). The gap junction between the oocyte and the follicle cells are extremely small in size, unlike the gap junctions between the granulosa cells (Espay and Stutts, 1972; Merk *et al.*, 1973). The presence of gap junctions between granulosa cells and between cumulus cells and the oocyte has suggested that these contact areas mediate the exchange of regulatory factors required for ovulation and luteinization (Linwin and Zampighi, 1980). Gap junctions have shown to be highly permeable to small molecules and to facilitate electrical coupling between cells (Sheridan, 1971; Gilula *et al.*, 1972; Albertini, 1992). Various electrophysiological and dye passage experiments supports this suggestion and demonstrate that ions and small molecular weight substances are freely exchanged through granulosa cell gap junctions.

The growth of the follicle is regulated by pituitary FSH and intra ovarian factors such as activins. The cells in the follicle like structure are connected by gap junctions formed of proteins called connexin 43 and 32. The reorganization of follicle is inhibited by inhibitors of gap junctions. Recently the effect of Lindane (g-hexachlorocyclohexane) an inhibitor of gap junction formation was studied (Li and Mather, 1997). Activin and FSH were added to primary follicle cultures along with diluted solution of Lindane. It was observed no change in the cell number and no formation of gap junctions. Gap junctions in the granulosa cells are believed to be involved in cell-to-cell attachment, communication and transport of hormones and nutrients within the follicle (Park-Sarge and Mayo).The membrane specializations of granulosa cells are also important due to the absence of vascular system within the granulosa cells. Zamboni (Anderson, 1979; Zamboni; Zambani, 1996) believed that the granulosa cell process and junctional devices in growing follicles ensure the isolation and stability of the follicular environment during critical phases of oocyte maturation. During follicular growth FSH stimulation preferentially promotes increases in surface gap junctions while internalization of surface junctions increased during the later phases of follicular growth (Burghardt and Matheson, 1982).

Gap junctions in granulosa cells are characterized by a range of packing order or P-face particles or E-face pits. Exposure of O.S.M. sucrose containing 1.8 m EGTA (Ethylene-bis-(β-aminoethyl ether)-NN′-tetraacetic acid) for 1 min results in a consistently close packing between granulosa cells. Initially gap junctions lose their regular shape and fragment into numerous tightly aggregates of P-face or E-pits repeated by unreplicated P-membrane (Bagovandoss and Midgley, 1981). Slab formation in granulosa cells of rat preovulatory follicles *in vitro* and *in vivo* was investigation. E_2* FSH primed adluminal granulosa cells showed blebs and 90 per cent were observed when culture was supplemented by serum. Fibronectin serum components act to induce micro-filament formation, cAMP inhibited bleb formation in granulosa cells both *in vivo* and *in vitro* (Campbell and Albertini, 1981). However, no difference in number of close junctions in perifollicular contractile tissue of rat during preovulatory period was observed (Capps *et al.*, 1981). Tight junctions of intertwining ridges were found on P-faces and complementary furrows on E-faces. Typical gap junctions in the form of clusters of particles on P-faces and complementary pits on E-faces were also observed associated with tight junctional elements. Sometimes, the junctions are the outermost elements of junctional complexes. These observations explain date previously obtained in tracer experiments with lanthanum (Riberio, 1983). The influence of altered carbohydrate structure on the surface, distribution, number and turnover of plasma membrane glycoprotein in chinese hamster ovary (CHO) cells by comparing the 3 lines that are resistant to the cytotoxic effects of wheat germ agglutinin (WGA) with parental CHO cells was studied (Fitzgerald *et al.*, 1984). Glycoproteins investigated were the members of a group of high MW acidic glycoproteins (HMWAG). One parental cell line represent the major surface component that became labelled by lactoperoxidase-catalyzed iodiation. They are only plasma membrane glycoprotein that bind to WGA.

The mutant likewise also possess indizable surface glycoprotein of high MW but these were less acidic. Cells generally do not bind to WGA. The surface glycoproteins of mutant cells had altered carbohydrate structure. Mouse antisera against the HMWAG, however, bound equally to both cell lines. By electron microscopic studies no change in topography of glycoprotein was observed in HMWAG (Fitzgerald *et al.*, 1984). The molecular alteration at genetic level need to be explored further to determine their specific significance in cell-to-cell interactions.

The ultrastructural studies on the extra oocyte cumulus components in hamster revealed that it consisted of granules and filaments of moderate electron density. The structural organization of zona pellucida revealed occasional small deposits on the oolema. The extra-cellular matrix observed between the cells of cumulus and corona layers extended into the outer one-third of zona. The granules and filamentous matrix was removed from the cumulus and corona, and pore of zona pellucida by brief treatment of hyaluronidase. Extra-cellular matrix components of oviductal oocyte-cumulus complex from hamsters and mice appear similar to oocyte follicle complexes removed from follicles of hamsters shortly before ovulation (Zaibot and Dicar, 1984). The metabolism of endogeneously labelled proteoglycans revealed that 90 per cent of the newly synthesized proteoglycans are transported to cell surface with a medium transit time (t/2) of 13 minutes. The membrane bound heparan sulphate proteoglycans (HSPG) is lost from the surface either by release into medium (30 per cent) (t/2 4h) or by internalization (t/2 4h) (40 per cent). Internalized HSPG which does not recycle to cell surface is degraded by two major pathways. In pathway I, 60 per cent of internalized MSPG migrates to lysosomes (t/2 30 minutes) where it is rapidly degraded releasing 30 g Chloroquine is inhibitory to this pathway. In pathway II, 40 per cent of internalized HSPG is first subjected to extensive proteolysis and limited endoglycosidic degradation yielding single haparan sulphate chain about one-third of original length (t/2 30 minutes). This is further degraded to 1/4 to 1/5 original HS (t/2 30–60 minutes). This step is inhibited by the chloroquine (Yanagishita and Hascall, 1984b). Dermatan sulphate DS-I is directly transported to cell surface from where it is released into medium (t/2 4–6 h). DS-II follows both pathways of glycolytic and internalization (Yanagishita and Hascall, 1984b). These cellular surface modifications need to be explored further so as to ascertain their specific role in cell-to-cell interact.

Granulosa cells produce a factor known as *kit*-ligand (KL) or *Steel factor* (SLF) from Steel (Sl) locus. The receptors for this KL *c-kit* are expressed by *While Spotting* (W) locus of the oocytes (Manova *et al.*, 1990; Manova *et al.*, 1993; Horie *et al.*, 1991). The production of KL is low in very small follicles but high in follicles with 3 layers of granulosa cells (Horie *et al.*, 1991). KL also appears within the oocytes of larger preantral follicles. Since oocytes do not produce KL mRNA soluble KL produced by granulosa cells may be internalized by receptor-mediated endocytosis. This *Steel* gene product KL of granulosa cells and its receptor *c-kit* from *White spotting* of oocyte represent an excellent paracrine regulating mechanism. Mutations in *steel* or *W* Loci result in a drastic decline in the number of germ cells in the ovary however small number of follicles apparently appearing normal have limited fertility (Brannan *et al.*, 1992). The *steel panda* mutation reduces the number of germ cells and arrests development of oocyte and follicles as well (Huang *et al.*, 1993). The production of KL is greatly reduced in these mutants oocyte growth is initiated in these ovaries, but stops before zona pellucida formation (Huang *et al.*, 1993). During early folliculogenesis, KIT together with KIT ligand controls oocyte growth and theca cell differentiation and protects the preantral follicles from apoptosis formation of an antral cavity requires a functional KIT-KIT ligand system. In large follicles the KIT-KIT ligand interaction modulates the ability of the oocyte to undergo cytoplasmic maturation and helps to maximize thecal androgen out put. Hence many steps of oogenesis and folliculogenesis appear to be at least in past, controlled by paracrine interactions between these two proteins (Zhao *et al.*, 2000). Recent studies suggest that intra follicular factors are involved in follicle development *in vitro* which especially at the early follicular stage plays a positive role in terms of growth as well as survival as analyzed by TEM (Pawshe *et al.*, 1998).

The Role of Granulosa Cells in Regulation of Oocyte Maturation

The fully grown mammalian oocytes on isolation from antral follicles undergo spontaneous germinal vesicle breakdown (GVB) but growing oocytes from the pre antral follicles fail to do so (Edwards, 1965; Pincus and Enzmann, 1935; Pincus and Saunders, 1939). The competence to undergo GVB is acquired after completion of growth (Sorensen and Wassarman, 1976; Szybek, 1972). The oocyte growth is largely dependent on the intercellular communication with the surrounding granulosa cells. However, the acquisition of meiotic competence is independent of gap-junctional association of granulosa cells as well as oocyte growth and is possibly an autonomous oocyte programme (Canipari *et al.*, 1984). The meiosis-arresting effect of the follicular environment is mediated by the transfer of substances from the granulosa cells to the oocyte through gap junctions (Dekel, 1988; Leibfried and First, 1980a; Racowsky and Baldwin, 1989; Tsafriri *et al.*, 1977). cAMP, hypoxanthine, guanosine, adenosine cGMP, a low molecular weight protein oocyte maturation inhibitor (OMI), β-endorphin and Mullarian inhibitory substance (MIS) have been implicated in the maintenance of meiotic arrest (Bornslager *et al.*, 1986; Cho *et al.*, 1974; Dekel and Beers, 1978; Magnusson and Hillensjo, 1977; Downs *et al.*, 1988; Eppig, 1985; Salustri *et al.*, 1988; Hubbard and Terranova, 1982; Tornell *et al.*, 1990; Tsafriri and Bar-Ami, 1982a and b; O.W.S., 1990; Takahashi *et al.*, 1986).

Since oocyte is devoid of LH-receptors, LH surge therefore cannot stimulate meiotic maturation directly obviously, granulosa cells-mediate the resumption of maturation in oocyte after gonadotrophin stimulation. The gonadotrophin-induce a decline in intercellular communication to the oocyte through gap junctions (Dekel, 1988; Leibfried and First, 1980a; Racowsky and Baldwin, 1989; Tsafriri and Channing, 1975) thus depriving the oocyte of the factors that maintain meiotic arrest (Dekel and Kraicer, 1978; Larsen *et al.*; Larsen *et al.*, 1987; Welt and Larsen, 1989) or *de navo* synthesis of factors that negate inhibitory milieu of the follicle (Downs *et al.*, 1988; Eppig and Downs, 1987). The morphometric observations on the cumulus-oocyte interacting surfaces support the hypothesis that LH-surge induced maturation is caused by a gap junction loss that result in a reduction in the meiosis arresting substances delivered to the oocyte (Wert and Larsen, 1990). Purines, c-AMP or both probably participate in granulosa cell-mediated meiotic arrest (Eppig and Downs, 1988; Schultz *et al.*; Fagbohun, 1991). The meiotic arrest is initially the result of autonomously within the oocyte, but later when oocyte germs and has become competent to resume meiosis, arrest is maintained by substances communicated by the gap junctions to the oocyte (Lawrence *et al.*, 1978). Two mechanisms have been proposed as how gonadotrophins induce resumption of meiosis. The first group of investigators stress on the reduction in gap junctional communication between the oocyte cumulus cell complex and membrana granulosa reducing the transmission of meiosis arresting substances (Dekel, 1988; Wert and Larsen, 1989; Wert and Larsen, 1990). The other group has proposed that gonadotrophins induce the generation of a positive maturation-inducing signal within the granulosa cells and transmission of this signal requires gap junctional communication (Downs *et al.*, 1988; Bilodeau *et al.*, 1993; Thibault *et al.*, 1987). The presence of granulosa cells is absolutely essential at the start of final oocyte maturation in sheep, bovine, pig and rabbit. These cells regulate the polypeptide synthesis which render the ooplasm competent to assume meiosis. Recently the role of gonadotrophins, steroid hormones, cyclic nucleotides, prostaglandins, GnRH, FF inhibitors, ions, electrolytes and various other growth factors have been implicated in meiotic arrest in oocyte. Oocyte meiotic inhibitor and purine are reversible in their action and act as physiological regulators of oocyte maturation (Colonna *et al.*, 1989; Eppig *et al.*, 1996), hypoxanthine is also present in bovine follicular fluids. There exists a synergism of c-AMP and OMI (Sirotkin *et al.*, 1988). They have studied the effect on of IGF-I and tyrosine kinase blockers on pig oocytes *in vitro* and found that IGF-I stimulated meiotic maturation in cumulus enclosed and free oocytes (Stock *et al.*, 1997). Inhibin A, and activin A secreted by granulosa cells play important roles during final stages of oocyte maturation (Eppig, 1989). The gonadotrophin induced interruption of cumulus-oocyte communicate decreases c-AMP concentration within the oocyte and trigger onset of maturation (Andreani *et al.*, 1997; Leibfried *et al.*, 1989). The decrease in cAMP is also attributed by enhanced phosphodiestrase activity and a decrease of adenylate cyclase activity.

The production of guanosine compounds by hypoxanthine rather than hypoxanthine alone increase concentration of cAMP within the oocytes the site of action of the purines is on the granulosa cells which transfer then on to the oocyte through intercytoplasmic channels which function as conduits for OMI, cAMP and other factors as well (Buccione *et al.*, 1990; El-Fouly *et al.*, 1977). 1 GF-I stimulates oocyte maturation in a dose-dependent manner, with maximal effect at a dose 100 mg/ml. A positive interaction of 1 GF-I and FSH in the presence or absence of granulosa cells was observed on meiotic maturation, DNA synthesis, protein synthesis and steroidogenesis is mainly caused by the increase of 1 GF-I receptors on granulosa cells by FSH 150.

The Role of Oocyte in Granulosa Cell Functions

The removal of oocyte from the follicle induced luteinization in pig and rabbit (Nekola and Nalbandov, 1971; Hilliensio *et al.*, 1981). However, subsequently studies have questioned the concept (Linder *et al.*, 1974; Muller and Urban, 1981) and it needs to be explored further. The oocyte-specific antiserum raised in isogenetic male mice against ovaries prevented granulosa-cell reaggregation *in vitro* thereby indicating that oocyte antigen might participate in follicular organization (Salustri *et al.*, 1989).

In the preovulatory follicles, the cumulus cells produce hyaluronic acid in response to gonadotrophic and initiate cumulus expansion or mucification (Dekel and Kraicer, 1978; Eppig, 1994; Bortolussi *et al.*, 1979), while mural granulosa cells undergo luteinization. The cumulus and mural cells differ in LH-receptors, m-RNA coding for LH-receptors and steroidogenic potential, and mRNAs for P450, IGF-I and Mullerian Inhibiting Substances (MIS) and other factors (Itahana *et al.*, 1996; Peng *et al.*, 1991; Sato and Ishibashi, 1977; Hiliensjo *et al.*, 1987; Zoller and Weisz, 1979; Zoller and Weisz, 1979b; Zlotkin *et al.*, 1986; Zhou *et al.*, 1991; Ueno *et al.*, 1989; Erickson *et al.*, 1985).

Cumulus expansion starts after preovulatory LH surge and it involves synthesis and secretion of hyaluronic acid (non-sulphated glycosaminoglycans) by the cumulus cells (Chen *et al.*, 1993). Hyaluronic acid after hydration expands and the intercellular spaces between cumulus cells increase and cells subsequently disperse in this matrix. This expansion is a must for ovulation, ovum uptake, sperm sequestering and fertilization (Mahi-Brown and Yanagimachi, 1983; Bedford and Kim, 1993; Salustri *et al.*, 1990). The LH, and FSH fail to stimulate cumulus expansion *in vitro* in the stripped cumuli oophori in the absence of oocytes, thereby suggesting the involvement of oocyte in cumulus expansion (Eppig *et al.*, 1993; Vanderhyden *et al.*, 1990). Medium conditioned by cumulus cell-free oocytes triggered the synthesis of hyaluronic acid and cumulus expansion *in vitro* in response to FSH (Vanderhyden *et al.*, 1992; Prochazka *et al.*, 1991; Takaoka *et al.*, 1985).

The cell division occurs more frequently in the granulosa cell population adjacent to the oocyte than in the distal cells (Hirshifield, 1986; Driancourt *et al.*, 2000). The extirpation of oocyte from the granulosa oocyte complex reduces the rate of proliferation, which can be restored by oocyte-conditioned medium. FSH reduces the cumulus proliferation as it promotes expansion (Takaoka *et al.*, 1985). The results indicate that oocyte produces granulosa proliferating factor that acts by paracrine regulatory mechanism and stimulates the granulosa cell proliferation (Lawrence *et al.*, 1978).

This oocyte-derived factor is referred as "cumulus expansion enabling factor" is heat labile protease sensitive having molecular weight less than 300 K Da (Vanderhyden *et al.*, 1990). This factor is expressed at the time of acquisition of meiotic competence in mouse (Vanderhyden *et al.*, 1992), and stops after fertilization. The somatic cell-conditioned medium also promotes the ability to secrete the cumulus expansion enabling factor (Prochazka *et al.*, 1991). The granulosa cells produce a paracrine factor that suppresses the synthesis and secretion of cumulus enabling factor by the oocyte until surge. This factor may be degraded or diluted to ineffective concentrations by increased follicular fluid volume after surge, thus stimulating expansion enabling factor by oocyte (Lawrence *et al.*, 1978). However, in pig cumuli oophori expand upon FSH stimulation even in the absence of oocyte (Takaoka *et al.*, 1985). The interaction of hyaluronic acid (HA) and proteins of inter-alpha-inhibitor family plays a critical role in organization and stabilization of the expanding cumulus extra cellular matrix (cECM) the cECM of COCs which expand within the intact follicles are more elastic and resistant to shear stress than those stabilized *in vitro.* Western blot analysis show that only the heavy chains of inter-alpha-inhibitor are incorporated into cECM and seems to be covalently linked to HA after stabilization *in vivo*, while intact inter-alpha-inhibitor is bound in the HA-enriched cECM by a non-covalent mechanism *in vitro* stabilized COCs. It is speculated that factor(s) secreted by granulosa cells within the follicle may catalyse a transestrification resulting in an exchange of chondroitin sulphate with HA at the heavy chain/chondrotin sulphate junction followed by release of chondroitin sulphate–bikunin into follicular fluid (Chen *et al.*, 1993). It becomes evident from the available information that granulosa-oocyte interaction are crucial for the success of oogenesis. Research on its molecular biology involving chronology of expression of specific GFs and their role in modulation of cell structure, interacting sites, general contour is imperative. The mechanism of coupling between granulosa-cells-oocyte its gating mechanism would be rewarding. Role of oocyte in 3D organization of different of granulosa cells and their specific sub populations need to be explored further. The cells forming the cumulus under the normal preovulatory stimuli undergo a change, which alters their ability to grow *in vitro*. The mechanism of this change is not known and further studies are required to analyzed as the cumulus cells apparently have a very definite life span after they are exposed to the influence of specific preovulatory hormones. During this phase the number of pycnotic cells increases and the oocytes produce soluble factors, which regulate a wide array of processes during follicular development, including cumulus expansion in the preovulatory period. These authors have discussed the similarities and differences in sequences, expression and function of the oocyte expressed TGF beta family members with respect to regulating folliculogenesis. The TGF-alpha and EGF receptor are expressed in the primordial to antral follicles, indicating a role of TGF-alpha in regulating follicular development through binding to the EGF receptor. Freeze thawing does not substantially charge immunoreactivities for TGF-alpha, EGF and EGF receptor in frozen ovarian tissue. These authors have not observed any significant difference in the immuno-histochemical staining for EGF receptor in ovarian tissue before and after cryopreservations. The results obtained are consistent with the suggestion that growth differentiation factor-9 (GDF-9) and GDF-9B may regulate human folliculogenesis in a manner specific to the ovary. Garridi *et al.* have demonstrated that the follicular environment is different in cases with endometriosis and suggested that infertility in patients with endometriosis may be related to changes within the oocyte, which in turn, result in embryos of lower quality, and with a reduced ability to implant.

The role of granulosa cell derived peptides in early embryogenesis and guiding further development need to be confirmed and established. Identification, purification of factors and their role on paracrine cells need to be reaffirmed on other cell types. Oocyte maturation, fertilization and early development largely depends on these interactions and an exhaustive research in these areas shall be directly affecting the production and thus economy

Summary

The association between germ cells and somatic granulosa/nurse cells is crucial to the growth, different maturation and fertilization of the oocyte as well as 3D differentiation of granulosa cells that ensures ultimate success of oogenesis. During early development, the survival, proliferation and migration of primodial germ cells 10 developing gonad is directed by fibronectin, TGF-β_1 leaokemia inhibiting factor, FGF and gene products of white spotting and steel from somatic cells formation of pregranulosa cells and their subsequent organization is largely dependent on oocyte-granulosa interactions through gap junction. The coupling in turn provide nutrients, metabolite and low MW GFs for normal development of oocyte, Gap junctions, annual nexuses, tight junctions spot and belt desmosomes. Na dependent ASC system and glycine dependent system are involved in this process. The cellular surface modifications endose their role in interaction. Granulosa cells produce kit ligend (KL) or steel factor (SLF) from steel (Sl) locus. The receptors for this KL-kit are expressed by white spotting (W) locus of oocytes. KIT-KIT ligend interactions are involved in initiation of follicle growth, and have antiapoptotic effect. The differentiation of granulosa cells and cytoplasmic maturation of oocyte both are influenced by this interaction. The granulosa cell derived factors like cAMP, hypoxanthine guanosine, cGMP, OMI β-endrophin and MIS affect the oocyte maturation. IGF-I, tyrosin kinase, inhibin A and activin A play important role in normal development growth and maturation of oocyte. Oocyte synthesizes and secretes factor/s that induce proliferation and three-dimensional organization of granulosa cells. Cumulus expansion enabling factors are also secreted by the oocyte. The physiological significance of these interactions has been discussed and future perspectives have been spelt out.

References

Albertini, D.F. and Anderson, E., 1974. *J. Cell Biol.*, 63: 234.

Albertini, D.F., 1980. In: *Biology of the Ovary Nijholt*, (Eds.) P.M. Motta and E.S.E. Hafez. The Hague, pp. 138–149.

Albertini, D.F., 1992. *Bioarrays*, 14: 97.

Albertini, D.F., Fawcett, D.W. and Olds, P.J., 1975., *Tiss. Cell.*, 7: 389.

Amsterdam, A. and Lindner, H.R., 1979. In: *Ovarian Felliculae Development and Function*, (Eds.) A.R. Midgley Jr and W.A. Sadler. Raven Press, New York, pp. 91–105.

Amsterdam, A. and Lindner, H.R., 1979. In: *Ovarian Felliculae Development and Function*, (Eds.) A.R. Midgley Jr and W.A. Sadler. Raven Press, New York, pp. 137.

Amsterdam, A., Josephs, R., Liebermann, E. and Lindner, H.R., 1976. *J. Cell. Sci.*, 21: 93.

Amsterdam, A., Kohen, F., Nimrod, A. and Linder, H.R., 1979a. In: *Ovarian Follicular and Corpus Luterum Function*, (Eds.) C.P. Channing, J. Marshand and W.A. Sadler. Plenum Press, Now York, pp. 69–75.

Amsterdam, A., Kooh, Y., Liebermann, E. and Lindner, H.R., 1975. *J. Cell. Biol.*, 67: 864.

Anderson, E. and Albertini, 1976. *J. Cell. Biol.*, 71: 680.

Anderson, E., 1979. In: *Ovarian Felliculae Development and Function*, (Eds.) A.R. Midgley Jr and W.A. Sadler. Raven Press, New York, pp. 91–105.

Andream, C.A., Diotallevi, L., Lazzarin, N., Pierro, E., Giannini, P., Lanzonc, A., Capitanio, P. and Mancuso, S., 1997. *Hum. Rep.*, 12: 89.

Bachvanrova, R., Baran, M.M. and Tehlum, A., 1980. *J. Exp. Zool.*, 211: 159.

Bagovandoss, P. and Midgley, A.R.I., 1981. *Tissue Cell*, 13: 669.

Bedford, J.M. and Kim, H.R., 1993. *J. Exp. Zool.*, 265: 321.

Billig, H., Tornell, J., Carlsson, B., Hillensjo, T. and Ahren, K., 1988. In: *Development and Function of the Reproductive Organs, Vol. 2: Ares*, (Eds.) M. Parvinen, I. Huhtaniemi and L.J. Pelliniemi. Serono Symposia, Rome, Italy, pp. 161: 170.

Bilodeau, S., Fortier, M.A. and Sirard, M.A., 1993. *J. Reprod. Fert.*, 97: 5.

Bjevsing, L., 1978. In: *The Vertebrate Ovary*, (Ed.) R.E. Jones. Academic Press, London, New York, pp. 303–309.

Bomslager, E.A., Matter, P. and Schultz, R.M., 1986. *Dev. Biol.*, 114: 453.

Bortolussi, M., Marini, G. and Reolon, M.L., 1979. *Cell and Tiss. Res.*, 197: 213.

Brannan, C.I., Bedell, M.A., Resnick, J.L., Eppig, J.J., Handel, M.A., Williams, D.E., Lyman, S.D., Donovan, P.I., Jenkins, N.A. and Coeland, N.G., 1992. *Gene. Dev.*, 6: 1832.

Brower, P. and Schultz, R., 1982. *Devl. Biol.*, 90: 144.

Buccione, R., Schroeder, A.C. and Eppig, J.J., 1990. *Biol. Reprod.*, 43: 543.

Burghardt, R.C. and Matheson, R.L., 1982. *Dev. Biol.*, 94: 206.

Byskov, A.G. and Hoyer, P.E., 1994. In: *The Physiology of Reproduction, Vol. 1*, (Eds.) E Knobil and J.D. Neill. Raven Press, New York, pp. 265.

Byskov, A.G., 1986. *Physiol. Rev.*, 66: 71.

Campbell, K.L. and Albertini, D.F., 1981. *Tissue Cell*, 13: 651.

Canipari, R., Palombi, F., Raminucci, M. and Mangia, F., 1984. *Devl. Biol.*, 102: 519.

Canipari, R., Palombi, F., Raminucci, M. and Mangia, F., 1984. *Dev. Biol.*, 102: 519.

Capps, M.L., Lawrence, J.E. (Jr) and Burden, H.W., 1981. *Tissue Res.*, 219: 133.

Chen, L., Russell, P.T. and Larsen, W.I., 1993. *Mol. Reprod. Dev.*, 34: 87.

Cho, W.K., Stem, S., Biggers, J.D., 1974. *J. Exp. Zool.*, 87: 383.

Colonna, R. and Mangia, F., 1983. *Biol. Reprod.*, 28: 797.

Colonna, R., Cecconi, S., Buccione, R. and Mangia, F., 1983. *Cell Biol. Int. Rep.*, 7: 1007.

Colonna, R., Cecconi, S., Tatone, C., Mangia, F. and Buccione, R., 1989. *Dev. Biol.*, 133: 305.

Colonna, R., Cecconi, S., Tatone, C., Mangia, R. and Buccione, R., 1989. *Dev. Biol.*, 133: 305.

Coons, L.W. and Espey, L.L., 1979. *J. Cell. Biol.*, 74: 321.

Cross, P.O. and Brinster, R.L., 1974. *Exp. Cell. Res.*, 86: 43.

Dekel, N. and Beers, W.H., 1978. *Proc. Natl. Acad. Sci.*, USA, 75: 4369.

Dekel, N. and Kraicer, P.F., 1978. *Endocrinology*, 102: 1797.

Dekel, N., 1988. In: *Cell to Cell Communication in Endocrinology*, (Ed.) F. Piva. Serono Symposia, Raven Press, New York, 82: 183–194.

Deno, S., Takahashi, M., Manganaro, T.F., Rangin, R.C. and Donahoe, P.K., 1989. *Endocrinology*, 124: 1000.

Downs, S.M., Daniel, S.A.J. and Eppig, J.J., 1988. *J. Exp Zool.*, 245: 89.

Driancourt, M.A., Reynaud, K., Corturindt, R. and Smitz, J., 2000. *Rev. Reprod.*, 5: 143.

Edwards, R.G., 1965. *Nature*, 208: 349.

Eicher, E.M. and Washburn, L.L., 1986. *Ann. Rev. Genet.*, 20: 327.

El-Fouly, M.A., Cook, B., Nekola, M. and Nalbandav, A.V., 1977. *Endocrinology*, 87: 288.

Eppi, J.J., Peters, A.H.F.M., Telfer, E.E. and Wigglesworth, K., 1993. *Mol. Reprod. Dev.*, 34: 450.

Eppig, J. and Downs, S.M., 1988. In: *Meiotic Inhibition Molecular Control of Meiosis: Progress in Clinical and Biological Research*, (Eds.) H. Haseltine and N.L. First. Liss, New York, pp. 103: 114.

Eppig, J.J. and Downs, S.M., *Dev. Biol.*, 119: 313.

Eppig, J.J. O'Brein M. and Wiggleswarth, K., 1996. *Mol. Reprod. Develop.*, 44: 260.

Eppig, J.J., 1979. *J. Exp. Zool.*, 209: 345.

Eppig, J.J., 1985. In: *Developmental Biology: A Comprehensive Synthesis, Vol. Isogenesis*, (Ed.) L.W. Browder. Plenum Press, New York, 13: 358.

Eppig, J.J., 1987. *Dev. Biol.*, 119: 313.

Eppig, J.J., 1989. *J. Reprod. Fertil.*, Suppl. 38: 3.

Eppig, J.J., 1994. *Sem. Develop. Biol.*, 5: 51.

Erickson, B.M., Magoffin, D.A., Dyer, C.A. and Hofeditz, C., 1985. *Endocr. Rev.*, 6: 371.

Espay, L.L. and Stutts, R.H., 1972. *Biol. Reprod.*, 6: 168.

Fagbohun, C.F. and Downs, S.M., *Biol. Reprod.*, 45: 851.

Fitzgerald, L.A. Denny, J.B., Baumbach, G.A., Ketcham, C.M., and Roberts R.M., 1984. *J. Cell Sci.*, 67: 1.

Gilula, N.B., Reeves, O.R. and Steinbach, A., 1972. *Nature*, 235: 262.

Gondos, B., 1970. In: *Gonadotrophins and Ovarian Development*, (Eds.) W.R. Bull. A.C. Crooke and M. Ryle. Livingstene, Edinburg, pp. 239.

Guraya, S.S., 1985. *Biology of Ovarian Follicles in Mammals*. Springer-Verlag, Heidelberg, Berlin, New York.

Guraya, S.S., 1998. Cellular and molecular biology of gonadal development and maturation in mammals. In: *Fundamentals and Biomedical Implications*. Narosa Publishing House, New Delhi.

Guraya, S.S., 2000. *Inter. Rev. Cytol.*, 199 (In press).

Haghigha, N. and Van Winkle, 1990. *J. Exp. Zool.*, 253: 71.

Heller, D.T., Cahill, D.M. and Schultz, R.M., 1981. *Dev. Biol.*, 84: 455.

Herlands, R.L. and Schultz, R.M., 1984. *J. Exp. Zool.*, 229: 317.

Herlands, R.L. and Schultz, R.M., 1984. *J. Exp. Zool.*, 229: 317.

Hiliensjo *et al.*, 1987.

Hilliensio, T., Magnusson, C., Svensson, U. and Lander, H., 1981. In: *Dynamics of Ovarian Function*, (Eds.) N.R. Schwartz and Hunjic Ker-Dunn. Raven Press, New York, pp. 105–110.

Hirshifield, A.N., 1986. *Biol. Reprod.*, 34: 229.

Horie, K., Takakura, K., Taii, S., Narimoto, K., Noda, Y., Nishikawa, S., Nakayama, H., Fujita, J. and Mori, T., 1991. *Biol. Reprod.*, 45: 547.

Huang, E.J., Manova, K., Packer, A.I., Sanchez, S., Bachvarova, R. and Besmer, P., 1993. *Dev. Biol.*, 157: 100.

Hubbard, C.J. and Terranova, P.F., 1982. *Biol. Reprod.*, 26: 628.

Itahana, K., Morikazu, Y. and Takeya, T., 1996. *Endocrinology*, 137: 5036.

Larsen, W.J., Wert, S.E. and Brunner, G.D. *Dev. Biol.*, 113: 517.

Larsen, W.J., Wert, S.E. and Brunner, G.D., 1987. *Dev. Biol.*, 122: 61.

Lawrence, T.S., Beers, W.H. and Gilula, N.B. *Nature*, 272: 501.

Lawson, K.A., 1994. *Ciba Foundation Symp.*, 182: 161.

Leibfried, L. and First, N.L., 1980a. *Biol. Reprod.*, 23: 699.

Leibfried, L. and First, N.L., 1980b. *Biol. Reprod.*, 23: 705.

Leibfried-Rutledge, M.L., Florman, H.M. and First, N.L., 1989. In: *Molecular Biology of Fertilization*, Eds.) H. Schatten and G. Schatten. Academic Press, New York, pp. 259: 101.

Li, R. and Mather, J.P., 1997. *Endocrinology*, 138: 4477.

Linda, H.R., Tsafriri, A., Lieberman, M.E., Zou, U., Koch, Y. and Barnea, B.S., 1974. *Prog. Hormone Res.*, 30: 79.

Linwin, P.N.T. and Zampighi, G., 1980. *Nature*, 283: 545.

Magnusson, C. and Hillensjo, T., 1977. *J. Exp. Zool.*, 201: 139.

Mahi-Brown, C.A. and Yanagimachi, R., 1983. *Gameli Res.*, 8: 1.

Makabe, S., Nagur, T., Nattola, S.A., Boreda, J. and Motta, P.M., 1991. In: *Ultrastructure of the Ovary*, (Eds.) G. Familiori, S. Makabe and P.M. Motta. Kluwer Academic Press, Boston, pp. 1.

Makabe, S., Naguro, T., Nottola, S.A., Pereda, J. and Motta, P.M., 1991. In: *Ultrastructure of the Ovary*, (Eds.) G. Familiori, S. Makabe and P.M. Motta. Kluwer Academic Press, Boston, pp. 1–28.

Manova, K., Huang, E.J., Angeles, M., Deleon, V., Sanchez, S., Pronovost, S.M., Besmer, S.M. and Bachvarova, R.F., 1993. *Dev. Biol.*, 157: 85.

Manova, K., Nocka, K., Besmer, P. and Bachvarova, R.F., 1990. *Development*, 110: 1057.

McGaughey, R., Racowsky, C., Rider, V., Baldwin, K., DeMarais, A.A. and Webster, S.D., 1990. *J. Electron Microsc. Tech.*, 16: 257.

McLaren, A., 1983. In: *Current Problems in Germ Cell Differentiation*, (Eds.) A. McLaren and C.C. Wylie. Cambridge Univ. Press, Cambridge, U.K., pp. 225.

McLaren, A., 1991. *Oxford Rev. Reprod. Biol.*, 13: 1.

McLaren, A., 1994. *Semi. Develop. Biol.*, 5: 43.

Merk, F.B.A., Bright, J.T. and Botticelti, C.R., 1973. *Anat Rec.*, 175: 107.

Moor, R.M., Smith, M.W. and Dawson, R.M.C., 1980. *Exp. Cell. Res.*, 126: 15.

Muller, U. and Urban, E., 1981. *Differentiation*, 20: 174.

Nagamine, C.M., Taketo, T. and Koo, G.C., 1987. *Differentiation*, 33: 212.

Nekola, M.V. and Nalbandov, 1971. *Biol. Reprod.*, 4: 154.

O.W.S. 1990. *Mol. Cell. Endocrin.*, 68: 181.

Oxbeny, B.A. and Greenwald, G.S., 1982. *Biol. Reprod.*, 27: 505.

Palmer, S.J. and Burgoyne, P.S., 1991. *Development*, 112: 265.

Park-Sarge, O.K. and Mayo, K.E. In: *Molecular Biology of Female Reproductive System*, (Ed.) J.K. Findlay. Academic Press, San Diego, pp. 153–203.

Pawshe, C.H., Rao, K.D and Totey, S.M., 1998. *Mol. Reprod. Dev.*, 49: 277.

Peng, X.R., Hsueh, A.J.W., Lapolt, P.S., Bjersing, L. and Ny, T., 1991. *Endocrinology*, 129: 3200.

Pincus, G. and Enzmann, E.V., 1935. *J. Exp. Med.*, 62: 655.

Pincus, G. and Saunders, B., 1939. *Anal Rec.*, 75: 537.

Prochazka, R., Nagyova, E., Riminucci, M., Nagai, T., Kikuchi, K. and Motlik, J., 1991. *J. Reprod. Fertil.*, 93: 569.

Racowsky, C. and Baldwin, K.W., 1989. *Dev. Biol.*, 134: 297.

Riberio, A.F., Ferronhna, M.H. and David-Ferreira, J.F., 1983. *J. Submicrose Cytol.*, 15: 415.

Salustri, A., Petrungaios, Oontti, M. and Stracusa, G., *Gamete Res.*, 11: 157.

Salustri, A., Yanagishita, M. and Hascall, V.C., 1989. *J. Biol. Chem.*, 264: 13840.

Salustri, A., Yanagishita, M. and Mascal, V.C., 1990. *Dev. Biol.*, 138: 26.

Sato, E. and Ishibashi, T., 1977. *J. Reprod. Fertil.*, 62: 22.

Schultz, R.M., Letourneau, G.E. and Wassarman, P.M. *J. Cell Sci.*, 30.

Sharma, R.K. and Guraya, S.S., 1990. *Acta Ernbr. Morph. Exp.*, 11: 107.

Sharma, R.K. and Sawhney, A.K., 1999. *Ind. J. Anim. Sci.*, 69: 109.

Sheridan, J.D., 1971. *Dev. Biol.*, 26: 6.

Sirotkin, A.V., Teradajnik, T.E., Makarevich, A.V. and Bulla, J., 1988. *Ann. Reprod. Sci.*, 51: 333.

Skinner, M. and Parrott, I.A., 1994. In: *Molecular Biology of the Female Reproductive System*, (Ed.) J.K. Findlay. Academic Press, Soni Diego, pp. 69–81.

Sorensen, R.A. and Wassarman, P.M., 1976. *Dev. Biol.*, 50: 531.

Stock, A.F., Woodruff, T.K. and Smith, L.C., 1997. *Biol. Reprod.*, 56: 1559.

Sutovsky, P., Felechon, J.E., Flechon, B., Motlik, J., Peynot, N., Patrick, C. and Heyman, Y., 1993. *Biol. Reprod.*, 49: 1277.

Szybek, K., 1972. *J. Endocrinol.*, 54: 527.

Takahashi, M., Koids, S.S. and Donahoe, P.K., 1986. *Mol. Cell. Endocr.*, 47: 225.

Takaoka, H., Satoh, H., Makinoda, S. and Khione, K., 1985. *Acta Obstct. Gynaecal. Japan*, 37: 92.

Thibault, C., Szollosi, D. and Gerard, M., 1987. *Reprod. Nutr. Develop.*, 27: 865.

Tornell, J., Carlsson, B. and Billig, H., 1990. *Endocrinology* 126: 1504.

Tsafriri A., Bar-Ami, S, and Dekel, N., 1982a and b.

Tsafriri, A. and Channing, C.P., 1975. *Endocrinology*, 96: 922.

Tsafriri, A., 1985. *Biology of Fertilization, Vol. 1,* (Eds.) C.B. Metz and A. Monroy. Academic Press, New York, 221: 22.

Tsafriri, A., 1988. In: *The Physiology of Reproduction, Vol. 1,* (Eds.) E. Kilobit and J.B. Nell. Raven Press, New York, 527: 566.

Tsafriri, A., Channing, C.P., Pomerantz, S.H. and Linder, H.R., 1977. *J. Endocrinol.*, 96: 922.

Van Winkle, U., 1988. *Biochem. Biophys Acta*, 941: 173.

Vanderhyden, B.C., Caron, P.J., Buccione, R. and Eppig, J.J., 1990. *Dev. Biol.*, 140: 307.

Vanderhyden, BC., Telfer, E.E. and Eppig, J.J., 1992. *Biol. Reprod.*, 46: 1196.

Wert, S.E and Larsen, W.J. 1990. *Tissue Cell.*, 22: 827.

Wert, S.E. and Larsen, W.J., 1989. *Gamete Res.*, 22: 143.

Yanagishita, M. and Hascall, V.C., 1984b. *J. Biol. Chem.*, 259: 10270.

Zaibot, P. and Dicar Cantorio, G., 1984. *Devel. Biol.*, 103: 159.

Zambani, 1996. In: *Ovulation in the Human*, (Eds.) P.G. Crosignani and D.R. Mishell. Academic Press, London, New York, pp. 1: 3.

Zamboni, L. and Upadhyay, S., 1983. *J. Exp. Zool.*, 228: 178.

Zamboni, L. In: *Endocrine Physiopathology of the Ovary*, (Eds.) R.I. Tozzini, G. Reeves and R.I. Lineda. Elsevier/North Holland Biomedical Press, 1980. Amsterdam New York, pp. 62.

Zamboni, L., 1974. Fine morphology of follicular wall and follicle cell-oocyte association. *Biol. Reprod.*, 10: 125–149.

Zhao, J., Dorland, M., Taverne, M.A., Van Der Weijden, G.C., Revers, M.M. and Van Den Hurk, R., 2000. *Mol. Reprod. Dev.*, 55: 65.

Zhou, J., Chin, E. and Bondy, C., 1991. *Endocrinology*, 129: 3281.

Zlotkin, T., Farkash, Y. and Orly, J., 1986. *Endocrinology*, 119: 2809.

Zoller, L.C. and Weisz, J., 1979. *Endocrinology*, 103: 303.

Zoller, L.C. and Weisz, J., 1979b. *Histochemistry*, 62: 125.

Chapter 16

Alterations in Certain Enzyme Activities in Muscles and Nerve Cord of *Periplanata americana* Induced by Sublethal Administration of Cypermethrin, Carbaryl and Monocrotophos

☆ *B.G. Kulkarni, Fatuma A. Mohammed and A.K. Pandey*

Abstract

Periplanata americana were administered with sublethal dose (less than the LD_{50} value for 96 hours) of cypermethrin (2.0×10^{-3} µg), carbaryl (5.0×10^{-2} µg) and monocrotophos (1.8×10^{-1} µg) for 24 and 98 hours. An elevation of aspartate aminotransferase (AAT) activity was noticed in the muscles and nerve cord of cockroaches treated with all the three insecticides. The observed increase in AAT activity was time-dependent with the exception of monocrotophos treated insects in which there was a decline in activity at 96 hours. However, the carbaryl-treated cockroaches showed pronounced increase in AAT activity as compared to cypermethrin and monocrotophos treated cockroaches. Elevated levels of alanine aminotranferase (ALAT) activity were also observed in cypermethrin and carbaryl treated cockroaches at 24 and 96 hours, however, a decline in ALAT activity was noted at 24 hours of monocrotophos exposure. A significant elevation in the acid phosphatase (ACP) and alkaline phosphatase (ALP) activities in both the tissues were observed in treated cockroaches. This elevation was time-dependent. However, after 96 hours of treatment, there was a slight decline in ACP activity in cypermethrin and monocrotophos treated cockroaches as compared to 24 hours of the treatments. Na^+/K^+-ATPase as well as Mg^{2+}-ATPase activities declined significantly in both the tissues of cockroaches treated with all the three pesticides. The present study indicates that cypermethrin, carbaryl and monocrotophos are potent inhibitors of Na^+/K^+-ATPase and Mg^{2+}-ATPase activities in the cockroach.

Keywords: *Cypermethrin, Carbaryl, Monocrotophos, Enzymatic changes, Periplanata americana.*

Introduction

Since enzymes control the formation of biochemical intermediates essential for all physiological functions, their assays have gained more importance in toxicological investigations. It has been established that organophosphate and carbamate insecticides exert their toxic effects by inhibiting acetylcholinesterase (AchE), the enzyme responsible for hydrolysis of acetylcholine in insects (Yu *et al.*, 1973; Fukuto, 1979; Bose, 1991). On the other hand, no specific target enzyme has been clearly established for other groups of insecticide such as organochlorine, cyclodienes, pyrethrins and synthetic pyrethroids. Pyrethrins, synthetic pyrethroids, DDT and related compounds are known for their interference in the sodium channel pumping mechanism. There exist reports that these compounds inhibit adenosine triphosphatase (ATPase), the enzyme responsible for immediate release of energy for physiological activities in both invertebrates and vertebrates (Cutkomp *et al.*, 1971; Desaiah and Cutkomp, 1973; Doherty *et al.*, 1981, Chandra and Poddar, 1990; Reddy *et al.*, 1992). However, effects of organophosphate and carbamate insecticides on other enzyme systems such as the aminotransferases and phosphatases in target organs of insects have not yet been clearly defined. Similarly, effects of the synthetic pyrethroid, cypermethrin on ATPase and other enzymes are not well known. Hence, the present investigation was undertaken to record the effects of cypermethrin, carbaryl and monocrotophos on activities of aspartate aminotransferase (AAT) or glutamic-oxaloacetic transaminase (GOT), alanine aminotransferase (ALAT) or glutamic–pyrivic transaminase (GTP), alkaline phosphatase (ALP), acid phosphatase (ACP) and adenosine triphosphatases (ATPases) in muscle and nerve cord of the cockroach for combating pesticidal stress.

Materials and Methods

Adult male cockroach, *Periplanata americana* (Linnaeus) (total length 3.92±0.08 cm, average weight 0.924±0.04 gm) were selected and acclimatized to the laboratory conditions for 4 days prior to exposure. They were kept in wired aluminum cages of 45 × 24 × 22 cm size and fed on roughage throughout the study period. Cockroaches of the experimental groups were administered sublethal dose (less than the LD_{50} value for 96 hours) of cypermethrin (2.0×10^{-3} μg), carbaryl (5.0×10^{-2} μg) and monocrotophos (1.8×10^{-1} μg) for 24 and 98 hours by topical application on the thorax using an aglamicrometer syringe (Kulkarni *et al.*, 2005). A set of control was also maintained for entire period of the treatment. The insects of both the groups were dissected at 24 and 96 hours in normal saline and the muscles of the thorax and coxae were carefully excised together with the ventral nerve cord. The tissues were blotted and homogenized in a motor-driven glass homogenizer with two volumes of chilled glass distilled water. Homogenates were centrifuged at 10,000 rpm for 10 minutes. The clean supernatants were used as an enzyme source. From the same enzyme source, protein content was estimated by the method of Lowry *et al.* (1951).

Aspartate Aminotransferase (AAT)

The AAT activity in the tissue homogenates of the cockroach was assayed as per the method of Bergmeyer and Bernt (1965a). The reaction mixture consisted of 1 ml sulphate buffer solution (0.1 M phosphate buffer, pH 7.4; 2×10^{-3} α-oxaloglutarate; 0.1 M L-aspartate) and 0.2 ml tissues homogenate. Incubation of the mixture was done at 37°C for 60 minutes and then 1 ml of ketone reagent (10^{-3} M 2,4-dinitrophenyl hydrazine) was added to it. After 20 minutes at room temperature, 10 ml of 0.4 N NaOH was added to the mixture. A control was prepared in the same manner except that the homogenate was added after addition of ketone reagent. Optical density (OD) was read after 5 minutes at 546 nm. One AAT unit was defined as μ mole of oxaloacetate formed per 60 minutes under assay conditions. Specific activity was defined as units of AAT per 60 minutes/mg of protein.

Alanine Aminotransferase (ALAT)

The method described by Bergmeyer and Bernt (1965b) was followed for assaying the ALAT activity in the tissue homogenates of the cockroach. The substrate-buffer solution consisted of 0.1 M phosphate buffer, pH 7.4; 0.2 M DL-alanine and 2×10^{-3} M α-oxaloglutarate. Mixture of the substrate-buffer (1 ml) and tissue homogenate (0.2 ml) was incubated at 37°C for 30 minutes. Rest of the procedure was exactly same as that followed for AAT. One ALAT unit was defined as μmole of pyruvate formed per 30 minutes under assay conditions. Specific activity was defined as units of ALAT per 30 minutes/mg of protein.

Alkaline Phosphatase (ALP)

Activity of alkaline phosphatase in the tissue homogenates of cockroach was determined by the procedure of Andersch and Sczcypinski (1947) with slight modification. To 1.1 ml of glycerine-NaOH buffer (0.05M, pH 10.5), 0.2 ml of p-nitrophenol phosphate (11×10^{-3} M) was added. To this reaction mixture, 0.2 ml of tissue homogenate was added and reaction started by incubating the test tubes at 37°C for 30 minutes. At end of this period, 4 ml of 0.2 N NaOH was added to stop the enzyme activity. The tubes were shaken well and the strong yellow colour developed of p-nitrophenol was read at 405 nm. A blank was prepared in which the homogenate was added after addition of 0.2 N NaOH. One phosphate unit (acid/alkaline) is defined as the amount of enzyme that liberates on n mole of p-nitrophenol under assay condition. Specific activity was defined as n moles of p-nitrophenol liberated per 30 minutes mg protein.

Acid Phosphatase (ACP)

Acid phosphatase activity in the tissue homogenates of cockroach was estimated by the method of Andersch and Szcypinski (1947) with slight modification. To 1 ml of citrate buffer (0.05 M, pH 4.5), 0.2 ml p-nitrophenol phosphate (11×10^{-3} M) was added. To this mixture, 0.2 ml of homogenate was added and the reaction started by incubating the tubes at 37°C for 30 minutes. At the end of the period, 4 ml of 0.2 N NaOH was added to stop the reaction. The tubes were shaken vigorously and the optical density was read at 405 nm. A blank was prepared in which the homogenate was added after addition of 0.2 N NaOH. The definition of specific activity of the enzyme was same as that of ALP.

Adenosine Triphosphatase (ATPase)

Changes in activities of ATPase were estimated as per method of Terri *et al.* (1973). Total ATPase activity was measured in a 2.0 ml reaction mixture containing 100 µmoles tris-HCl buffer (pH 7.4; 0.1M), 20 µmoles of $MgCl_2$, 25 µmoles of NaCl-KCl, 8 µmole of disodium salt of ATP and 0.5 ml homogenate. The Mg^{2+}-ATPase reaction mixture contained 100 µmoles tris-HCl buffer (pH 7.4; 0.1M), 25 µmoles of $MgCl_2$, 20 µmoles NaCl-KCl, 8 µmoles of disodium salt of ATP, 10 µmoles of ouabain and 0.5 ml of extract. The contents were incubated at 37°C for 15 minutes. The reaction was stopped by adding 2 ml of ice-cold 10 per cent TCA. The contents were centrifuged and 1 ml of the filtrate was taken for the estimation of organic phosphorus content (Fiske and Subbarow, 1925). The Na^+/K^+-ATPase activity was obtained by subtracting the Mg^{2+}-ATPase from the total ATPase. Units of ATPase activity were expressed as µmoles of Pi liberated/mg of protein/hour.

Results and Discussion

Aspartate Aminotransferase

Aspartate aminotransferase (AAT) activity in thoracic muscle, coxal muscle and nerve cord of control cockroach were 6.26±0.87, 4.13±0.23 and 1.94± 0.19 µmole of oxaloacetate formed/mg protein/minute. However, significant elevation in AAT activity was noticed in muscles and nerve cord of the cockroaches treated with all the three insecticides. The observed increase in AAT activity was time-dependent with exception in monocrotophos treated cockroaches where there was a decline in activity at 96 hours. Interestingly, carbaryl treated cockroaches showed more pronounced increase in AAT activity on both the exposure periods as compared to those treated with cypermethrin and monocrotophos (Table 16.1).

Alanine Aminotransferase

Alanine aminotransferase (ALAT) activity in thoracic muscle, coxal muscle and nerve cord of control cockroach were found to be 3.87±0.25, 3.74±0.24 and 1.67± 0.05 µmole of pyruvate formed/mg protein/minute. Elevated levels of ALAT activity was observed in cypermethrin and carbaryl treated cockroaches at 24 and 96 hours. However, a decline in ALAT activity was recorded after 24 hours in cockroaches treated monocrotophos followed by significant elevation at 96 hours (Table 16.2).

Alkaline and Acid Phosphatases

Alkaline phosphatase (ALP) activity in thoracic muscle, coxal muscle and nerve cord of control cockroach were 4.06±0.13, 2.73±0.05 and 2.15± 0.09 nmole of pnp liberated/mg protein/minute whereas the corresponding values for acid phosphatases (ACP) were 4.06±0.13, 2.73±0.05 and 1.85±0.11 nmole of pnp liberated/

Table 16.1: AAT Activity (μ mole of oxaloacetate formed/mg protein/min) in Muscles and Nerve Cord of *P. americana* Exposed to the Pesticides

Treatment	*Control*	*Experimental*	
		24 hours	*96 hours*
		Cypermethrin	
Thoracic muscle	6.26 ± 0.87	8.74 ± 0.78[a] (+40)	16.48 ± 2.25[b] (+163)
Coxal muscle	4.13 ± 0.23	8.71±1.26[b] (+111)	9.75±0.04[b] (+136)
Nerve cord	1.94 ± 0.19	4.50 ± 0.37[b] (+132)	4.41 ± 0.20[b] (+127)
		Carbaryl	
Thoracic muscle	6.26 ± 0.87	11.34 ± 1.77[a] (+81)	12.33 ± 0.55[b] (+97)
Coxal muscle	4.13 ± 0.23	11.29 ± 0.61[b] (173)	13.44 ± 0.24[b] (+225)
Nerve cord	1.94 ± 0.19	7.86 ± 0.99[b] (+305)	11.16 ± 1.89[b] (+475)
		Monocrotophos	
Thoracic muscle	6.26 ± 0.87	8.91 ± 0.96[a] (+42)	3.77 ± 0.19[b] (-40)
Coxal muscle	4.13 ± 0.23	9.95 ± 0.93[b] (+141)	2.80 ± 0.08[b] (-32)
Nerve cord	1.94 ± 0.19	3.20 ± 0.16[b] (+65)	2.03 ± 0.03 (+5)

Values are mean ± S. D. of 5 animals.

Significant responses: [a] $P < 0.01$, [b] $P < 0.001$.

Numbers in parentheses indicate per cent decrease over control.

Table 16.2: ALAT Activity (μ mole of pyruvate formed/mg protein/min) in Muscles and Nerve Cord of *P. americana* Exposed to the Pesticides

Treatment	*Control*	*Experimental*	
		24 hours	*96 hours*
		Cypermethrin	
Thoracic muscle	3.87 ± 0.25	22.58 ± 2.55[b] (+483)	18.84 ± 1.71[b] (+387)
Coxal muscle	3.74 ± 0.24	37.65 ± 3.90[b] (+907)	24.34 ± 2.20[b] (+551)
Nerve cord	1.67 ± 0.05	16.42 ± 2.47[b] (+883)	5.13 ± 0.34[b] (+207)
		Carbaryl	
Thoracic muscle	3.87 ± 0.25	68.02 ± 7.48[b] (+1658)	65.92 ± 9.47[b] (+1603)
Coxal muscle	3.74 ± 0.24	30.85 ± 4.02[b] (+725)	33.48 ± 2.35[b] (+795)
Nerve cord	1.67 ± 0.05	16.61 ± 1.24[b] (895)	20.01 ± 3.00[b] (+1098)
		Monocrotophos	
Thoracic muscle	3.87 ± 0.25	2.22 ± 0.42[a] (-43)	7.50 ± 0.15[b] (+94)
Coxal muscle	3.74 ± 0.24	3.92 ± 0.31 (+5)	6.20 ± 0.04[b] (+66)
Nerve cord	1.67 ± 0.05	1.49 ± 0.05 (-11)	2.11 ±0.23[b] (+26)

Values are mean ± S. D. of 5 animals.

Significant responses: [a] $P < 0.01$, [b] $P < 0.001$.

Numbers in parentheses indicate per cent decrease over control.

mg protein/minute. Significant elevations in the ALP and ACP activities in the tissues of treated cockroaches were observed at both the exposure periods. Though this elevation was time-dependent, there was a slight depletion in ACP activity in cypermethrin and monocrotophos treated cockroaches at 96 hours as compared to 24 hours of the corresponding treatments (Tables 16.3–16.4).

Adenosine Triphosphatase

Total adenosine triphosphatase (ATPase) activity in thoracic muscle, coxal muscle and nerve cord of control cockroach were 8.99±0.24, 6.11±0.25 and 11.48±0.52 μmole of Pi liberated/mg protein/hour. However, Na^+/K^+-ATPase activity in thoracic muscle, coxal muscle and nerve cord of control cockroach were found to be 4.24±0.38, 3.45±0.23 and 6.61±0.40 μmole of Pi liberated/mg protein/hour whereas the corresponding activity of Mg^{2+}-ATPase in different tissues were 4.70±0.05, 2.66±0.25 and 4.87±0.21 μmole of Pi liberated/mg protein/hour. Total ATPase, Na^+/K^+-ATPase and Mg^{2+}-ATPase activities were depleted significantly in the muscles and nerve cord of the cockroaches treated with all the three pesticides (Tables 16.5–16.7). Carbaryl was found to be a potent inhibitor of total ATPase activity (Table 16.5).

The treatment of *Periplanata americana* with sublethal dose of cypermethrin, carbaryl and monocrotophos brought about variations in enzymatic activities in their muscles and nerve cord. In general, the insecticidal stress elevated the activities of transaminases and phosphatases in the muscles and nerve cord whereas the ATPases activity was declined in experimental cockroaches. The findings in present study are in agreement with those of other investigators who have recorded similar observations in fishes, crustaceans and molluscs exposed to various insecticides (Kabeer, 1979; Rao *et al.*, 1980).

Amino acid metabolism in animals, especially in invertebrates, has been directly implicated in energy metabolism (Adiyodi and Nayar, 1966; Chen, 1985; Pandey and Mathur, 1990). Of the enzymes involved in protein metabolism, the most important are transaminases. Although the relationship between transaminase activity and protein synthesis is not clear, these enzymes are believed to play an important role in protein metabolism. However, an inverse relationship between amino transferases activity and protein content of the tissues in cockroaches exposed to the insecticides has been observed in the present study (Tables 16.1 amd 16.2). The probable reasons for heightened activities of aminotransferases in the cockroach tissues treated with the insecticides may be attributed to amino acid degradation by gluconeogenesis. This degradation is of physiological importance during starvation and stress because insects utilize a number of amino acids, particularly proline, as supplementary fuel.

Alkaline phosphatase activity is associated with the maintenance of the orthophosphate pool, the transfer of phosphoryl groups, the hydrolysis and esterification of metabolites moving across membranes within the cell and between the cells and extracellular spaces (Saev, 1963; Morton, 1965). Acid phosphatase has been identified with other hydrolases in lysosomes (Novikoff, 1961). Changes in the activity of phosphatases in the tissues are associated with growth, differentiation and lysis of the cells (de Duve and Wattaux, 1966; Varute and Sawant, 1971; Singh and Pandey, 1987). Koundinya and Ramamurthi (1982) attributed the increase in

Table 16.3: ALP Activity (μmole of pnp liberated/mg protein/min) in Muscles and Nerve Cord of *P. americana* Exposed to Pesticides

Treatment	*Control*	*Experimental*	
		24 hours	*96 hours*
		Cypermethrin	
Thoracic muscle	4.06 ± 0.13	13.83 ± 2.69 [b] (+241)	23.33 ± 2.72 [b] (+475)
Coxal muscle	2.73 ± 0.05	6.99 ± 0.68 [b] (+156)	12.44 ± 0.66 [b] (+356)
Nerve cord	2.15 ± 0.09	11.12 ± 1.20 [b] (+417)	9.59 ± 1.06 [b] (+342)
		Carbaryl	
Thoracic muscle	4.06 ± 0.13	17.56 ± 1.64 [b] (+337)	17.47 ± 1.02 [b] (+330)
Coxal muscle	2.73 ± 0.05	6.83 ± 0.45 [b] (+150)	8.17 ± 0.30 [b] (+199)
Nerve cord	2.15 ± 0.09	7.09 ± 0.80 [b] (+230)	13.07 ± 0.5 [b] (+508)
		Monocrotophos	
Thoracic muscle	4.06 ± 0.13	5.43 ± 0.21 [b] (+34)	6.79 ± 1.27 [b] (+67)
Coxal muscle	2.73 ± 0.05	4.19 ± 0.11 [b] (+53)	3.13 ± 0.19 [a] (+15)
Nerve cord	2.15 ± 0.09	4.52 ± 0.20 [b] (+110)	4.52 ± 0.46 [b] (+110)

Values are mean ± S. D. of 5 animals.

Significant responses: [a] $P < 0.01$, [b] $P < 0.001$.

Numbers in parentheses indicate per cent decrease over control.

Table 16.4: ACP Activity (μmole of pnp liberated/mg protein/min) in Muscles and Nerve Cord of *P. americana* Exposed to the Pesticides

Treatment	*Control*	*Experimental*	
		24 hours	*96 hours*
		Cypermethrin	
Thoracic muscle	4.06 ± 0.13	4.22 ± 0.25 (+4)	4.67 ± 0.90 [a] (+15)
Coxal muscle	2.73 ± 0.05	3.33 ± 0.16 [b] (+22)	3.04 ± 0.75 (+11)
Nerve cord	1.85 ± 0.11	4.03 ± 0.65 [b] (+118)	1.60 ± 0.14 (-14)
		Carbaryl	
Thoracic muscle	4.06 ± 0.13	8.06 ± 1.67 [b] (+99)	7.42 ± 0.87 [b] (+83)
Coxal muscle	2.73 ± 0.05	3.42 ± 0.22 [b] (+25)	5.34 ± 0.31 [b] (+96)
Nerve cord	1.85 ± 0.11	3.28 ± 0.49 [b] (+77)	8.50 ± 0.46 [b] (+359)
		Monocrotophos	
Thoracic muscle	4.06 ± 0.13	6.20 ± 0.33 [b] (+53)	6.00 ± 1.40 [b] (+48)
Coxal muscle	2.73 ± 0.05	6.21 ± 0.32 [b] (+127)	2.59 ± 0.28 (-5)
Nerve cord	1.85 ± 0.11	3.89 ± 0.08 [b] (+110)	3.57 ± 0.47 [b] (+93)

Values are mean ± S. D. of 5 animals.

Significant responses: [a] $P < 0.01$, [b] $P < 0.001$.

Numbers in parentheses indicate per cent decrease over control.

Table 16.5: Total ATPase Activity (μmole of Pi liberated/mg protein/hour) in Muscles and Nerve Cord of *P. americana* Exposed to the Pesticides

Treatment	*Control*	*Experimental*	
		24 hours	*96 hours*
		Cypermethrin	
Thoracic muscle	8.99 ± 0.24	6.16 ± 0.58[b] (-31)	4.79 ± 0.10[b] (-47)
Coxal muscle	6.11 ± 0.25	5.15 ± 0.09[a] (-16)	4.13 ± 0.11[b] (-32)
Nerve cord	11.48 ± 0.52	9.54 ± 0.54[b] (-18)	7.79 ± 0.43[b] (-32)
		Carbaryl	
Thoracic muscle	8.99 ± 0.24	6.21± 0.62[b] (-31)	2.33 ± 0.08[b] (-74)
Coxal muscle	6.11± 0.25	2.60 ± 0.22[b] (-57)	1.91 ± 0.05[b] (-69)
Nerve cord	11.48 ± 0.52	7.97 ± 0.77[b] (-31)	3.79 ± 0.14[b] (-67)
		Monocrotophos	
Thoracic muscle	8.99 ± 0.24	5.98 ± 0.12[b] (-33)	5.39 ± 0.22[b] (-40)
Coxal muscle	6.11± 0.25	5.19 ± 0.16[b] (-15)	4.86 ± 0.14[b] (-20)
Nerve cord	11.48 ± 0.52	9.47 ± 1.14[b] (-18)	8.83 ± 0.58[b] (-23)

Values are mean ± S. D. of 5 animals.

Significant responses: [a] $P < 0.01$, [b] $P < 0.001$.

Numbers in parentheses indicate per cent decrease over control.

Table 16.6: Na^+/K^+-ATPase Activity (μmole of Pi liberated/mg protein/hour) in Muscles and Nerve Cord of *P. americana* Exposed to the Pesticides

Treatment	*Control*	*Experimental*	
		24 hours	*96 hours*
		Cypermethrin	
Thoracic muscle	4.24 ± 0.38	2.73 ± 0.40[c] (-36)	3.27 ± 0.14[b] (-23)
Coxal muscle	3.45 ± 0.23	3.38 ± 0.18 (-2)	2.75 ± 0.13[c] (-20)
Nerve cord	6.61 ± 0.40	5.42 ± 0.78[b] (-18)	4.76 ± 0.60[c](-29)
		Carbaryl	
Thoracic muscle	4.24 ± 0.38	2.82 ± 0.73[c] (-33)	1.52 ± 0.13[b] (-68)
Coxal muscle	3.45± 0.23	0.925 ± 0.19[b] (-73)	1.45 ± 0.05[b] (-55)
Nerve cord	6.61± 0.40	5.33 ± 0.52[b] (-19)	2.19 ± 0.15[b] (-55)
		Monocrotophos	
Thoracic muscle	4.24 ± 0.38	1.99 ± 0.06[c] (-53)	1.99 ± 0.28[c] (-53)
Coxal muscle	3.45 ± 0.23	3.24 ± 0.16 (-6)	3.11 ± 0.22[a] (-10)
Nerve cord	6.61± 0.40	4.94 ± 0.77[c] (-25)	5.53 ± 0.31[b] (-16)

Values are mean ± S. D. of 5 animals.

Significant responses: [a] $P < 0.05$, [b] $P < 0.001$, [c] $P < 0.01$.

Numbers in parentheses indicate per cent decrease over control.

Table 16.7: Mg^{2+}-ATPase Activity (µmole of Pi liberated/mg protein/hour) in Muscles and Nerve Cord of *P. americana* Exposed the Pesticides

Treatment	*Control*	*Experimental*	
		24 hours	*96 hours*
		Cypermethrin	
Thoracic muscle	4.70 ± 0.05	3.43 ± 0.23[b] (-27)	1.55 ± 0.09[b] (-67)
Coxal muscle	2.66 ± 0.25	1.77 ± 0.12[b] (-33)	1.38 ± 0.02[b] (-48)
Nerve cord	4.87 ± 0.21	4.11± 0.29[a] (-16)	3.08 ± 0.38[b] (-37)
		Carbaryl	
Thoracic muscle	4.70 ± 0.05	3.36 ± 0.20[b] (-29)	1.52 ± 0.13[b] (-68)
Coxal muscle	2.66 ± 0.25	1.67 ± 0.06[b] (-37)	1.45 ± 0.05[b] (-45)
Nerve cord	4.87 ± 0.21	2.64 ± 0.31[b] (-46)	2.19 ± 0.15[b] (-55)
		Monocrotophos	
Thoracic muscle	4.70 ± 0.05	3.99± 0.15[b] (-15)	3.39 ± 0.17[b] (-28)
Coxal muscle	2.66 ± 0.25	1.95 ± 0.17[b] (-27)	1.75 ± 0.11[b] (-34)
Nerve cord	4.87 ± 0.21	4.54 ± 0.41 (-7)	3.30 ± 0.16[b] (-32)

Values are mean ± S. D. of 5 animals.

Significant responses: [a] $P < 0.01$, [b] $P < 0.001$.

Numbers in parentheses indicate per cent decrease over control.

alkaline phosphatase activity of sumithion-treated fishes with increased rates of glycogenolysis. Since oxidative phosphorylation of glucose is an energy-requiring process, increase in phosphatase activity which catalyses the liberation of inorganic phosphate from phosphate esters like glycerophosphate, phenyl phosphate etc is justifiable. Since ACP is a lysosomal enzyme, its elevation in insecticide treated cockroaches may be useful for removing absorbed insecticides in the cells. Further, lysis in the body tissues may be responsible for elevated levels of AAT, AALT, ACP and ALP. This view is in agreement with other workers who have reported similar results in different animals (Kabeer, 1979; Biswas, 1986; Ghosh, 1989; Bose, 1991).

The impact of insecticidal stress has created an inhibitory effect on ATPases. The inhibition pattern was more or less similar in muscles and nerve cord in all the exposures. Both Mg^{2+} and Na^+/K^+-dependent ATPases were inhibited in muscles and nerve cord of insecticide-treated cockroaches. Therefore, the present study indicates that cypermethrin, carbaryl and monocrotophos are potent inhibitors of Mg^{2+} and Na^+/K^+-ATPase. The inhibition of these two ATPases both *in vivo* and *in vitro* with chlorinated hydrocarbons such as DDT, allethrin, chloraden etc have been reported by a number of investigators (Cutkomp *et al.*, 1971; Desaiah and Cutkomp, 1973; Kulkarni, 1988; Chandra and Poddar, 1990; Reddy *et al.*, 1992). ATPases are known to play a strategic role in regulating oxidative phosphorylation, ionic transport and several other membrane transport phenomena. Therefore the observed depletion in ATPase in treated cockroaches may lead to interferences in these functions.

Acknowledgements

One of the authors (FAM) is grateful to the Muslim Education and Welfare Association, Mombasa, Kenya for sponsoring her study in India. We are thankful to Prof. S.A. Suryawanshi, former Director, Institute of Science, Mumbai for providing necessary facilities to carry out the work.

References

Adiyodi, K.G. and Nayar, K.K., 1966. Haemolymph proteins and reproduction in *Periplanata americana*. *Curr. Sci.*, 35: 587–588.

Andersch, M.A. and Szcypinski, A.J., 1947. Estimation of alkaline and acid phosphatase activities in normal human serum by using p–nitrophenol phosphate. *Am. J. Clin. Pathol.*, 17: 571–574.

Bergmeyer, H.U. and Bernt, E., 1965a. Glutamate–oxaloacetate transminase. In: *Methods of Enzymatic Aanalysis* (Ed.) Bergmeyer, H.U. Academic Press, New York, p. 837–847.

Bergmeyer, H.U. and Bernt, E., 1965b. Glutamate-pyruvate transminase. In: *Methods of Enzymatic Analysis* (Ed) Bergmeyer, H.U. Academic Press, New York, p. 847–853.

Bose, C., 1991. Malathion–induced biogenic amine levels and acetylcholinesterase activity in the cockroach, *Periplanata americana*. *Curr. Sci.*, 60: 707–709.

Chandra, M. and Poddar, M.K., 1990. *In vivo* and *in vitro* effects of aldrin on rat brain synaptosomal Mg^{2+} and Na^{+}/K^{+}–adenosine triphosphatase. *Biochem. Pharmacol.*, 40: 1449–1456.

Chen, P.S., 1985. Amino acid and protein metabolism. In: *Comparative Insect Physiology: Biochemistry and Pharmacology. Vol. 10* (Eds.) Kerkut, G.A. and Gilbert, L.I. Pergaman Press, Oxford, New York, Toronto and Sydney, p. 177–217.

Cutkomp, L.K., Yap, H.H., Vea, E.V. and Koch, R.B., 1971. Inhibition of oligomycin–sensitive (mitochondrial) Mg^{2+}–ATPase by DDT and selected analogs in fish and insect tissue. *Life Sci.*, 10: 1201–1209.

de Durve, C. and Wattaux, R., 1966. Functions of lysosomes. *Physiol. Rev.*, 28: 435–492.

Desaiah, D. and Cutkomp, L.K., 1973. The effect of pyrethins on ATPase in cockroach and bluefish. *Pyrethrum Post.*, 12(2): 70–75.

Doherty, J.D., Salem N. Jr, Lauter, C.J. and Trams, E.G., 1981. Mn^{2+} and Ca^{2+} ATPases in lobster axon plasma membranes and their inhibition by pesticides. *Comp. Biochem. Physiol.*, 69C: 185–190.

Fiske, C.H. and Subbarow, V., 1925. The colorimeteric determination of phosphorus. *J. Biol. Chem.*, 66: 375–400.

Fukuto, T.R., 1979. Effect of structure on the interaction of organophosphorus and carbamate esters with acetylcholinestersae. In: *Neurotoxicology of Insecticides and Pheromones* (Ed.) Narahashi, T. Plenum Press, New York and London, p. 277–295.

Kabeer, A.A., 1979. Studies on some aspects of protein metabolism and associated enzyme systems in the freshwater teleost, *Tilapia mossambica* (Peters) subjected to malathion exposure. *Ph.D. Thesis*. S.V. University, Tirupati, India.

Koundinya, P.R. and Ramamurthi, R., 1982. Effect of organophosphorus pesticide sumithion (Fenitrothion) on alkaline phosphatase activity of freshwater teleost, *Sarotherodon mossambicus* (Peters). *Curr. Sci.*, 50: 191–192.

Kulkarni, B.G., 1988. ATPase activity in the tissues and blood of the crab, *Scylla serrata* (Forskal) exposed to malathion. *Nat. Acad. Sci. Letters*, 11: 197–199.

Kulkarni, B.G., Mohammad, F.A., Kupekar, S.S. and Pandey, A.K., 2005. Acute toxicity studies of cypermethrin, carbaryl and monocrotophos to the cockroach, *Periplanata americana* (Linnaeus). *J. Natcon.*, 17: 299–306.

Lowry, O.H., Rosenbrough, N.J., Farr, A.L. and Randall, R.J., 1951. Protein measurement with folin–phenol reagent. *J. Biol. Chem.*, 193: 265–275.

Morton, R.K. (1965) Phosphatases. In: *Comparative Biochemistry. Vol. 16.* (Eds.) Florkin, M. and Statz, E.H. Elsevier, New York, p. 55–84.

Novikoff, A.B., 1961. Lysosomes and related particles. In: *The Cell. Vol. 2.* (Ed.) Mirsky, A. Academic Press, New York and London, p. 423–428.

Pandey, R. and Mathur, Y.K., 1990. Amino acid alterations in insecticidal treated insects. *Indian J. Entomol.*, 52: 197–202.

Rao, S.P., Prasad, S. and Rao, V.R., 1980. Sublethal effect of methyl parathion on tissues proteolysis of the freshwater mussel, *Lamellidens marginalis* (Lamarck). *Proc. Indian Natl. Sci. Acad.*, 46B: 164–167.

Reddy, A.N., Venugopal, N.B.R.K. and Reddy, S.L.N., 1992. Effect of endosulfan 35 EC on ATPases in the tissues of fresdhwater field crab, *Barytelphusa guerini. Bull. Environ. Contam. Toxicol.*, 48: 216–222.

Saev, G.K., 1963. Some concepts of intracellular functions of alkaline phosphatases based on investigations of the mechanism of their action. *Enzymologia*, 26: 169–175.

Singh, N. and Pandey, R.S., 1987. Acid phosphatase activity in the life stages of *Dicrisia oblique* Walker. *Indian J. Entomol.*, 49: 114–117.

Terri, R., Lagerspetz, K.Y.H. and Kohenson, J., 1973. Temperature dependence of ATPase activities in brain homogenates during post–natal developments of the rat. *Comp. Biochem. Physiol.*, 44: 473–480.

Varute, A.T. and Sawant, S.K., 1971. β-glucoronidase during embrogenesis of the larvae of blowfly, *Crysomigia ruriferacis Comp. Biochem. Physiol.*, 38: 211–223.

Yu, C.C., Park, K.S. and Metcalf, R.L., 1973. Correlation of toxicity and acetylcholinesterase (AchE) inhibition in 2–alkyl substitute 1,3–benzodioxyl–4–N–methylcarbamates and related compounds. *Pestic. Biochem. Physiol.*, 4: 178–184.

Plate 17.1: Showing Some Aquatic Birds in Mansar Lake

Insectivores were further categorized into 7 sub-groups *i.e.* Aerial Insectivores (AI), Shore Insect Plovers (SIP), Aquatic Insectivore (AqI), Understorey Insectivores (UI), Canopy Insectivores (CI), Terrestrial Insectivore (TI) and Trunk or Bark Feeders (T/BF). Out of these, 5 species belonged to AI category, 4 species represented TI, 2 species each by UI and T/BF and 1 species each by CI and AqI (Figure 17.2) (Table 17.2.

Table 17.1: Showing the Number of Bird Species Under Each Category of Feeding Guild

Feeding Guild	*Number of Species*
Insectivorous	15
Carnivorous	15
Grainivorous	5
Frugivorous	3
Omnivorous	2
Herbivorous	2
More than one feeding guild	21
Total	63

Table 17.2: Showing the Number of Bird Species Under Each Category of Insectivorous Type of Feeding Guild

Insectivorous Guilds Feeding	*Number of Birds*
Aerial Insectivores	5
Canopy Insectivores	4
Terrestrial Insectivores	2
Aquatic Insectivores	2
Understorey insectivores	1
Trunk or Bark feeder	1
Total	15

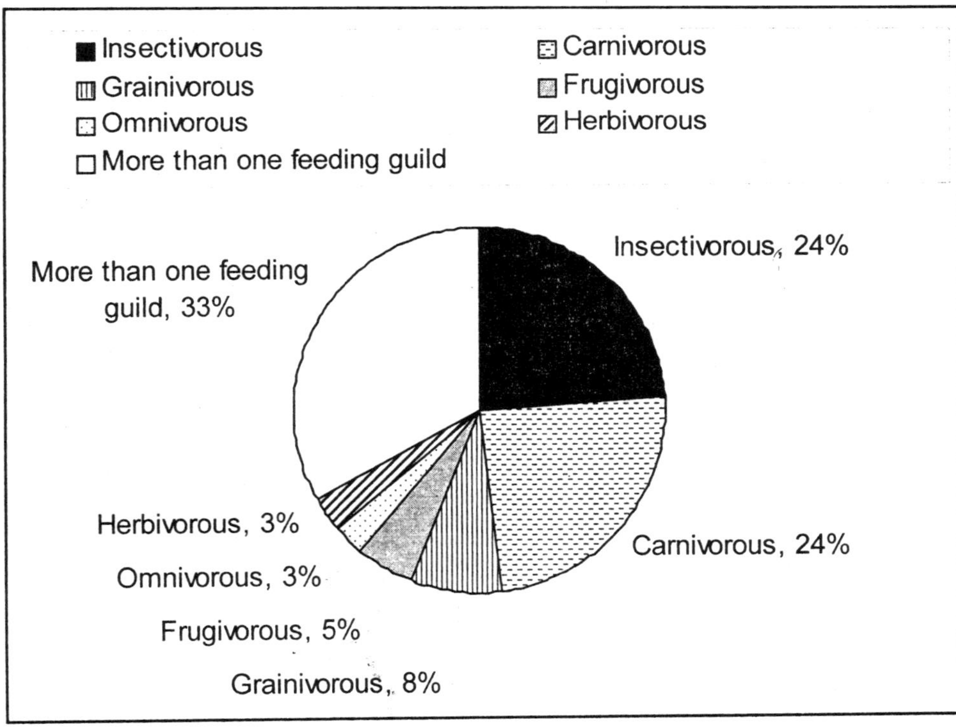

Figure 17.1: Showing the Percentage of Bird Species Belonging to Different Feeding Guilds Utilized by them

Carnivorous were categorized into 4 sub-groups *i.e.* Wading Carnivore (WC), Diving Carnivore (DC), Arboreal Terrestrial Carnivore (ATC), Arboreal Aquatic

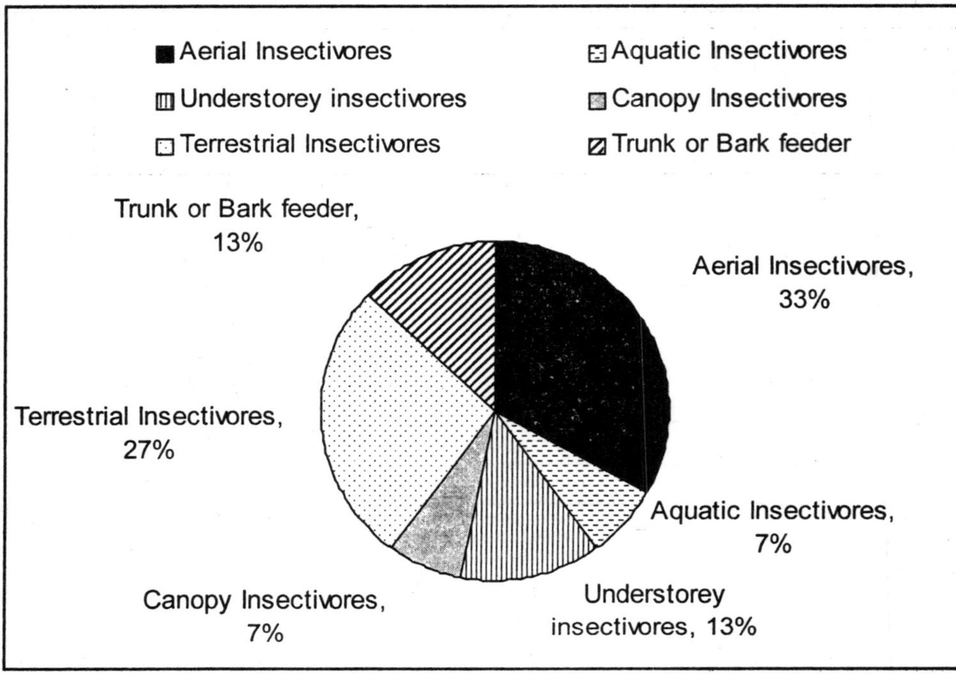

Figure 17.2: Showing the Percentage Proportion Constituted by Different Feeding Guilds Utilized by Birds Under Insectivorous Feeding Category

Carnivore (AAqC). Out of this, 8 species constitute the category of WC, 1 specie DC, 4 species ATC and 2 species AAqC (Figure 17.3) (Table 17.3).

A view of above data showed that overall highest population is of insectivorous as well as carnivorous followed by grainivorous. Frugivorous feeding guild lies at next position followed by omnivorous and herbivorous feeding guild. Little Grebe was observed to be a diving carnivore. This observation goes well with that of Kumar (2005).

Table 17.3: Showing the Number of Bird Species Under Each Category of Carnivorous Type of Feeding Guild

Carnivorous Feeding Guilds	*Number of Bird Species*
Wading Carnivore	8
Diving Carnivore	1
Arboreal terrestrial carnivore	4
Arboreal aquatic carnivores	2
Total	15

As far as association of the birds is concerned, different categories was made in order to show a particular type of association. It includes solitary feeding, feeding in groups of 2–3 birds, feeding in small parties of 3–7 birds and feeding in small groups or in loose parties. As far as substrate preference of a particular type of species is concerned, it consists of aquatic, ground, ground as well as aquatic, arboreal feeders, aerial feeders, ground as well as arboreal feeders.

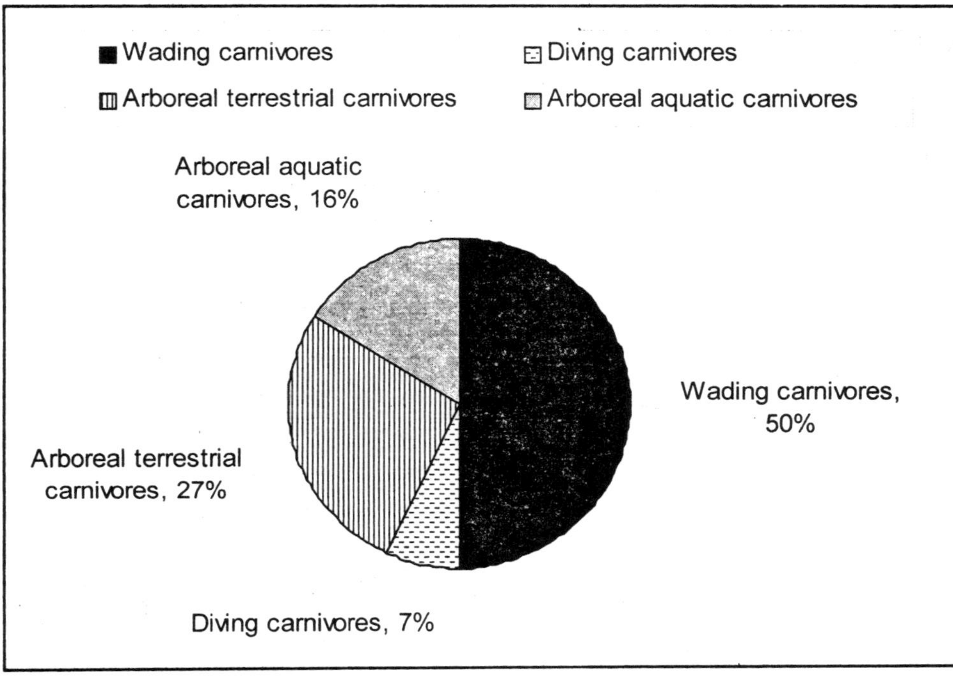

Figure 17.3: Showing the Percentage Proportion Constituted by Different Feeding Guilds Utilized by Birds Under Carnivorous Feeding Category

The bird community in the light of trophic relations presents a specific pattern of food resources. The proportion of the birds occupying the broad food categories like insectivorous, carnivorous, frugivorous, nectarivorous and omnivorous has been discussed by many workers (Dhindsa and Saini, 1994; Pramod, 1995; Makuloluwa *et al.*, 1997; Srinivasulu *et al.*, 1997; Bhatt and Sharma, 2000 and Sahi and Sharma, 2006). Bhatt and Sharma (*op cit.*) has tried to break-up major food categories into various guilds for more specific approach towards the feeding behaviour of species. A similar attempt has been made in the present study with regard to avifauna of Mansar, Jammu.

To conclude it can be said that the avifauna of Mansar exhibit a wide variety of feeding habits. All these depend upon a particular type of substrate. Moreover the avifauna of Mansar shows a characteristic type of association during feeding.

Feeding Guilds

Insectivore: Feeding on insects; *Carnivore*: Feeding on animal matter like fishes, amphibians, reptiles; *Herbivore*: Feeding on vegetable matter; *Frugivore*: Feeding on fruits; *Omnivore*: Feeding on all types of food including vegetable-matter, fruits, insects and other animal matter included in carnivore category; *Grainivore*: Feeding on grains; *Nectarivore*: Feeding on nectar of flowers.

Table 17.4: Showing Various Feeding Guilds of Avian Fauna at Mansar

Sl.No.	Name of the Bird	Feeding Guild
1.	Little Grebe	DC
2.	Little Cormorant	WC
3.	Darter	WC
4.	Indian Pond Heron	WC
5.	Night Heron	WC
6.	Grey Heron	WC
7.	Black Bittern	WC
8.	Little Egret	WC
9.	Cattle Egret	WC
10.	Pariah Kite	ATC
11.	Steppe Eagle	ATC
12.	Black Winged Stilt	SIP/AqC
13.	Red Wattled Lapwing	SIP/TI
14.	Indian Moorhen	WC/SIP
15.	Indian Coot	DH
16.	Indian White Breasted Waterhen	WC/SIP/TI
17.	Blue Rock Pigeon	G
18.	Little Brown Dove	G
19.	Indian Spotted Dove	G
20.	Alexandrine Parakeet	F
21.	Rose Ringed Parakeet	F
22.	Northern Spotted Owlet	ATC
23.	Small Blue Kingfisher	AAqC
24.	White Breasted Kingfisher	ATC/AAqC
25.	Pied Kingfisher	AAqC
26.	Indian Small Green Bee-eater	AI
27.	Indian Koel	F/I
28.	European Hoopoe	G/UI
29.	Blue Throated Barbet	F
30.	Lesser Golden Backed Woodpecker	T/BF
31.	Yellow Crowned Woodpecker	T/BF
32.	Golden Oriole	F/I
33.	Black Drongo	AI
34.	Indian Myna	G/F/I
35.	Brahminy Myna	G/F/I
36.	Indian Jungle Crow	O

Contd...

Table 17.4–Contd...

Sl.No.	Name of the Bird	Feeding Guild
37.	Rufous Tree pie	O
38.	Red Vented Bulbul	F/I
39.	White Cheeked Bulbul	F/I
40.	Jungle Babbler	UI
41.	Common Babbler	UI
42.	Paradise Flycatcher	AI
43.	Indian Tailor bird	CI/TI
44.	Himalayan Blue Whistling Thrush	O/UI
45.	Magpie Robin	TI
46.	Indian Robin	TI
47.	Brown Rock Chat	TI
48.	Pied Bush Chat	TI
49.	White Capped Water Redstart	AqI
50.	Indian White Eye	CI
51.	Purple Sunbird	N/I
52.	Indian House Sparrow	G/I
53.	Spotted Munia	G
54.	Grey Tit	C/I/F
55.	Crested Bunting	G
56.	Red Rumped Swallow	AI
57.	Wire Tailed Swallow	AI
58.	Indian White Wagtail	SIP/TI
59.	Large Pied Wagtail	SIP/TI
60.	Grey Wagtail	SIP/TI
61.	Indian Grey Shrike	ATC
62.	Common Grey Hornbill	F/I
63.	Mallard	DH

Insectivores and Carnivores are further categorized according to the place a species occupy during feeding. Each of the feeding guild is designated as per the below stated abbreviation

Insectivores

AI: Aerial Insectivore

CI: Canopy Insectivore

TI: Terrestrial Insectivore

SIP: Shore Insect Plover

AqI: Aquatic Insectivore

UI: Understorey Insectivore
T/BF: Trunk or Bark Feeder

Carnivores

WC: Wading Carnivore
DC: Diving Carnivore
ATC: Arboreal Terrestrial Carnivore
AAqC: Arboreal Aquatic Carnivore

Others

F: Frugivore
G: Grainivore
O: Omnivore
N: Nectarivore
DH: Diving Herbivore.

References

Ahmed, A., 2004. Diversity and community structure of the birds of Tehsil Doda, Jammu. *M.Phil. Dissertation*, University of Jammu, Jammu.

Altmann, J.C., 1974. Observational study of behaviour: Sampling method. *Behaviour*, 49: 227–285.

Bhatt, D. and Sharma, R., 2000. Diversity, status and feeding ecology of avifauna in Motichur area of Rajaji National Park, India. *Ann. For.*, 8(2): 179–191.

Dhindsa, M.S. and Saini, H.K., 1994. Agriculture ornithology: An Indian perspective. *J. Biosci.*, 19(4): 391–402.

Kumar, S., 2006. Diversity of avian fauna of District Kathua, J&K. *Ph.D. Thesis*. University of Jammu, Jammu.

Macdonald, M., 1960. Communal nesting in Babblers. *J. Bomb. Nat. Hist. Soc.*, 56(1): 132–134.

Makuloluwa, S. Alagoda, T. and Santiapillai, C., 1997. Avifaunal diversity in the highlands of Sri Lanka. *Tiger Paper*, 24(3): 17–2.

Parmod, P., 1995. Ecological studies of bird communities of Silent Valley and neighbouring forests. *Ph.D. Thesis*, Calicut Univ., pp. 99.

Rana, B.D., 1970. Some observations on the food of Jungle Babbler, *Turdoides striatus* and the Common Babbler, *Turdoides caudatus* in the Rajasthan desert, India. *Pavo*, (8): 35–44.

Sahi, D.N. and Sharma, B., 2006. Diversity status and feeding ecology of avifauna of Ramnagar Wildlife Sanctuary, Jammu. In: *Perspectives in Animal Ecology and Reproduction*, (Eds.) V.K. Gupta and A.K. Verma. Daya Publishing House, New Delhi, 3: 244–264.

Sharma, B., 2003. Faunal diversity of Ramnagar Wildlife Sanctuary, Jammu. *M.Phil. Dissertation*, University of Jammu, Jammu.

Snow, D.W., 1962. A field study of the black and white manakin, *Manacus manacus*, in Trinidad. *Zoological*, New York, 47: 65–104.

Srinivasulu, B., Srinivasulu, C., Rao, V.V., Koteshwarulu, C. and Nagulu, V., 1997. Avian use of Paddy agroecosystem. *Pavo*, 35: 75–84.

Chapter 18

Penguins: The Amazing Creatures

☆ *Anil K. Verma and B.L. Koul*

Introduction

Penguins are unusual amazing, delightful and wonderful creatures. They walk upright like clowns in a circus and do not fly and their tidy plumage makes it easy to believe that they are dressed in a suit. They are much loved across the world thanks to various TV channels and are some of the most animal favourites in zoos across the world.

Although now among the most popular of birds yet they are relatively new to us in the northern hemisphere. It was not until the 15th century that sailors venturing into southern waters discovered penguins (Figures 18.1–18.2). These first penguin/ mankind interactions were pretty brutal. Sailors regarded penguins, who were perhaps the only birds they had ever seen which showed no fear of man, as a source of easy food. All they had to do was a walk into a colony and club them to death. Later they were killed not just for food but for oil.

However, it is unfortunate that oil spills in South Africa and farming in South America, as well as farming and a collection of introduced predators in New Zealand are threatening them. Yellow-eyed Jackass, Humboldt's and Little penguins are all suffering declining populations. Fortunately, no penguin species have been so far, driven to extinction by this process and now that we have learned more respect for the world around us, we are free to admire their sometimes comical but always amazing existence.

Specific Distribution

There are 17 species of penguins all of which have a southerly distribution ranging from the Antarctic itself to the Galapagos Islands (Figure 18.3) *viz.*, 1. Emperor Penguin, 2. Adelie Penguin, 3. King Penguin, 4. Royal Penguin, 5. Gentoo Penguin,

Figures 18.1–18.2: Showing Penguins in Ventral View (1) and Lateral View (2)

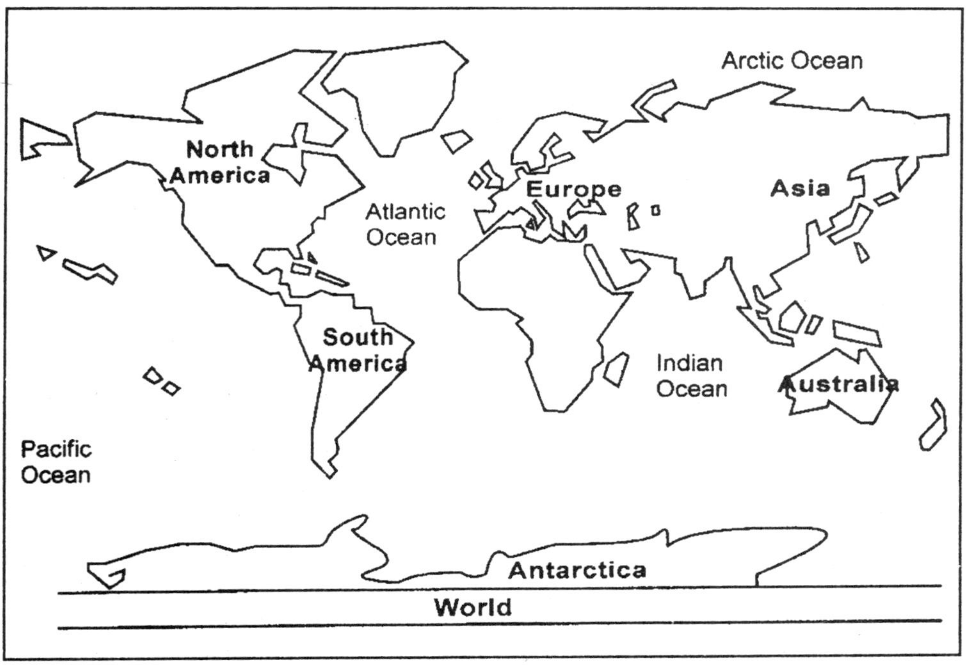

Figure 18.3: Showing Distribution of Penguins Around the World

6. Chinstup Penguin, 7. African Penguin, 8. Magallenic Penguin, 9. Humboldt Penguin, 10. Yellow-eyed Penguin, 11. Rockhopper Penguin, 12. Macaroni Penguin, 13. Little Penguin, 14. Galapagos Penguin, 15. Fiordland Penguin, 16. Snares Island Penguin, 17. Erect crested Penguin.

Only two species, Emperor and Adelie, have entirely Antarctic distributions. The rest live more northerly lives to varying degrees, 5 species being sub-antarctic, 6 southern temperate and 4 sub-tropical. Though one species, the Galapagos Penguin (currently the rarest species), lives in the equatorial band it is protected to some extent form the heat by the cold Antarctic currents which bath the islands. Penguins are not found in the northern hemisphere. Penguins are highly adapted to marine life and some species spend up to 80 per cent of their life at sea.

Europeans have on several occasions attempted to introduce Penguins of various species to the northern hemisphere, and although some of the introduced specimens survived for a number of years none were observed to breed. Breeding for a number of species has been achieved successfully in zoos however.

Penguins as Excellent Divers

Penguins do not fly and walk only slowly, though they can outdistance a running man while tobogganing on their bellies. Penguins look awkward, even comical, on land, they are however elegant when in the sea. They have evolved to live in the freezing southern waters. They are excellent swimmers using their strengthened and modified wings to fly through the water. Their whole bodies are designed to make them a success in their chosen environment. Propelled by their wings and steering by their feet. Penguins can reach speeds of up to 13 miles per hour.

Penguins are excellent divers and the Emperor is the record holder for both dive duration (18 minutes) and dive depth (175 ft.). The King Penguin comes in second with recorded dives of 783 ft. However, dives like this are exceptional, normally penguins dive much more shallowly and far less deeply. A survey reported in 1995 by the scientists (TG and GL Kooyman, 1995) indicate that on average Emperor Penguins dive to depths of between 25 to 40 metres and for times of 4 to 5 minutes.

Special Adaptations

Many Penguins live in cold environments, under extreme conditions. Male Emperor Penguins can endure temperatures below –20°C for several months without food while incubating their single egg during the Antarctic winter. To achieve this remarkable feat they obviously have had to evolve some special adaptations. Penguin's first layer of defence is the ubiquitous and remarkably efficient feathers which form a double layer of protection incorporating layer of highly insulating air between them. This feather barrier supplies 84 per cent of a Penguin's thermal insulation.

A Penguin's second layer of defence is a thick layer of blubber immediately beneath the skin. Blubber is basically fatty oils and is a bad conductor of heat as well as a valuable store of energy. During the breeding season some of the more southerly Penguins may be as much as 32 per cent blubber by weight. This blubber is invaluable in saving Penguin lives.

Penguin's thermal insulation is so efficient that penguins living on the ice shelves of the Antarctic overheat if they spend too long in the sea. Water freezes at 0°C, if it is liquid it must be warmer than this (usually about 4°C). This can be much warmer than the surrounding air which gets as low as –10°C near the sea. So penguins standing around on the ice are not just enjoying the scenery, many of them are actually cooling down before getting back into much warmer sea to look for food.

Not all Penguins however live in the frozen water of the Antarctic, coastal Africa, South including the Galapagos, are home to Penguins. Needless to say many of these habitats are considerable warmer than mainland Antarctica and in the more northerly ones this excellent insulation that protects more southerly living species can become a problem, causing overheating. Penguins respond to overheating in several ways all of which are designed to increase heat loss.

Foraging and Dietary Habits

Penguins hunt for their food in the seas and oceans of the Southern Hemisphere. They consume considerable amounts of fish, krill and squid. The amount of time penguins spend at a sea foraging depends on the time of year. Some species are prone to longer trips with Emperor Penguins spending 60 to 75 days at sea at a time while Gentoo Penguins make foraging trips of only 4 to 12 hours during the breeding season. Foraging during the breeding season obviously puts a heavy burden on the prey populations near the nesting site as penguins are then grouped together in large numbers in breeding grounds called rookeries. The rookeries are like mad houses where the inmates (Penguins) make deafening noises.

Table 18.1: Showing Breeding Traits in Emperor Penguin and King Penguin

Common Name	*Scientific Name*	*Height*	*Weight*	*Home*
Emperor	*Aptenodytes fosteri*	120 cm. 4 ft.	23-45 Kg.	Antarctic
Nest	*Eggs*	*Young*	*Season*	*Incubation*
No Nest	1 white egg	1	Oct. to Nov.	60 days
Common Name	*Scientific Name*	*Height*	*Weight*	*Home*
King	*Aptenodytes patagonicus*	90 cm. 3 ft.	14-18 Kg.	Antarctic and Sub-Antarctic Islands
Nest	*Eggs*	*Young*	*Season*	*Incubation*
No Nest	1 egg	1	Nov. to Mar.	54 days

Breeding Traits

Penguins are monogamous and breed in a diverse variety of habitats ranging from the frozen water of the Antarctic to the scorching larva flows of the Galapagos Islands. Most Penguins congregate in large dense colonies to nest (except Yellow-eyed and Fiordland Penguins). These colonies range in size from 200 to 300 birds for

Gentoo Penguins through to 600 000 or more birds in Chinstrap, King and Macaroni Penguins. Most species nest on the surface with simple unlined nests though some will nest in burrows and or caves.

Clutch Size and Parental Care

Most species of Penguin lay two eggs, the exceptions to this being the two larger species Emperor and King Penguins which both lay only one egg. Both parent play an active role in incubating the egg and raising the young, though in emperor Penguins the male does all the incubating and the female all the caring before creching. Young Penguins aggregate in creches after a certain period of time. This frees both parents to go foraging to feed the hungry young and prevents them from getting lost on their own (Table 18.1). It also helps protect them against predators.

Predators and Threats to Penguins

The primary predators of eggs and chicks are Skuas and Gulls for most species of Penguins except the Emperors which have to deal with Giant Petrels instead. However, since he coming of human beings introduced predators such as Dogs, Cats, Pigs, Ferrets and Rats have become increasingly more important as predators. At sea Penguins have to deal with a different selection of predators. Leopard Seals, Fur Seals, Sea Lions and Elephant Seals are all considerable predators of Penguins at sea. Often these predators will be waiting in the sea near nesting colonies to prey on adults coming to and from foraging trips and on young birds entering the sea for the first time.

References

http :/www.earthlife.net/birds/sphenicidae.html.

Lyod, S. Davis and Renner, M., 1995. *Penguins London.* T&AD Poysa.

Sparks, I. and Soper, T., 1987. *Penguins.* David and Charles, London.

Williams, T.D., 1995. *The Penguins (Bird Families of the World).* Oxford University Press.

Chapter 19

Importance of New Systematics in Parasitological Taxonomy

☆ *P.L. Koul and M.K. Raina*

Introduction

By applying the correlation coefficient formula the present study is an attempt to introduce a new element in Parasitological Taxonomy which is mostly based on "*The classical methods*" (Crites, 1962) although the main criterion in Parasitological Taxonomy is yet not absolute in itself especially regarding speciation. The need therefore is to support the classical methods of speciation by the "

" *i.e.* creating new species and verifying the validity of already existing ones on the basis of reproductive behaviour of inbreeding populations and reproductive isolation, chromosome counts, histochemical and biochemical analyses, DNA studies and the application of correlation coefficient formula.

Materials and Methods

Studies were conducted on some genera of ACANTHOCEPHALA Rudolphi, 1809 which constitutes one of the most interesting groups of parasites commonly reffered to as "the spiny headed worms" which are found in fishes, ducks and many wild animals. These are endoparasitic dioecious pseudocoelomate vermiform bilateria without a digestive tract (Hyman, 1951) and with a retractile proboscis as the most obvious and characteristic feature (Crompton, 1970) which is armed with hooks. The parasites of the genera ACANTHOSENTIS, NEOECHINORHYNCHUS and HEBESOMA collected from the hosts pressed and fixed in Carnoy's fluid, GURR 1962 stained with acetoalum carmine, dehydrated and permanent slides prepared in Canada balsam.

Classical method of taxonomy was used but in addition the correlation coefficient formula, given under, was used to determine the closeness of the genera and species.

$$r_{xy} = \frac{\sum x_i y_i - n\overline{x}\overline{y}}{(n-1)s_x s_y} = \frac{n\sum x_i y_i - \sum x_i \sum y_i}{\sqrt{n\sum x_i^2 - \left(\sum x_i\right)^2}\sqrt{n\sum y_i^2 - \left(\sum y_i\right)^2}}$$

Observations

The hook size was used for determining the correlation coefficient and the hook size of the genera and the species is as under.

Table 19.1

Sl.No.	Genus.	Sp.	Male	Female
1.	*Acanthosentis*	*Umai*	1st circlet 35, 37, 37, 38, 42, 42	38, 40, 42, 43, 46, 50
			2nd circlet 22, 24, 24, 27, 28, 33	22, 23, 26, 27, 30, 31
			3rd circlet 18, 21, 21, 21, 22, 22	16, 24, 25, 25, 25, 26
2.	*Acanthosentis*	*Vancleavei*	1st circlet 40, 45, 50, 50, 52, 58	50, 55, 55, 55, 58, 58
			2nd circlet 29, 29, 30, 32, 34, 36	31, 31, 31, 33, 33, 35
			3rd circlet 22, 22, 25, 27, 27, 30	24, 24, 28, 33, 33, 35
3.	*Acanthosentis*	*Kawi*	1st circlet 42, 46, 46, 48, 50, 50	43, 44, 45, 51, 54, 55
			2nd circlet 25, 32, 33, 33, 42, 42	28, 31, 33, 33, 35, 38
			3rd circlet 25, 26, 26, 31, 32, 34	23, 24, 27, 31, 33, 33
4.	*Hebesoma*	*Guptai*	1st circlet 20, 20, 24, 24, 26, 34	20, 25, 25, 27, 27, 31
			2nd circlet 15, 17, 24, 24, 25, 28	17, 26, 26, 31, 31, 34
			3rd circlet 15, 17, 18, 26, 27, 28	15, 20, 20, 24, 26, 28
5.	*Neoechino-rhynchus*	*Ladakhensis*	1st circlet –	60, 60, 60, 63, 63, 69
			2nd circlet –	15, 20, 21, 27, 27, 30
			3rd circlet –	15, 27, 27, 30, 33, 36

In determining the correlation coefficient *r* in the formula used

n: No. data points
ΣXY: Sum of XY products
$\Sigma X \Sigma Y$: Sum of X series multiplied sum of Y series
ΣX^2: Sum of the square of X series
$(\Sigma X)^2$: Sum of X series whole square
ΣY^2: Sum of the square of Y series
$(\Sigma Y)^2$: Sum of Y series whole square
ΣX: Sum of X series
ΣY: Sum of Y series

If the value of the coefficient is 1 it denotes a very close relationship between the specimens and when the value is less than one it indicates lesser relationship and when the value is less than 0.6 meaning lesser relationship.

Example

Acanthosentis kawi	*Neoechinorhynchus ladakhensis*			
Male	*Female*			
X	*Y*	*XY*	X^2	Y^2
1st circlet				
42	60	2520	1764	3600
46	60	2760	2116	3600
46	60	2760	2116	3600
48	63	3024	2304	3969
50	63	3154	2500	3969
50	69	3450	2500	4761
282	375	17664	13300	23499
$(\Sigma X)^2$	$\Sigma X \Sigma Y$	$(\Sigma Y)^2$		
79524	105750	140625		
n=6				

Calculations (We need to actually show, multiplications and subtraction)

Hence,

$$r = \frac{6 \times 17664 - 105750}{\sqrt{[6 \times 13300 - 79524][6 \times 23499 - 140625]}}$$

$$r = \frac{105984 - 105750}{\sqrt{[79800 - 79524][140994 - 140625]}}$$

$$r = \frac{234}{\sqrt{[278][369]}}$$

$$r = \frac{234}{319.1}$$

$$r = 0.733$$

Acanthosentis kawi	*Neoechinorhynchus ladakhensis*			
Male	*Female*			
X	*Y*	*XY*	*X²*	*Y²*
2nd Circlet				
25	15	2520	1764	3600
32	20	2760	2116	3600
33	21	2760	2116	3600
33	27	3024	2304	3969
42	27	3150	2500	3969
42	30	3450	2500	4761
207	140	17664	13300	23499
42849	28980	19600		
r = 0.88				
3rd circlet				
25	15	375	625	225
26	27	702	676	729
26	27	702	676	729
31	30	930	961	900
32	33	1056	1024	1089
34	36	1224	1154	1296
174	168	4989	5116	4968
30276	29232	28224		
r = 0.84				

By applying the formula the following observations were made:

Acanthosentis umai	Male	0.94, 0.91, 0.96
	Female	
A. vancleavei	Male	0.88, 0.96, 0.97
	Female	
A. kawi	Male	0.91, 0.94, 0.95
	Female	
A. umai	Male	0.91, 0.97, 0.71
A. vancleavei	Female	
A. umai	Male	0.78, 0.96, 0.67
A. kawi	Female	
A. umai	Male	0.94, 0.82, 0.77
A. kawi	Female	
A. vancleavei	Male	0.90, 0.94, 0.62
A. kawi	Female	
A. vancleavei	Male	0.88, 0.91, 0.94
A. kawi	Female	
A. vancleavei	Male	0.96, 0.78, 0.95
A. kawi	Female	
A. umai	Male	0.80, 0.92, 0.90
A. ladakhensis	Female	
A. vancleavei	Male	0.83, 0.91, 0.83
N. ladakhensis	Female	
A. kawi	Male	0.73, 0.88, 0.84
N. ladakhensis	Female	

18 point basis

Hebesoma gupti	Male	0.778
	Female	
H. gupti	Male	0.419
A. umai	Female	
H. guptai	Male	0.404
A. vancleavei	Female	
H. guptai		0.577
A. kawi		
H. guptai	Male	0.3877
N. ladakhensis	Female	

Discussion

Expressing his concern in his presidential report 2006 Prof. David Cutler president Linneaen society London said "I hope we can work together on determining the way forward and tackle the root problems in the apparent decline in taxonomy and in the number of taxonomists. Taxonomy plays a key role in our understanding; conservation and management of the natural world; we need effective policies to ensure that a succession of new taxonomists is trained and all major groups of organisms are covered." According to Crites (1962) species have so far been created just on 'classical methods' which although the main criterion in parasitological taxonomy is not absolute in itself especially regarding speciation. It is therefore desirable that the so called classical methods of speciation be supported by the new systematics (Crites, 1962) in parasitological taxonomy *i.e.* creating the new species and verifying the validity of the already existing ones on the basis of reproductive behaviour, inbreeding populations and reproductive isolation; histochemical and biochemical analyses; chromosome counts and DNA studies. This will generate new enthusiasm in the field of taxonomical studies and establish it on modern footing.

Daskalov (1965) reports the establishment of *H. contortus* and *H. placei* on reproductive isolation. Debates are being conducted as to whether the 'future classification be exclusively DNA based' by the Linneaen society London. On 30th November 2006 Dr. Alfried Vogler in a debate argued strongly that DNA based techniques produce results that are unbiased by character interpretations and clearly define species and their relationships. The more we learn about the DNA of the organisms the less relevant morphology becomes to define them.

Although DNA may not provide the "silver bullet" to taxonomy yet it can be used as a strong supportive tool for modern taxonomy.

In the present write up an attempt has been made to introduce yet another element in new systematics especially in parasitological taxonomy in general and acanthocephalan taxonomy in particular.

When a correlation coefficient was obtained for the hooks between the male and female of the same species mostly a value more than 0.9 was obtained indicating a very close relationship but when the value was obtained for different species of the same genus it was slightly less indicating lesser relationship but when the value for different genera was obtained it was less than 0.6 indicating poor relationship hence justification for separate genera and species. It is hoped that the suggestion to use correlation coefficient as a supportive method shall receive the attention of helminth taxonomists.

References

Crites, J.L., 1962. Morphology as a basis of identification and classification of parasites. *J. Parasitol.*, 48(5): 652–655.

Crompton, D.W.T., 1969. On the environment of *Polymorphus minutus* in ducks. *Parasitology*, 58: 19–28.

Cutler, David, 2006. President's report. *The Linnean Society of London, Annual Report*, pp. 3.

Daskalov, P., 1965. The reproductive isolation between *Haemonchus contortus* and *H. placei. IZV Tsent. Khelmint. Lab. Sof,* 10: 11–17.

Gurr, E., 1962. *Staining Animal Tissue Practical and Theoretical.* Leonard Hill (Books) Ltd., The Tower Brooke, Green Rd, London, W6, pp. 1–631.

Hyman, L.H., 1951. *The Invertebrate, Vol. 3: Acanthocephala, Aschelminthes and Entoprocta The pseudocoelomate bilateria.* McGraw Hill Book Company Inc., New York, Toronto.

Vogler, Alfried, 2006–07. *Linnean Tercentenary 2007 Newsletter and Proceedings of the Linnean Society of London,* 23(2): 8–9.7.

This humble write up is dedicated to my alma mater my respected teachers and the alumni.

P.L. Kaul

Chapter 20

Insect Pests Associated with the Medicinal Plants in Shivalik Region of Jammu, J&K State

☆ *Madhu Sudan**

ABSTRACT

The present study was conducted in foot hills of shivaliks in Rajouri region of J&K. During the periodical surveys, as many as 24 species of insect pests were recorded from the medicinal plants. The pests were identified and studied with respect to their taxonomic status, mode of infestation and nature and extent of damage done to the plants. *Henosepilachna vigintioctopunctata, Paramecops farinosa, Papilio polytes, Papilio polyctor, Anosia chrysippus, Nezara viridula, Spilostethus pandurus* were the pests of economic importance, recorded in the study.

Keywords: *Medicinal plants, Pests, Infestation, Damage, Economic loss etc.*

Introduction

India being the largest producer of medicinal herbs is appropriately called the "Botanical Garden of the World" (Seth and Sharma 2004) and it has one of the richest plant medical cultures in the world. Indian people have a tremendous passion for medicinal plants and use them for a wide range of health related applications from common cold to memory improvement and treatment of snake bites to cure for muscular dystrophy and the overall enhancement of general body immunity. Medicinal plants

*E-mail: madhu.ento@gmail.com/sudan.madhu92@yahoo.in.

as a group constitute approximately 8000 species and account for around 50 per cent of all the higher flowering plant species of India.

Jammu and Kashmir State is well known for its rich variety of medicinal plants. It has a rich medicinal plant wealth and more than 2,500 plant species are used by "Vaids" and "Hakims" (persons having knowledge of Unani medicine). The local Hakims attribute some medicinal property to almost every plant species. At least 45 to 50 plant species used in the Indian system of medicine and found in Jammu and Kashmir are on the brink of extinction, according to a working group on medicinal project (c.f. Internet). Despite this entomofauna associated with medicinal flora of Jammu and Kashmir has not been paid due attention.

Workers like Gould (1960), Popov (1971,1972, 1973a,b), Neubauer (1974), Tiwari and Joshi (1974), Verma *et al.* (1978), Ali (1980), Reddy *et al.* (1981), Sagar (1981), Spiridonova and Nosyrev (1981), Bhutani (1982), Campbell and Pike (1985), Sagar and Reddy (1987a, b), Ramji and Sagar (1990a,b,c) , Ramzan *et al.* (1990), Joshi *et al.* (1991), Swamy and Rajagopal (1995a, b), McPartland (1996), Murugesan *et al.* (1997), Rai and Singh (1997), Tosic *et al.* (1997) Tripathi *et al.* (1997), Meshram and Garg (1999), Mitra and Biswas (2002), Arif and Kumar (2003), Chandra (2004), Meshram (2005), Sharma and Verma (2005), Banjo *et al.* (2006) and Venkatesha (2006) have done some work on various aspects of insect pests of medicinal plants in different parts of the country and world.

Hardly any detailed information is available on the survey, bio-ecology and control measures of the insects infesting the medicinal plants, though some sporadic records of insect pests on medicinal and aromatic plants from this region are available (Mathur and Srivastava, 1967, Srivastava and Saxena, 1976 and Bhagat, 2004). The recent records of insect pests of medicinal plants available from district Rajouri are that of Sharma *et al.* (2007, 2008a, b) and Sudan (2008).

Above authors have just casually reported individual insect pests but nothing is known about their bio-ecological interactions, life cycles, modes of infestations, and symptoms of infestations which result in heavy losses to medicinal plants.

So keeping in view, the existing lacunae, the present work was carried out to investigate the diversity, distribution, morphology, bionomics, ecology and floral association of insects in District Rajouri of Jammu region.

Material and Methods

The present work has been carried out in District Rajouri of Jammu division which falls in Pir Panjal Range of the J&K State. Field studies were conducted during the period from March 2007 to February, 2008 mainly in plain and kandi belt of the district *viz.*, Sunderbani, Nowshera, Kalakote and Rajouri

Rajouri district with an area of 2,630 sq. kms is situated between 70° to 74° 4' East Longitude and 32° 58' and 33° 35' North Latitude. It is located in the foothills of Pir Panjal Range. It is flanked by district Poonch in the North, Jammu in the South, Udhampur in the East and Pakistan in the West. The district has some peculiar physical features. The Dhaula-Dhar range runs across the north eastern part of the district. While topography of Rajouri, Budhal, Thanna Mandi and Darhal hillsis subtemperate, the tehsils of Sunderbani, Nowshera and Kalakote, mostly plain and kandi situated in South is relatively warmer and sub-tropical. The average rainfall in the district is 500 mm and average temperature varies from a minimum of 7.42°C to a maximum of 37.4°C.

Medicinal plants from all the four study stations (Figure 20.1) were screened randomly for insect pests. The insects along with their immature stages *i.e.*, eggs, larvae, pupae and adults were collected from the fields by various methods such as hand picking, stem beating and also with the help of entomological nets and were preserved were preserved by methods (Sudan, 2008) for further studies. Observations with regards to various stages of the insect pests, their feeding behaviour and mode of damage caused by each insect species were made both in the field as well as in the laboratory.

In the field, symptoms of damage by feeding of larvae and adults were recorded by visual observations. In the laboratory, insects were reared both in clear plastic jars with a porous lid as well as in insect rearing cages with wire meshed windows for aeration and provided fresh tender leaves of their host plants and their mode of feeding was recorded. General morphological studies were made under different magnifications of the stereoscope microscope. Photographs have been taken with Digital Still Camera, SONY Cyber-Shot (Model No.: DSC-T10) with 5x optical zoom having 7.2 effective mega pixels with inbuilt macro function for extreme closeup shots.

Observations and Discussion

During the period of observations, species belonging to 22 genera under 12 families of 4 insect orders have been recorded. Out of 24 species, 11 species belong to order Coleoptera; 7 species belong to order Lepidoptera; 5 species belong to order Hemiptera and 1 species belong to order Orthoptera. Thus, order Coleoptera contributes to the maximum with highest number of insects and order Orthoptera being the minimum.

A total of 24 insect species belonging to 12 families were recorded from the following medicinal plants *viz. Calotropis procera* belonging to family Asclepiadaceae, *Cannabis sativa* belonging to family Cannabaceae, *Justicia adhatoda* belonging to family, *Mentha* sp. belonging to family Labiatae (Lamiaceae), *Ocimum sanctum* belonging to family Labiatae (Lamiaceae), *Solanum nigrum* belonging to family Solanaceae, *Vitex negundo* belonging to family Verbenaceae and *Zanthoxylum armatum* belonging to family Rutaceae for the first time from District Rajouri of Jammu province of Jammu and Kashmir State (Table 20.1).

These plants are affected during several stages, including flowering, fruiting and also during the vegetative growing stage. Based on all the observations, it can be concluded that like any other agricultural crop or plant, medicinal plants (both wild as well as cultivated) too have to bear the devastating attacks of injurious insect pests.

Out of 12 families of insects, 6 insects species in 6 families were recorded from *Calotropis procera*; 5 insect species in 4 families from *Cannabis sativa*; 2 insect species

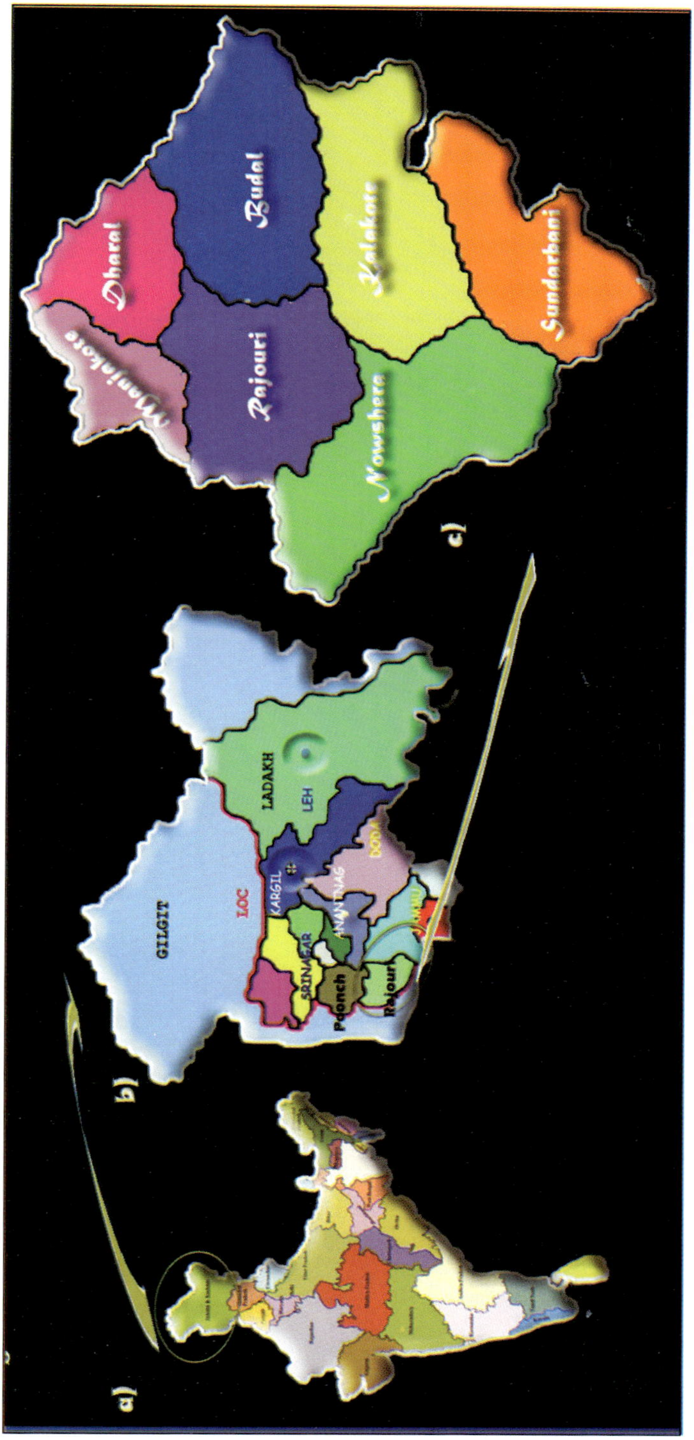

Figure 20.1: Map Showing Study Stations

Figure 20.2

Table 20.1: List of Insects Recorded on some Important Medicinal Plants Investigrated during Studies

Sl.No.	Plant Species	Insects Recorded	Family	Order	Uses of Plant
1.	*Calotropis procera*	*Corynodes peregrines*	Chrysomelidae	Coleoptera	Exhibits anti-cancer, antispasmodic, anti-inflammatory, antihelminthic and anti-blood coagulating activity. Also used in Cholera, leprosy, spleenomegaly, painful rheumatic joints, swellings, sores and wounds
		Paramecops farinose	Curculionidae	Coleoptera	
		Anosia chrysippus	Nymphalidae	Lepidoptera	
		Spilostehus pandurus	Lygaeidae	Hemiptera	
		Dolycoris baccarum	Pentatomidae	Hemiptera	
		Poekilocerus pictus	Pyrgomorphidae	Orthoptera	
2.	*Cannabis sativa*	*Oxycetonia versicolor*	Scarabaeidae	Coleoptera	Antihelminthic, diuretic, expectorant and aphrodisiac analgesic for cancer and AIDS patients undergoing chemotherapy. Increases heart frequency and peripheral vasodilation. Also used in diarrhoea, dysentery, insomnia, neuralgia, haemorrhoids, neurological disorders and skin diseases.
		Amblyrhinus poricollis	Curculionidae	Coleoptera	
		Helicoverpa armigera	Noctuidae	Lepidoptera	
		Nezara viridula	Pentatomidae	Hemipetra	
		Dolycoris baccarum	Pentatomidae	Hemiptera	
3.	*Justicia adhatoda*	*Nezara viridula*	Pentatomidae	Hemiptera	Exhibits strong respiratory stimulatory activity, moderate hypotensive activity and cardiac depressant effect. Used in dyspepsia, anorexia, haemoptysis, cold, allergic cough, whooping cough, chronic bronchitis, asthma and fever.
		Eysarcoris ventralis	Pentatomidae	Hemipetra	
4.	*Mentha longifolia*	*Chrysolina exanthematica*	Chrysomelidae	Coleoptera	Antibacterial, antiviral, antifungal, antiparasitic, antispasmodic, antiasthmatic, antinausea, antiseptic, diuretic, analgesic, nervine, antiemetic, antihysteric, carminative and stimulant. Used in common cold, cough, bronchitis, inflammation of mouth and pharynx and for liver and gall bladder complaints, stomachache, colic, diarrhoea, cardalgia, catarrh and migraine and externally for myalgia and neurologic ailments. Acts as a digestive tonic, relieves muscle spasms, relaxes peripheral blood vessels and stimulates bile secretion.
		Chrysolina bella	Chrysomelidae	Coleoptera	
		Altica cyanea	Chrysomelidae	Coleoptera	
		Monolepta signata	Chrysomelidae	Coleoptera	
		Oxycetonia versicolor	Scarabaeidae	Coleoptera	
		Trichoplusia ni	Noctuidae	Lepidopetra	
		Dolycoris baccarum	Pentatomidae	Hemiptera	
		Eysarcoris ventralis	Pentatomidae	Hemiptera	

Contd...

Table 20.1–Contd...

Sl.No.	Plant Species	Insects Recorded	Family	Order	Uses of Plant
5.	*Ocimum sanctum*	*Monolepta signata* *Pericalli ricini* *Nezara viridula* *Eysarcoris ventralis*	Chrysomelidae Arctiidae Pentatomidae Pentatomidae	Coleoptera Lepidoptera Hemipetra Hemiptera	Exhibits antifungal, antibacterial, antipyretic and antistress property. Specific for alleviating anorexia, parasitic infections, rhinitis, catarrh, cold, cough, asthma, fever, spleenic affections, toxicosis and skin eruptions. Stimulates digestion, cures dysurea, vitiation of blood, urticaria, chronic skin diseases, piles, parasitic infections, vomiting, earache, conjunctivitis, post parturition pain, chest diseases; pharynx, bronchial and lung infections.
6.	*Solanum nigrum*	*Henosepilachna vigintioctopunctata* *Eysarcoris ventralis* *Nezara viridula*	Coccinellidae Pentatomidae Pentatomidae	Coleoptera Hemiptera Hemipetra	Analgesic, antispasmodic, anti-inflammatory, antiperiodic, antiphlogistic, antiseptic, diaphoretic, diuretic, emollient, febrifuge, narcotic, purgative, sedative and vasodilator. Specific for cough, asthma, rhinitis, anorexia and parasitic infections, diabetes, hepatitis, skin lesions, oedema, cough, insomnia and muscular atrophy. Also in the treatment of cancerous sores, boils, leucoderma, wounds, cirrhosis of liver, viral hepatitis, as an adjuvant to hepatotoxic drugs and for inflammation of joints.
7.	*Vitex negundo*	*Hyblaea puera* *Erthesina fullo*	Hyblaeidae Pentatomidae	Lepidoptera Hemiptera	Exhibits antibacterial, anti-inflammatory, antihelminthic, analgesic, febrifugal, expectorant and diuretic properties. Used in dyspepsia, rheumatism, catarrh, cough, asthma also for boils and piles.
8.	*Zanthoxylum*	*Colasposoma semicostatum* *Monolepta signata* *Platymycterus himalayanus* *Papilio polytes* *Papilio polyctor* *Erthesina fullo* *Nezara viridula*	Chrysomelidae Chrysomelidae Curculionidae Papilionidae Papilionidae Pentatomidae Pentatomidae	Coleoptera Coleoptera Coleoptera Lepidopetra Lepidoptera Hemiptera Hemipetra	Exhibits antiseptic antihelminthic antispasmodic carminative, stomachic hypoglycemic activity. Used as a pain relieving drug for discomfort and congestion in chest, cardiac region, piles and tumours, circulatory stimulant and to treat arthritis, leg ulcers and chronic pelvic inflammatory affections, diseases of mouth and throat.

in 1 family from *Justicia adhatoda;* 8 species in 4 families from *Mentha longifolia;* 4 insect species in 3 families from *Ocimum sanctum;* 3 insect species in two families from *Solanum nigrum;* 2 insect species in 2 families from *Vitex negundo and* 7 insect species in 4 families from *Zanthoxylum armatum* (Figures 20.3 and 20.4).

It is found that majority of insects are polyphagous like *Nezara viridula, Dolycoris indicus, Eysarcoris ventralis, Erthesina fullo, Oxycetonia versicolor, Monolepta signata, Henosepilachna vigintioctopunctata* etc. while few are oligophagous like the lepidopteran caterpillars of *Danaus chrysippus* and nymphs and adults of *Spilostethus pandurus* which feed on the leaves, buds, flowers and seed pods of *Calotropis procera*. The *Calotropis procera* produces a cardiac glycoside 'Calotropin' permitting only chemically adapted insects to feed on it. The lepidopteran larva has adapted itself marvelously to feed on the leaves containing this glycoside for imbibing it for saving itself from predators and such adaptations have been of great interests to scientists.

In nutshell, the data recorded includes 4 insect orders, 12 families, 22 genera and 24 identified species from some economically important medicinal plants. Out of 4 orders of insects recorded, Coleoptera contributes to the maximum (11 species), Lepidoptera (7 species), Hemiptera (05 species) and Orthoptera with 01 species only (Figure 20.5). the species belonging to order Coleoptera (*viz., Chrysolina exanthematica, Chrysolina bella, Colasposoma semicostatum, Corynodes peregrines, Altica cyanea, Monolepta signata, Oxycetonia versicolor, Henosepilachna vigintioctopunctata, Paramecops farinosa, Amblyrrhinus poricollis, Platymycterus himalayanus*), Lepidoptera (*viz., Danaus chrysippus, Papilio polytes, Papilio polyctor, Trichoplusia ni, Helicoverpa armigera, Pericallia ricini, Hyblaea puera*) and Orthoptera (*Poekilocerus pictus*) are defoliators where

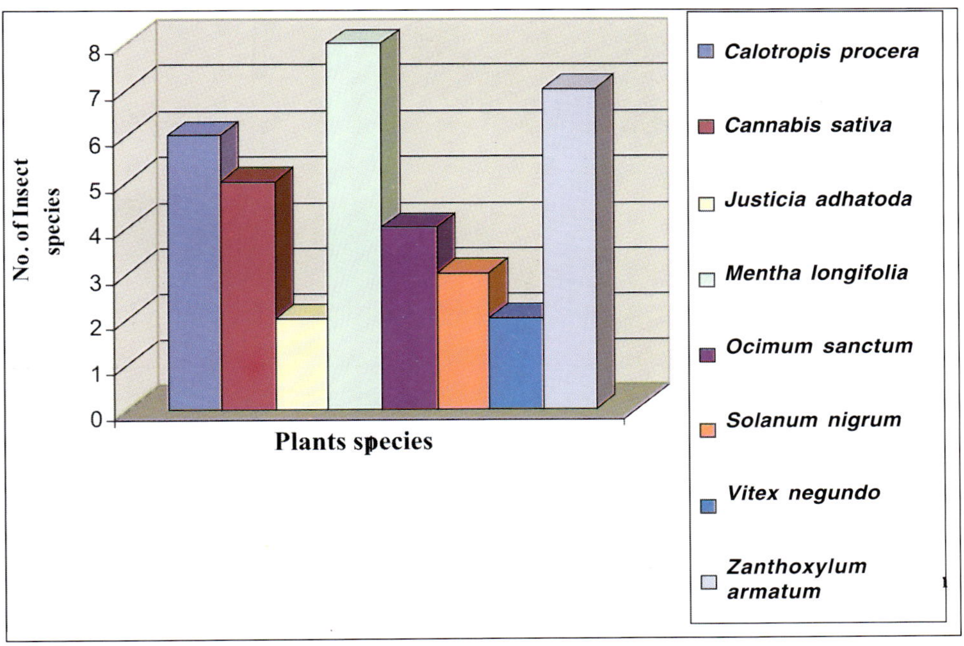

Figure 20.3: No. of Insect Species Found on Each Plant

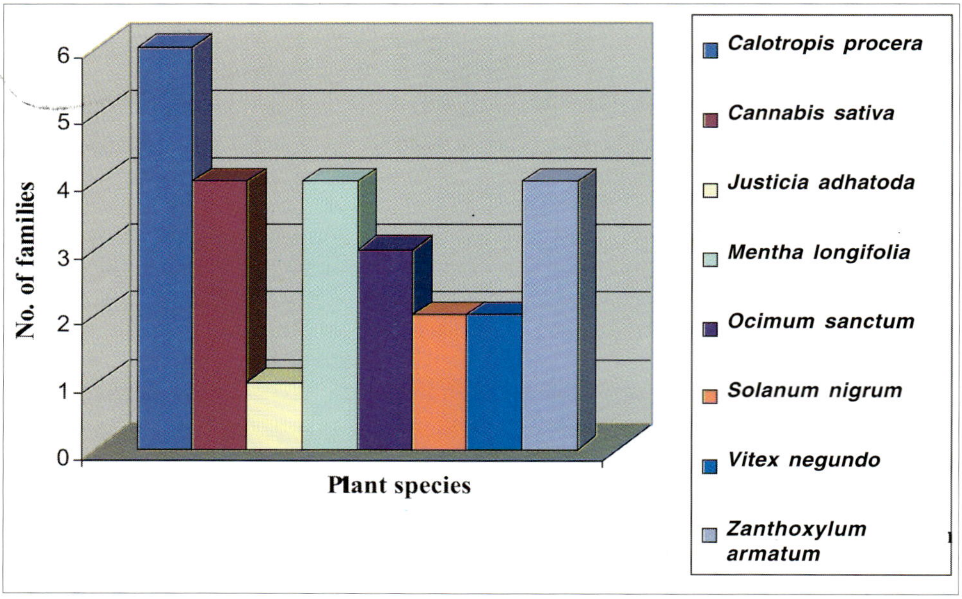

Figure 20.4: Showng No. of Families on Each Plant Species

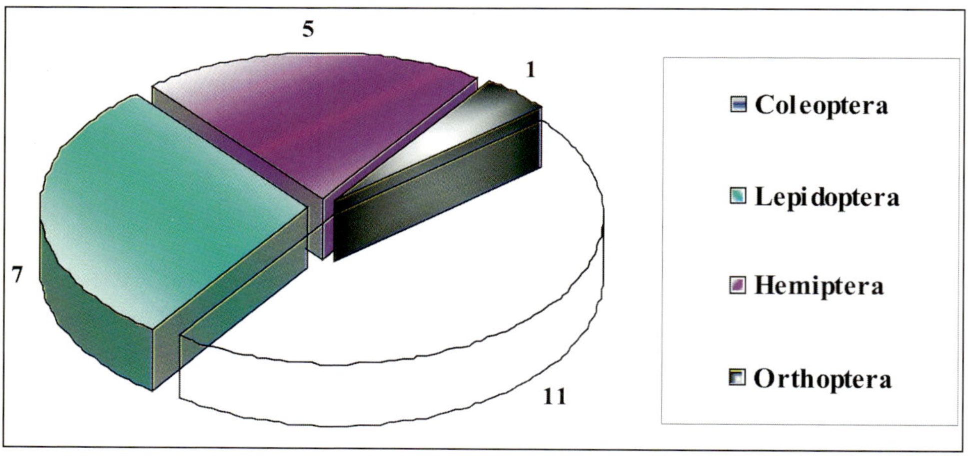

Figure 20.5: Contribution of Different Orders

as Hemipterans (*viz., Nezara viridula, Dolycoris baccarum, Erthesenia fullo, Spilostethus pandurus*) are found to be sap suckers.

Among defoliators, *Henosepilachna vigintioctopunctata* is a serious pest of *Solanum nigrum* causing a typical and distinctive "lace pattern damage" to the leaves of the plant. Both adults and larvae feed on leaves by scraping away chlorophyll from epidermal layers or surface cells between the main veins leaving the veins and veinlets causing irregular shaped holes or stripes and giving them a characteristic skeletonized or lace like appearance or forming ladder like windows in the leaves. Later, the affected areas on leaves become dry and fall off and damage appear in the form of holes in the leaves.

Both adults and larvae of *Chrysolina bella* and *Chrysolina exanthematica* are voracious feeders leading to complete defoliation in severe infestation. Adults of *Corynodes peregrinus* feed gregariously on the leaves of *Calotropis procera.* They start eating the leaf from its margin gnawing it completely leading to the total defoliation of plants where as adults of *Monolepta signata, Colasposoma semicostatum* and flea beetles, *Altica cynaea* make small holes often causing characteristic "shot holing" damage by completely perforating with numerous small holes in the foliage on which they feed.

Adults of *Oxycetonia versicolor* were found feeding exclusively on the flower buds of *Mentha arvensis, Mentha longifolia, Mentha piperita* and inflorescences of *Cannabis sativa* destroying the flowers as well as leaves. They are principally pollen feeders damaging the pollens and sucking nectar and sap. They are recorded as pests for they destroy the anthers and pistil of flowers in their quest for nectar and pollens.

Among weevils, *Paramecops farinosus* and *Amblyrhinus poricollis* were observed feeding on the leaves, shoots and inflorescences of *Calotropis procera* and *Cannabis sativa* respectively. Damaged leaves wither; turn pale yellowish resulting in premature abscission. Tops of plants lose their moisture content, become dry and fall off.

Caterpillar of Plain Tiger, *Papilio polytes, Papilio polyctor, Trichoplusia ni, Helicoverpa armigera* and *Pericallia ricini* are the most destructive pests feeding voraciously on the host plant leaves leading to total defoliation of the plants. The larval stages, particularly the third and fourth stages feed vigorously, consuming sufficient amount of food. The caterpillars generally feed on leaves but sometimes also feed on flowers and inflorescences of host plants. In severe infestations, whole of the branches are eaten up there by causing complete defoliation of the plant. Damaged leaves often become brown, wither, curl up and fall out prematurely.

Poekilocerus pictus is a serious pest causing extensive damage to *Calotropis procera.* Both adults and nymphs feed voraciously on leaves of the plant causing severe defoliation. They eat the leaves form their margins giving them irregular shaped cuts. Sometimes they eat the leaves creating small holes in the leaf lamina. The leaves are skeletonized and in severe infestations the bark is also chewed by these insects. Damaged leaves turn yellow become dry and fall out prematurely.

Among sap suckers *i.e.*, bugs, the nymphs and adults of *Spilostethus pandurus, Nezara viridula, Erthesina fullo,* and *Eysarcoris ventralis* were found feeding by sucking the sap by piercing plant tissue with needle like stylets. They suck the plant sap (juice) from leaves, flowers, fruits and seed pods leading to the necrosis, especially of young parts of the plant. In severe infestations damage caused by them resulted in retarded growth, withering and premature abscission of leaves and fruits. Feeding on stems or twigs may cause dwarfing or wilting. Leaf feeding results in a characteristic spotting or browning or curling and wilting of the leaves. As a result damaged leaves, flowers and seed pods become dry and shrivelled and fall out prematurely.

From the above, it is evident that insect pests are responsible for widespread damage to both commercially and economically important medicinal plants and it is clear that the wider gap that exists in our knowledge on the insect fauna associated with medicinal plants of the Jammu region need to be brought in notice, managed scientifically, in particular taxonomically, biologically and ecologically to provide the state with health security system, source of livelihood herbal industry and income.

Though India has a rich biodiversity, the growing demand is putting a heavy strain on the existing resources. While the demand for medicinal plants is growing, some of them are increasingly being threatened in their natural habitat due to outbreak of insect pests and their over exploitation. This poses a definite threat to the genetic stocks and to the diversity of medicinal plants if biodiversity is not sustain ably used. Therefore, cultivation of medicinal plants tolerant to the insect pests attack has to be encouraged for meeting the future needs.

Acknowledgements

The authors are grateful to Prof. Baldev Sharma, former Head, Department of Zoology, University of Jammu for providing the necessary facilities to conduct the work. The junior author is also thankful to Dr. V.V. Ramamurthy, Principal Scientist, Entomological Division, IARI, New Delhi. for the identification of Insects.

References

Ali, S., 1980. Record of damage to certain vegetable and medicinal plants by Bombay locust, *Patanga succincta* Linn. *Journal of Science Research*, 2(3): 227–229.

Arif, M. and Kumar, N., 2003. *Papilio machon asiatica* Men: Severe infestation on medicinal plant (*Ammi majus* L.) at 1676 altitude in Central Himalayas. *Journal of Applied Zoological Researches*, 14(2): 207–208.

Banjo, A.D., Lawal, O.A. and Aina, S.A., 2006. Insects associated with some medicinal plants in South–Western Nigeria. *World Journal of Zoology*, 1(1): 40–43.

Bhagat, K.C., 2004. Mango mealy bug, *Drosicha mangiferae* (Green) (Margarodidae: Hemiptera) on Ashwagandha– a medicinal plant. *Insect Environment*, 10(1): 14.

Butani, D.K., 1984. Spices and pest problems: Mint. *Pestology*, 8(11): 24–28.

Campbell, C.L. and Pike, K.S., 1985. Life history and biology of *Prussia orphisalis* Walker (Lepidoptera: Pyralidae) on mint in Washington. *Pan Pacific Entomologist*, 61(1): 42–47.

Chandra, R., 2004. Status of medicinal plants with respect to infestation of insect pests in and around Chitrakoot, District–Satna (M.P.). *Flora and Fauna, Jhansi*, 10(2): 88–92.

Gould, G.E., 1960. Problems in the control of Mint insects. *Journal of Economic Entomology*, 53(4): 526–531.

Joshi, K.C., Meshram, P.B., Sambath, S., Usha, K., Humane, S. and Kharkwal, G.N., 1991. Insect pests of some medicinal plants in Madhya Pradesh. *Indian Journal of Forestry*, 15(1): 17–26.

Mathur, A.C. and Srivastava, J.B., 1967. Record of Insect pests of Medicinal and Aromatic Plants in Jammu and Kashmir. *Indian Forester*, 93: 663–671.

McPartland, J.M., 1996. *Cannabis* pests. *Journal of the International Hemp Association*, 3(2): 49, 52–55.

Meshram, P.B., 2005. New reports of defoliator *Psilogramma menephron* on *Rauvolfia serpentine* and white grub *Holotrichia serrata* on *Withania somnifera*. *Indian Forester*, 131(7): 969–970.

Meshram, P.B. and Garg, V.K., 1999. A report on the occurrence of *Scutellera nobilis* Fab. on *Emblica officinalis* Gaertn. *Indian Forester*, 125(5): 536.

Mitra, B. and Biswas, B., 2002. Insects of Ashwagandha in South 24 Parganas, West Bengal. *Insect Environment*, 8(3): 122.

Murugesan, S., Kumar, S. and Sundararaj, R., 1997. Blister bettles as a threat to medicinal/ ornamental plants of arid and semi–arid regions. *Indian Forester*, 123(4): 341–344.

Neubauer, S., Kral, J. and Klimes, K., 1974. Insect pests of peppermint. *Nase Liecive Rastliny*, 11(2): 38–41.

Popov, P., 1971. Insect pests of medicinal plants in Bulgaria. Part II: Homoptera. *Pharmazie*, 26(7): 424–431.

Popov, P., 1972. Insect pests of medicinal crops in Bulgaria. Part III: Coleoptera. *Rasteniev"dni Nauki*, 9(5): 165–175.

Popov, P., 1973a. Insect pests on medicinal plants in Bulgaria. Part I: Hemiptera. *Rasteniev"dni Nauki*, 10(1): 157–164.

Popov, P., 1973b. The thrips on medicinal plants in Bulgaria. *Rastitelna Zashchita*, 21(9): 28–29.

Rai, S.N. and Singh, J., 1997. Population trend in *Brevipalpus phoenicis* on Ashwagandha in relation to weather factors in Varanasi. *Pestology*, 21(3): 38–43.

Ramji and Sagar, P., 1990a. Development of Cabbage semilooper larvae on different species of Mint in Punjab. *Indian Perfumer*, 34(2): 130–132.

Ramji and Sagar, P., 1990b. Population abundance of *Thysanoplusia* orichalcea (Fab.) on different species/cultivars of Mint at Ludhiana. *Indian Perfumer*, 34(1): 26–29.

Ramji and Sagar, P., 1990c. Seasonal history of Cabbage semilooper, *Thysanoplusia orichalcea* on Japanese mint, *Mentha arvensis* sub sp. *Haplocahyx* var. *piperascens* in Punjab. *Indian Journal of Agricultural Sciences*, 60(12): 852–853.

Ramzan, M., Singh, D., Singh, G., Mann, G.S. and Bhalla, J.S., 1990. Comparative development and seasonal abundance of Hadda beetle, *Henosepilachna vigintioctopunctata* (Fabr.) on some Solanaceous host plants. *Journal of Research, Punjab Agricultural University*, 27(2): 253–262.

Reddy, D.N.R., Puttaswamy and Hedge, N.S., 1981. Pests infesting *Catharanthus roseus* (L.) – a medicinal plant and their control. *Lal Baugh*, 26(2): 49–51.

Sagar, P., 1981. Comparative population abundance of Leaf Webber, *Syngamia abruptalis* Walker (Pyralidae: Lepidoptera) on three species of Mint at Ludhiana, Punjab. *Indian Perfumer*, 27(3&4): 200–202.

Sagar, P. and Reddy, V.O., 1987a. Nature and extent of damage of Mentha leaf webber, *Syngamia abruptalis* Walker, a pest of Japanese mint in the Punjab. *Indian Perfumer*, 31(4): 379–382.

Sagar, P. and Reddy, V.O., 1987b. *Syngamia abruptalis* Walker and its population abundance on five hosts in the Punjab. *Indian Perfumer*, 31(4): 335–339.

Seth, S.D. and Sharma, B., 2004. Medicinal plants in India. *Indian Journal of Medical Research*, July 2007.

Sharma, B., Tara, J.S. and Sudan, M., 2007. A report on insect pest, *Epilachna vigintioctopunctata* Fabricius (Coleoptera: Coccinellidae), of *Solanum nigrum* in Jammu region. Abstract published in "18th All India Congress of Zoology and National Seminar on "Current Issues on Applied Zoology and Environmental Sciences with Special Reference to Eco–restoration and Management of Bioresourses" held at University of Lucknow, Lucknow, SCIAZE, December 7–9, 2007, OS–156, P–65.

Sharma, B., Tara, J.S. and Sudan, M., 2008a. A checklist of insect pests attacking medicinal plants in various parts of district Rajouri, Jammu (J&K). Abstract published in " 3rd J&K State Science Congress" held at University of Jammu, February 26–28, 2008, zol–47, P–165.

Sharma, B., Tara, J.S. and Sudan, M., 2008a. *Henosepilachna vigintioctopunctata* Fabricius (Coleoptera: Coccinellidae), a serious pest of *Solanum nigrum* (Solanaceae), an important medicinal plant in Jammu region. Abstract published in "National Seminar on Technical Advances in Environment Management and pplied Zoology" held at University of Kurukshrtra, Kurukshetra, January 2325, 2008, S. No. 10, P–12.

Sharma, N. and Verma, T.D., 2005. Life stages and development of *Danaus chrysippus* Linn. Infesting commercially cultivated medicinal plants of mid hill regions of Himachal Pradesh. *Journal of Hill Research*, 18(1): 33–34.

Spiridonova, V.P. and Nosyrev, V.I., 1981. Pests of the medicinal plant crops in the Potava region. *Entomologicheskoe Obozrenie*, 60(3): 570–576.

Srivastava, J.B. and Saxena, B.P., 1976. Host preference of Solanaceous medicinal plants by *Epilachna vigintioctopunctata* F. (Coccinellidae: Coleoptera) and fate of Solasodine in the damaged leaves of *Solanum aviculare* Forst. *Science and Culture*, 42(4): 125–126.

Sudan, M., 2008. Survey of Insect pests infesting some medicinal plants in district Rajouri (J&K). M. Phil. Dissertation, University of Jammu, Jammu.

Swamy, B.C.H. and Rajagopal, D., 1995a. Insect pests of Vasaka – a medicinal plant. *Current Research University of Agricultural Sciences Bangalore*, 24(7): 129.

Swamy, B.C.H. and Rajagopal, D., 1995b. Insect pests of *Gloriosa superba* Linn. – an Indian medicinal plant. *Indian Journal of Forestry*, 18(2): 158–160.

Tiwari, K.C. and Joshi, P., 1974. A record of some insect pests attacking medicinal plants at Ranikhet. *Indian Journal of Pharmacy,* 36(5): 111–112.

Tosic, D., Spasic, R. and Petrovic, O., 1997. A study on the insect fauna on medicinal plants in Serbia. *Quatrieme Conference Internationale sur les Ravageurs en Agriculture, 6–7–8 janvier, 1997, le–corum, Montpellier, France Tome–2.* 1997: 531–540.

Tripathi, J., Singh, D. and Mathur, S., 1997. Outbreak of *Spodoptera litura* on bramhi – an important medicinal plant. *Insect Environment,* 2(4): 134.

Venkatesha, M.G., 2006. Seasonal occurrence of *Henosepilachna vigintioctopunctata* (F.) (Coleoptera: Coccinellidae) and its parasitoids on Ashwagandha in India. *Journal of Asia Pacific Entomology,* 9(3): 265–268.

Verma, A., Srivastava, S.K. and Sinha, T.B., 1978. Seasonal occurrence in the population of different pests of *Calotropis procera. Indian Journal of Entomology,* 40(2): 204–210.

Previous Volumes

— Volume 1 —

1996, xvi+332p., figs., pls., tab., 25 cm Rs. 700

ISBN 81-7035-156-1

Section I: Fish and Limnology

21. **Population Structure and Seasonal Succession of Zooplankton of Lake Surinsar, Jammu (India)**

    M.K. Jyoti & H.S. Sehgal

Section II: Wildlife

22. **On the Habitat and Behaviour, Maturity, Size and Sexual Dimorphism in a Population of Freshwater Emydid Turtles of Jammu, J & K State**

    P.L. Duda, Anil K. Verma & D.N. Sahi

23. **Status and Management of Wildlife in Jammu & Kashmir State**

    B.L. Kaul & Indu Kanwal

24. **Eco-geographical Distribution and Present Status of Herptiles in Kashmir Himalayas**

    D.N. Sahi & P.L. Duda

25. **Shrinking Wetlands of India**

    S.K. Chadha

26. **Ecology and Status of Wildlife in Ladakh**

    B.L. Kaul

27. **On the Freshwater Chelonian Fauna of Jammu and Kashmir**

    D.N. Sahi & Anil K. Verma

28. **Threatened Wildlife Habitat in Kashmir Himalayas**

    B.L. Kaul

29. **Role of Gastropods in Trematode Transmission Among Herptiles - Part I: Amphibia A Numerical Analysis**

    Anil K. Verma & P.L. Duda

— Volume 2 —

1999, xvi+231p., fig., plts., 25 cm **Rs. 600**

ISBN 81-7035-205-3

Section I: Fish and Limnology

1. **Role of Thyroid Gland in the Regulation of Metabolic Rate in Fishes with Special Reference to Indian Teleosts**

   B.N. Pandey

2. **Alcohol Dehydrogenase Isozyme Expression in the Air-breathing Fish, *Clarias batrachus* and *Heteropneustes fossilis* of North Eastern India**

   Alka Prakash & Sant Prakash

3. **The Ecological Role of Algal Weeds, Charophytes in Particular in Fisheries Water**
Usha Moza

4. **Importance of Fish Food Organisms (Live Food) in Aquaculture Practice**
Seem Langer, K. Gupta & R. Gandotra

5. **Morphological Studies of Alimentary Canal of Fishes of Lake Mansar**
Arunk K. Gupta, Seema Langer & S.C. Gupta

6. **Transgenic Fish: Production and Improvement of Fish Resources**
Anil K. Verma & B.L. Kaul

7. **Sewage Fed Fisheries: A Biotechnological Application**
Y.R. Malhotra, Seema Langer & S. Raina

8. **The Histopathology of *Pallisentis jagani* and *Pomphorhynchus bulbocolli* Infection in *Channa striatus* and *Schizothorax sinuatus***
P.L. Kaul & M.K. Rana

9. **Female Reproductive System of *Pallisentis jagani***
P.L. Kaul, M.K. Raina & Usha Zutshi

10. **Bacterial Microflora: Their Distribution and Relationship with Fish and Its Environment: A Review**
J.P. Sharma & V.K. Gupta

11. **A Comparison of the Feeding Rates of *Streptocephalus torvicornis* and *Chirocephalus diaphanus* (Crustacea : Anostraca) on Rotifers**
S.S.S. Sarma and K.R. Dierckens

12. **Population Growth of *Brachionus calyciflorus* Pallas (Rotifera) in Relation to Algal (*Dictyosphaerium chlorelloides*) Density**
S.S.S. Sarma, E.D. Fiogbe & P. Kestemont

13. **Ecological Crisis in Lake Mansar Jammu, J & K State**
B.L. Kaul & Anil K. Verma

14. **Zooplankton Composition, Abundance and Dynamics in a Lentic Habitat (Kalika Pond, Dhar, M.P.)**
R.K. Dave, M.M. Prakash & N.K. Dhakad

15. **Impact of Nutrient Influx on Water Quality Trends of a Vindhyan Lake**
S. Pani & A. Wanganeo

16. **Seasonal Variations in Biochemical Composition of Muscle During the Annual Ovarian Cycle of Female *Channa gachua* (Ham.)**
K. Gupta, Sujata Raina, R. Gandotra & S. Langer

Section II: Wildlife

— Volume 3 —

2004, xix+317p., figs., tabls., ind., 25 cm Rs. 990

ISBN 81-7035-327-0

Section I: Fish and Limnology

Section II: Wildlife

— Volume 4 —

2007, xxii+259p., 12 col. plts., figs., tabls., ind., 25 cm Rs. 1200

ISBN 81-7035-517-6

Section I: Fish and Limnology

Index

F

G